Scheduling: Control-Based Theory and Polynomial-Time Algorithms

Applied Optimization

Volume 43

Series Editors:

Panos M. Pardalos
University of Florida, U.S.A.

Donald Hearn
University of Florida, U.S.A.

The titles published in this series are listed at the end of this volume.

Scheduling: Control-Based Theory and Polynomial-Time Algorithms

by

Konstantin Kogan

Department of Computer Science,
Center for Technological Education,
Holon, Israel

and

Eugene Khmelnitsky

Department of Industrial Engineering,
Tel-Aviv University,
Tel-Aviv, Israel

SPRINGER-SCIENCE+BUSINESS MEDIA, B.V.

A C.I.P. Catalogue record for this book is available from the Library of Congress.

ISBN 978-1-4613-7116-8 ISBN 978-1-4615-4675-7 (eBook)
DOI 10.1007/978-1-4615-4675-7

Originally published by Kluwer Academic Publishers in 2000
Softcover reprint of the hardcover 1st edition 2000

CONTENTS

PART II: ONE-MACHINE, ONE-PRODUCT-TYPE SCHEDULING

Preface

This book presents a first attempt to systematically collect, classify and solve various continuous-time scheduling problems. The classes of problems distinguish scheduling by the number of machines and products, production constraints and performance measures. Although such classes are usually considered to be a prerogative of only combinatorial scheduling literature, the scheduling methodology suggested in this book is based on two mathematical tools - optimal control and combinatorics. Generally considered as belonging to two totally different areas of research and application, these seemingly irreconcilable tools can be integrated in a unique solution approach with the advantages of both. This new approach provides the possibility of developing effective polynomial-time algorithms to solve the generic scheduling problems.

This book is aimed at a student audience - final year undergraduates as well as master and Ph.D. students, primarily in Operations Research, Management, Industrial Engineering and Control Systems. Indeed, some of the material in the book has formed part of the content of undergraduate and graduate courses taught at the Industrial Engineering Department of Tel-Aviv University, the Logistics Department of Bar-Ilan University and the Technology Management Department of Holon Center for Technological Education, Israel.

The book is also useful for practicing engineers interested in planning, scheduling and optimization methods. Since the book addresses the theory and design of computer-based scheduling algorithms, applied mathematicians and computer software specialists engaged in developing scheduling software for industrial engineering and management problems will find that the methods developed here can be embedded very efficiently in large applications.

The comments of many colleagues and students have been helpful during the writing of the book. We would especially like to thank Profs. Sheldon Lou (California State University San Marcos), Toshihide Ibaraki (Kyoto University), Eugene Levner (Holon Center for Technological Education), Oded Maimon (Tel-Aviv University), Avi Shtub (Technion, Haifa), for fruitful discussions of various topics underlying the book. Finally, we are much obliged to Arthur Kemelman who read the text and corrected the English phrasing with great care.

K. Kogan, E. Khmelnitsky

PART 1

INTRODUCTION

1

INTRODUCTION TO CONTROL-BASED SCHEDULING

1.1 SCHEDULING APPROACHES

Modern production systems are characterized by resources and the products these resources are designed to manufacture in response to market demand. When demand for products at a time horizon is given, scheduling generally means assigning individual resources to individual products in order to trace the demand along the horizon. Since typically, a large number of possible schedules can be constructed, the problem of scheduling is to select the schedule which ensures optimal functioning of the production system in terms of a certain performance measure. Due to the great variety of production systems, of performance measures as well as of production demand forms, a vast number of scheduling models has been developed to mathematically formalize the scheduling problem.

Historically, the first approach to formalizing and solving the scheduling problem was combinatorial. In the beginning of the 20th century, simple techniques were used to allocate production tasks to machines in rapidly growing industrial plants. These techniques resulted in a sequence of tasks to be carried out. The sequence was illustrated in a graphic form known as a *Gantt chart*. The advantage of Gantt charts is that they provide control over the machines and work-in-process and are easy to implement on the shop floor. However, enumerating the charts is likely to become very tedious if, among all possible sequences, the most profitable ones are sought. This is especially true, when a tight deadline is set for all tasks so that the enumeration cannot be stopped until a chart which does not violate the deadline is found. To overcome tediousness of the chart-based techniques, scientists began developing combinatorial optimization algorithms for sequencing tasks. The idea was to model the tasks as distinct, indivisible jobs or products characterized by production times. Since production times are considered to be fixed, the task sequencing unambiguously determines their timing. As with chart-based techniques, selection of the best schedule is reduced to finding the permutation of products which satisfies an optimality condition. This approach to modeling and solving scheduling problems thus falls in the mathematical area of combinatorics. First, the combinatorial approach was successfully applied to proving optimal rules for non-preemptive scheduling of multiple tasks on a single machine under various performance measures. These measures were mainly related to either completion times of the tasks or the due dates before which the tasks are to be completed. Efficient algorithms have been also developed for non-preemptive scheduling of two tandem machines with the objective of minimizing maximal completion time (makespan). In addition, *NP*-hardness of a number of problems has been proved. This indicates that for such problems no polynomial time algorithm can be developed. Therefore, a great number of combinatorial heuristics have been suggested to tackle complex scheduling problems. Results accumulated in the area of combinatorial scheduling led researchers to systematize their studies by the detailed classification of problems with respect to the number of machines and products, configuration of the production system and the performance measure. In turn, this led to elaboration of the combinatorial theory of scheduling.

Beginning in the 1950s, mathematical programming enriched the combinatorial theory for those cases when multiple machines and multiple units of each product are involved in the production process. This is accomplished by modeling production times of each product on a machine as decision variables. Thus, timing no longer results from sequencing, rather both sequencing and timing are sought simultaneously. Specifically, most scheduling problems can be modeled as mixed integer programming problems. In some cases, where tasks are allowed to be preempted, these models transform into a linear programming problem known to be efficiently solvable. The theory of mathematical programming suggests its own methods, some of which take advantage of combinatorial techniques, for example Branch and Bound method, which underlies mixed integer programming.

The other approach to modeling and solving scheduling problems is due to the rapid development of control theory in the 1970s. The approach suggests to formulate the problems as control models. This is accomplished by considering accumulated product units as state variables and production rates, rather than production times, as decision or control variables. These variables are now functions of time. As a result, the control model consists of a number of differential equations, which describe dynamic state of production processes over an interval of time. Such a representation leads to developing either an optimal off-line control with respect to a given performance measure, or a stable on-line feedback policy to maintain desired parameters of the system. By applying the optimal control theory, which is based on the maximum principle, some scheduling problems can be solved analytically, for example, one-machine, one-product-type problem. The others are solvable only numerically.

Similar to combinatorial scheduling theory, the control model induces a classification of the scheduling problems. This book suggests classification and methods based on the idea of identifying classes of continuous-time scheduling problems which can be solved by polynomial-time algorithms either exactly or with a given accuracy. The methodology is, first, to analyze each problem class in order to obtain analytical properties of the optimal schedules. This is accomplished by taking advantage of the maximum principle which yields the necessary optimality conditions for control problems. Second, an algorithm is developed to construct the schedule which satisfies the analytical properties. Since construction of such schedules takes advantage of both mathematical programming and combinatorial theory, this book presents an approach which integrates control, combinatorics and mathematical programming to classifying and solving scheduling problems. It also includes the worst-case complexity analysis of the algorithms and the accuracy analysis of the obtained solutions when the algorithm approximates the optimum.

1.2 DETERMINISTIC CONTROL-BASED SCHEDULING

In general, scheduling problems considered in this book are characterized by the number of machines, M, and the number of products, N. Each machine m, $m=1,...,M$, is defined by maximum production rate U_{nm} with respect to product n, $n=1,...,N$. Each product is characterized by demand known in advance, $D_n(t)$, which is the number of product n units to be delivered by time t. The variables of the scheduling problems are: the rate of producing product n on machine m at time t, $u_{nm}(t)$, and the number of product n units accumulated in the production system by time t, $X_n(t)$. The former is referred to as the *control variable*, while the latter is referred to as the *state variable*.

Scheduling means determining production rates of each machine on each product at each point of time $u_{nm}(t)$ over a planning horizon $[0,T]$ in order to track the demand under imposed constraints. This definition supposes that the decision $u_{nm}(t)$ can be made at every point of time which implies that production of any product unit can be preempted at any time and restarted later at no cost. Similar to classical combinatorial approach, such schedules are referred to as *preemptive*.

In contrast to the classical combinatorial scheduling approach, a machine is allowed to produce more than one product at a time. Although this simplifies scheduling, such a generalization often results in an optimal solution in which each machine produces only one product at a time. Another generalization arises from stating the demand for products. The classical combinatorial approach supposes that demand is given as a unit of product to be produced by a time point called due date. The above definition of scheduling implies that an arbitrary amount of product units can be demanded at each point of time along the planning horizon. As a result, demand is presented as a function of time, a special case of which is the due date statement.

There are two general constraints in control-based scheduling. The first constraint describes the dynamics of the production process

$$\dot{X}_n(t) = \sum_{m=1}^{M} u_{nm}(t) - d_n(t), \quad X_n(0) = X_n^0, \quad n=1,\dots,N, \tag{1.1}$$

where $d_n(t) = \frac{d}{dt} D_n(t)$ is the demand rate of product n at time t and X_n^0 is the number of product n units at the beginning of the planning horizon. Equation (1.1) implies that the change in the accumulated level of product n units is equal to the difference between the total production rate of the machines producing product n and the demand rate for this product. If $X_n(t) \geq 0$, the units of product n compose the surplus which is stored in a buffer for this product. Otherwise, if $X_n(t) < 0$, the units of product n are backlogged. Throughout the book notation $X_n^+(t) = \max\{X_n(t), 0\}$ will stand for inventory surplus and $X_n^-(t) = \max\{-X_n(t), 0\}$ for inventory backlog of product n at time t.

The second constraint is due to the limited capacity of the machines

$$0 \leq u_{nm}(t) \leq U_{nm}, \; n=1,\dots,N, \quad m=1,\dots,M, \tag{1.2}$$

$$\sum_{n=1}^{N} \frac{u_{nm}(t)}{U_{nm}} \leq 1, \; m=1,\dots,M. \tag{1.3}$$

Generally, other constraints feature characteristics of modern production environments, e.g. multiple resource and budget constraints, as well as the setup times required to reconfigure the machines from one type of production to another. In this book, we assume there are no other constraints than resource constraints (1.2)-(1.3). It is also assumed that setup times and associated with them setup costs are negligible.

The following cost parameters characterize the production system:

c_{xn}^+ - product n surplus cost coefficient;

c_{xn}^- - product n backlog cost coefficient;

c_{unm} - machine m production cost coefficient for product n;

c_T - planning horizon cost coefficient.

To evaluate feasible schedules, the following performance measures will be used to minimize the total schedule cost:

$$L_x = \int_0^T \sum_{n=1}^{N} c_{xn}^{+} X_n^{+}(t) + c_{xn}^{-} X_n^{-}(t)dt$$ - linear inventory cost;

$$Q_x = \int_0^T \sum_{n=1}^{N} c_{xn}^{+} X_n^{+}(t)^2 + c_{xn}^{-} X_n^{-}(t)^2 dt$$ - quadratic inventory cost;

$$L_u = \int_0^T \sum_{n=1}^{N} \sum_{m=1}^{M} c_{unm} u_{nm}(t)dt$$ - linear production cost;

$$Q_u = \int_0^T \sum_{n=1}^{N} \sum_{m=1}^{M} c_{unm} u_{nm}(t)^2 dt$$ - quadratic production cost;

$c_T T$ - planning horizon cost;

and the following parameters will be used to maintain stable inventory of the production system

$\overline{X}_n$ - desired level of product n.

Measures L_x and Q_x present the inventory related cost incurred over the planning horizon. Measures L_u and Q_u present machine-related production costs incurred over the planning horizon. Measure $c_T T$ presents planning horizon related costs when the planning horizon is variable. Finally, measure $\overline{X}_n$ presents the case when the objective is to maintain stable inventory of products rather than to minimize costs.

For the linear cost measures, L_x and L_u, the meaning of coefficient c_{xn}^{+} is the cost of carrying a unit of product n for a unit of time; the meaning of c_{xn}^{-} is the cost of delay in shipping a unit of product n for a unit of time; the meaning of c_{unm} is the cost of production of a unit of product n on machine m. For the quadratic measures Q_x and Q_u, the meaning of the cost coefficients changes. Now they can be calculated by regression analysis of inventory and production cost data of a specific plant. The quadratic cost measures are consistent with the notion that costs increase as inventory level or production quantity approach the plant capacity.

So far we have discussed the general constraints and performance measures for cases when the demand rate of product n, $d_n(t) = \frac{d}{dt} D_n(t)$ is given as a function of time. If, however, this is not the case, that is the demand $D_n(t)$ is given as the number of product n units to be delivered by time t, the dynamics of the production process is described as

$$\dot{X}_n(t) = \sum_{m=1}^{M} u_{nm}(t), \qquad X_n(0) = X_n^0, \qquad n=1,\dots,N, \tag{1.4}$$

where $X_n(t)$ now presents the number of product n units produced by time t, rather than the buffer level in (1.1). The buffer level in this case is determined as the difference between the cumulative production and demand, i.e. $Z_n^+(t) = \max\{X_n(t) - D_n(t), 0\}$ is the inventory surplus level and $Z_n^-(t) = \max\{D_n(t) - X_n(t), 0\}$ is the inventory backlog level. The capacity constraints (1.2) and (1.3) remain unchanged.

The change in the production dynamics results in the following change in the inventory-related performance measures:

$L_x = \int_0^T \sum_{n=1}^{N} c_{xn}^+ Z_n^+(t) + c_{xn}^- Z_n^-(t) dt$ - linear inventory cost;

$Q_x = \int_0^T \sum_{n=1}^{N} c_{xn}^+ Z_n^+(t)^2 + c_{xn}^- Z_n^-(t)^2 dt$ - quadratic inventory cost.

1.3 STOCHASTIC CONTROL-BASED SCHEDULING

The previous section defines scheduling problems characterized by a demand function known in advance. However, in practice, knowledge about future demand is generally incomplete. This implies that at the beginning of the planning horizon, we may predict with a certain probability a number of products to be demanded by time t. As a result, inventory levels $X_n(t)$ also become uncertain at every point of time.

1.3.1 DETERMINISTIC REPRESENTATION OF STOCHASTIC DEMAND

Generally, a stochastic demand process is completely determined by its multivariable distribution function. However, in practice, this function is rarely available. Therefore, various types of series expansions and quasi-deterministic forms are often employed to represent the process. In such a case, the process is described by a number of deterministic functions with randomized parameters. Widely used trigonometric series is an example of random process representation by deterministic trigonometric functions with random amplitude. This representation is very useful if the stochastic process is periodic and stationary, because then the random amplitudes become uncorrelated. If this is not the case, more general orthogonal representations are utilized.

The well-known Korhunen-Loeve type of series can be adopted to illustrate the reduction of the stochastic problem to a deterministic one for the most general random demands which are not necessarily stationary and periodic. If accuracy of such a representation is specified, then stochastic demands can be expanded in the following K-element Korhunen-Loeve series

$$d_n(t) = \sum_{k=1}^{K} \xi_{nk} \pi_{nk}(t), \; t \in [0, T] \tag{1.5}$$

where $\pi_{n1}(t), \pi_{n2}(t), \ldots, \pi_{nK}(t)$ are orthonormal deterministic sample functions, and ξ_{nk} are uncorrelated random coefficients with expected values μ_{nk}, variances σ_{nk}^2 and density functions $\varphi_{nk}(\xi)$. In the following lemma we prove that the stochastic scheduling problem can be reduced to a deterministic one if inventory and backlog cost coefficients are equal. The proof is conducted for the quadratic case of the cost function. The linear case is easier and accomplished in a similar way.

LEMMA 1.1. Given $c_{xn}^{+} = c_{xn}^{-} = c_{xn}$, the problem of minimizing total expected cost $J = E\left[\int_0^T \sum_n c_{xn} X_n(t)^2 + \sum_n \sum_m c_{unm} u_{nm}(t) dt\right]$ subject to constraints (1.1)-(1.3) is equivalent to the following deterministic problem

$$J = \int_0^T \sum_n c_{xn} X_n(t)^2 + \sum_n \sum_m c_{unm} u_{nm}(t) dt$$

subject to constraints (1.2), (1.3) and

$$\dot{X}_n(t) = \sum_m u_{nm}(t) - \sum_k \mu_{nk} \pi_{nk}(t) .$$

PROOF. Introduce new variables $\hat{X}_n(t)=\int_0^t \sum_m u_{nm}(\tau)d\tau$ and denote by $\hat{\pi}_{nk}(t)=\int_0^t \pi_{nk}(\tau)d\tau$. Then, the objective is

$$J=E\left[\int_0^T \sum_n c_{xn}\left(\hat{X}_n(t)-\sum_k \xi_{nk}\hat{\pi}_{nk}(t)\right)^2 dt+\int_0^T \sum_n \sum_m c_{unm}u_{nm}(t)dt\right]=$$

$$=\int_0^T\left(\sum_n \sum_m c_{unm}u_{nm}(t)+\sum_n c_{xn}\int_{\{\xi_{nk}\}}\left(\hat{X}_n(t)-\sum_k \xi_{nk}\hat{\pi}_{nk}(t)\right)^2 \prod_k \varphi_{nk}(\xi_{nk})d\xi_{nk}\right)dt. \quad (1.6)$$

The objective now can be simplified by opening parentheses and replacing integrals of the uncorrelated random coefficients multiplied by their density functions with the corresponding moments. The resulting terms are again collected into the following final functional

$$\int_0^T\left(\sum_n \sum_m c_{unm}u_{nm}(t)+\sum_n c_{xn}\left(\hat{X}_n(t)-\sum_k \mu_{nk}\hat{\pi}_{nk}(t)\right)^2\right)dt+$$

$$+\int_0^T \sum_n c_{xn}\left(\sum_k \sigma_{nk}^2\hat{\pi}_{nk}^2(t)+2\sum_{k'\neq k''}\mu_{nk'}\mu_{nk''}\hat{\pi}_{nk'}(t)\hat{\pi}_{nk''}(t)\right)dt.$$

By taking into account that the second term of the last expression is constant, which does not influence the optimization, and denoting $\hat{X}_n(t)-\sum_k \mu_{nk}\hat{\pi}_{nk}(t)=X_n(t)$, we conclude the proof.

■

Thus, if the surplus cost is equal to the backlog cost, then scheduling problems with stochastic demand can be reduced to deterministic scheduling problems which are treated in this book.

1.3.2 INVENTORY CONTROL POLICIES

Inventory theory suggests a number of control policies which are used when demand is a stochastic process. The policies are generally classified with respect to the way the inventory is reviewed, either continuously or periodically.

In *continuous review*, a control effort is changed at each point of time depending on the current inventory level $X_n(t)$, in order to maintain a desired inventory level $\overline{X}_n(t)$ over an infinite planning horizon. In control theory, such policies are referred to as *feedback control*. This book covers a number of feedback policies and discusses stability of inventories when the policies are implemented in various production environments.

In *periodic review*, the inventory level is checked at discrete times and control decisions are made only at these times in the form of a time-dependent function over a period. The control function is chosen in order to minimize the expected value of a certain performance measure. In control theory, such policies are referred to as *optimal control*. The periodic review implies that the optimal control is calculated off-line at the beginning of each period. It is exercised over this period regardless of which surplus or shortage of inventory is really accumulated over the period. Thus, at each review point, parameters of the scheduling problem are updated and a new optimal solution is generated for the next period. Such re-scheduling continues over infinitely *rolling* planning horizon and ensures inventory stability under stochastic demand by means of optimization. Since this book develops polynomial time algorithms, the periodic review policy can be efficiently implemented even if the review period is short.

1.4 DEMAND CLASSIFICATION

The form of the demand function significantly influences the statement and tractability of scheduling problems. The most general form of demand is the number of product units required per time unit at point *t*. This demand rate function was denoted in the previous section as $d_n(t)$. The cumulative demand $D_n(t)$ is the total number of product *n* units required at time interval $[0,t]$. We distinguish between two forms of the cumulative demand: the differentiable one, $d_n(t) = \frac{d}{dt} D_n(t)$, and the due date

$$D_n(t) = \begin{cases} D_n, & \text{if } t \geq t_{nd}, \\ 0, & \text{otherwise,} \end{cases}$$

where D_n is the amount of the product *n* to be produced by the due date t_{nd}. Note that if D_n=1, i.e. if only a single item of each product is required, the demand function becomes equivalent to that used in the combinatorial approach. In other words, the control-based scheduling in these cases assigns machines to the products.

We further distinguish between some special shapes of the demand functions. For the due date form of the demand, a common due date case is of a special interest. In this case, all products are required by the same time point $t_d = t_{nd}$ for all n.

For the demand rate function form, we introduce two classifications. The first one relates demand to the system capacity and outlines four special classes of interest:

- *weak demand* rate, $d_n(t) \le U_{nm}$ for all m, n, and $\sum_n \frac{d_n(t)}{\sum_m U_{nm}} \le 1$ along the planning horizon[0,T]. Weak demand implies that the total capacity of the production system is sufficient to produce strictly according to the demand at each point of time.
- *moderate demand* rate, $\frac{1}{t}\int_0^t d_n(\tau)d\tau \le U_{nm}$ for all m, n, and $\frac{1}{t}\int_0^t \sum_n \frac{d_n(\tau)}{\sum_m U_{nm}} d\tau \le 1$ for all $0 < t \le T$. Moderate demand implies that the total capacity of the production system is sufficient to produce all products without any backlog along the planning horizon.
- *pressing demand* rate, $\frac{1}{t}\int_0^t d_n(\tau)d\tau \ge U_{nm}$ for all m, n, and $\frac{1}{t}\int_0^t \sum_n \frac{d_n(\tau)}{\sum_m U_{nm}} d\tau \ge 1$ for all $0 < t \le T$. Pressing demand implies that the total capacity of the production system is insufficient to produce all products without any backlog along the planning horizon.
- *strong demand* rate, $d_n(t) \ge \sum_m U_{nm}$ along the planning horizon. Strong demand implies that the total capacity of the production system is insufficient for a product n to be produced strictly according to its demand at each point of time.

The second classification of the demand functions, concerned with the rate of change in the demand at an interval of time (t_1, t_2), outlines four special classes of interest:

- *monotone demand* rate, non-decreasing if $d_n(t_x) \le d_n(t_y)$ for all t_x and t_y, $t_1 < t_x < t_y < t_2$; non-increasing if $d_n(t_x) \ge d_n(t_y)$ for all t_x and t_y, $t_1 < t_x < t_y < t_2$.

- *non-lumpy demand* rate, $\left|\dot{d}_n(t)\right| \le A$, where A is a given parameter. Non-lumpiness implies that the demand rate is differentiable for $t_1 < t < t_2$ and the derivative is uniformly bounded.

- *constant demand* rate, $d_n(t_x) = d_n(t_y)$ for all t_x and t_y, $t_1 < t_x < t_y < t_2$. Constant demand rate is evidently monotone and non-lumpy with A=0.

- *periodic demand* rate, $d_n(t_x) = d_n(t_x + T)$ for all t_x, $t_1 < t_x < t_2 - T$, where T is the demand period. Periodic demand is usually considered on an infinite planning horizon that often implies periodicity of the control $u_{nm}(t)$ and state $X_n(t)$ variables.

When applying this classification to the entire planning horizon, one may divide the horizon into K intervals on each of which the demand rate meets a specific class. Then, we distinguish K-level *piece-wise monotone*, *piece-wise non-lumpy* and *piece-wise constant* demand. Finally, if there exists a point of time $0 < \hat{t} < T$ such that the demand rate is non-decreasing before $\hat{t}$ and non-increasing after $\hat{t}$, the demand function is referred to as *single-mode*.

1.5 CLASSIFICATION OF SCHEDULING PROBLEMS

A great variety of scheduling problems can be formulated by considering the various characteristics of different production environments such as a number of products and machines, limited or unlimited capacity, finite or infinite planning horizon, demand form and performance measure. This motivates the introduction of systematic notation as a basis for a classification scheme. Such a notation of problem types would greatly facilitate the presentation and discussion of scheduling problems.

The notation we suggest consists of five fields $\alpha|\beta|\gamma|\delta|\varepsilon$. The first two fields denote respectively, the number of parallel machines, α =1,...,M and the number of product types, β=1,...,N. The third field presents constraints imposed on the control and state variables and includes at most three parameters, $\gamma = \gamma_1, \gamma_2, \gamma_3$. Parameter $\gamma_1 \in \{u(t) \in [0,\infty], u(t) \in [0,U]\}$ characterizes the production capacity:

$\gamma_1 = (u(t) \in [0,\infty])$ - unlimited production capacity,

$\gamma_1 = (u(t) \in [0,U])$ - production capacity is limited by U product units per time unit.

The second parameter $\gamma_2 \in \{\varnothing, X(t) \ge 0\}$ indicates the possibility of backlogging:

$\gamma_2 = \varnothing$ - backlog is allowed,

$\gamma_2 = (X(t) \geq 0)$ - backlog is disallowed.

The third parameter $\gamma_3 \in \{\varnothing, X(0) = X(T), X(T) = X_T\}$ presents boundary conditions for the state variables:

$\gamma_3 = \varnothing$ - no terminal condition is imposed on the state variables,

$\gamma_3 = (X(0) = X(T))$ - periodicity is required for the state variables, i.e. the level of surplus (backlog) of product units must be the same at time points t and t+T over an infinite planning horizon,

$\gamma_3 = (X(T) = X_T)$ - terminal condition is imposed on the state variables, i.e. X_{nT} units of product n must be accumulated by the end of the planning horizon.

The fourth field characterizes planning horizon and demand function $\delta = \delta_1, \delta_2$. The first parameter $\delta_1 \in \{\varnothing, d(t), D\}$ describes the demand function:

$\delta_1 = \varnothing$ - no demand is specified,

$\delta_1 = d(t)$ - demand is specified as a rate function,

$\delta_1 = D$ - demand is specified in the due date form.

The second parameter $\delta_2 \in \{\varnothing, T\}$ indicates whether the planning horizon is finite or not:

$\delta_2 = \varnothing$ - the planning horizon is unspecified or infinite,

$\delta_2 = T$ - the length of the planning horizon is equal to T.

The fifth field defines the objective (performance measure) $\varepsilon = \varepsilon_1, \varepsilon_2, \varepsilon_3, \varepsilon_4, \varepsilon_5, \varepsilon_6$.

The first parameter $\varepsilon_1 \in \{\varnothing, L_u, Q_u\}$ describes production cost:

$\varepsilon_1 = \varnothing$ - no production cost, $c_{unm} = 0$

$\varepsilon_1 = L_u$ - linear production cost,

$\varepsilon_1 = Q_u$ - quadratic production cost.

The second parameter $\varepsilon_2 \in \{L_x, Q_x\}$ describes inventory cost:

$\varepsilon_2 = L_x$ - linear inventory cost,

$\varepsilon_2 = Q_x$ - quadratic inventory cost.

The third parameter $\varepsilon_3 \in \{\varnothing, T\}$ indicates whether the cost of the planning horizon is included in the performance measure or not:

$\varepsilon_3 = \varnothing$ - the cost of the planning horizon is not included because its length is fixed,

$\varepsilon_3 = T$ - the cost of the planning horizon is included because its length is variable.

The fourth parameter $\varepsilon_4 \in \{\varnothing, c_x^+ = 0\}$ indicates whether the surplus inventory cost is incurred or not:

$\varepsilon_4 = \varnothing$ - the surplus inventory cost is incurred with rate c_x^+,

$\varepsilon_4 = (c_x^+ = 0)$ - no surplus inventory cost is incurred.

The fifth parameter $\varepsilon_5 \in \{\varnothing, c_T = 0\}$ indicates whether the planning horizon cost is incurred or not for the case of the variable planning horizon:

$\varepsilon_5 = \varnothing$ - the planning horizon cost is incurred with rate c_T,

$\varepsilon_5 = (c_T = 0)$ - no planing horizon cost is incurred.

The last parameter $\varepsilon_6 \in \{\varnothing, \overline{X}\}$ indicates whether the inventory should be maintained stable with respect to desired level $\overline{X}$:

$\varepsilon_6 = \varnothing$ - there is no specific requirement for maintaining stable inventory,

$\varepsilon_6 = \overline{X}$ - the inventory is to be maintained stable.

The use of this notation is illustrated by the following examples.

Example 1.1

Problem $1|N|u(t) \in [0, \infty]|d(t), T|\, L_u, L_x$ reads as follows: Given a demand rate function over a finite planning horizon and unlimited machine capacity, schedule N product types on a single machine in order to minimize the total linear cost of inventory and production. The mathematical statement of the problem is

$$J = \int_0^T \sum_n \left[c_{xn}^+ X_n^+(t) + c_{xn}^- X_n^-(t) + c_{un} u_n(t)\right] dt \rightarrow \min,$$

$$X_n^+(t) = \max\{X_n(t), 0\}, \; X_n^-(t) = \max\{-X_n(t), 0\}$$

subject to

$$\dot{X}_n(t) = u_n(t) - d_n(t), \qquad X_n(0) = 0,$$

$$u_n(t) \geq 0.$$

■

Example 1.2

Problem $1|1|u(t) \in [0, U]|d(t)|\overline{X}$ reads as follows: Given an infinite planning horizon, limited machine capacity, and the desired inventory level, determine a control policy $u(X(t), \overline{X})$ which schedules single-product-type units on a single machine in order to maintain stable inventory under an unknown arbitrary demand rate. The problem is stated as follows

$$\sup_{0\le t<\infty} |X(t)| < \infty ,$$

$$\dot{X}(t) = u(t) - d(t) , \qquad X(0) = 0 ,$$

$$0 \le u(t) \le U .$$

■

Example 1.3
Problem $1|N|\, u(t) \in [0, U]\, |D,T|\, Q_x$ reads as follows: Given the due-date form of demand over a finite planning horizon and limited machine capacity, schedule N product types on a single machine in order to minimize the quadratic cost of inventory. The mathematical statement of the problem is

$$J = \int_0^T \sum_n \left[c_{xn}^+ Z_n^+(t)^2 + c_{xn}^- Z_n^-(t)^2 \right] dt \to \min ,$$

$$Z_n^+(t) = \max\{X_n(t) - D_n(t), 0\} , \; Z_n^-(t) = \max\{D_n(t) - X_n(t), 0\}$$

subject to

$$\dot{X}_n(t) = u_n(t) , \qquad X_n(0) = 0 ,$$

$$0 \le u_n(t) \le U_n ,$$

$$\sum_n \frac{u_n(t)}{U_n} \le 1 ,$$

where

$$D_n(t) = \begin{cases} D_n, \text{if } t \ge t_{dn}, \\ 0, \text{otherwise.} \end{cases}$$

■

Example 1.4
Problem $M|N|\, u(t) \in [0, U] , X(t) \ge 0 , X(0) = X(T)\, |d(t)|\, L_u , L_x$ reads as follows: Given periodic demand rate over an infinite planning horizon, limited machine capacity and backlogs are disallowed, schedule N product types on M parallel machines in order to minimize the total linear cost of inventory and production. The mathematical statement of the problem is

$$J=\int_0^T \sum_n \left[c_{xn}^{+} X_n(t) + \sum_m c_{unm} u_{nm}(t) \right] dt \to \min,$$

subject to

$$\dot{X}_n(t)=\sum_m u_{nm}(t)-d_n(t), \quad X_n(0)=X_n(T), \quad X_n(t)\ge 0,$$

$$0 \le u_{nm}(t) \le U_{nm},$$

$$\sum_n \frac{u_{nm}(t)}{U_{nm}} \le 1.$$

■

Example 1.5

Problem $1|M|\, u(t)\in[0,U], X(T)=X_T \,\|\, L_u, L_x, T$ reads as follows: Given limited machine capacity, variable planning horizon and the number of product units to be accumulated by the end of the planning horizon, schedule N product types on a single machine in order to minimize the total linear cost of inventory, production and planning horizon. Since $X_n(t)$ does not decrease along the planning horizon, no backlog can occur. The mathematical statement of the problem is

$$J=\int_0^T \sum_n \left[c_{xn}^{+} X_n(t) + c_{un} u_n(t) \right] dt + c_T T \to \min,$$

subject to

$$\dot{X}_n(t)=u_n(t), \quad X_n(0)=0, \quad X_n(T)=X_{Tn},$$

$$0 \le u_n(t) \le U_n,$$

$$\sum_n \frac{u_n(t)}{U_n} \le 1.$$

■

2

MATHEMATICAL BACKGROUND

The methodology suggested in this book is based on three mathematical tools - optimal control, combinatorics and mathematical programming - which are traditionally related to separate areas of research and application. The methodology involves, first, investigating a continuous-time problem with the aid of the maximum principle and then reducing it to a discrete combinatorial or mathematical programming problem solvable in polynomial time. The following sections present selected combinatorics, the maximum principle and a constructive approach for integrating both mathematical tools.

2.1 SELECTED COMBINATORIAL METHODS

This section addresses three combinatorial methods which are extensively used in the scheduling algorithms developed in this book.

2.1.1 SORTING

Given an array $a[1,...,N]$ of N numbers, sorting is a procedure which arranges N elements of the array in non-decreasing order: $a[n+1] \geq a[n]$, n=1,...,N-1. A great variety of sorting algorithms has been developed. We present here the *mergesort* algorithm which is based on the divide-and-conquer approach. For the sake of simplicity, let N be a power of 2.

Algorithm 2.1 (*mergesort*)

Step 1:

Assign level s=1 and array size $h = 1$. Divide array $a[1,...,N]$ into N arrays $u_{ns}[1]$ such that $u_{ns}[1]$=$a[n]$, n=1,...,N.

Step 2:

Assign r=1, p=2 and q=1.

Step 3:

Do Subroutine $merge\left(h, u_{rs}[1,\ldots,h], u_{ps}[1,\ldots,h], u_{qs+1}[1,\ldots,2h]\right)$.

Step 4:

If $p<\frac{N}{h}$, then assign $r=r+2$, $p=p+2$ and $q=q+1$. Go to Step 3.

Otherwise, assign $s=s+1$, $h=2h$ and go to Step 5.

Step 5:

If $s<\log N+1$, then go to Step 2. Otherwise STOP.

Subroutine merge $\left(h, u_{rs}[1,\ldots,h], u_{ps}[1,\ldots,h], u_{qs+1}[1,\ldots,2h]\right)$ ***to Algorithm 2.1***

(i) Assign i=1, j=1, k=1.

(ii) If $u_{rs}[i] \le u_{ps}[j]$, then assign $u_{qs+1}[k]=u_{rs}[i]$, i=i+1;
otherwise, assign $u_{qs+1}[k]=u_{ps}[j]$, j=j+1;

(iii) If $i \le h$ and $j \le h$, then assign k=k+1 and go to (ii). Otherwise go to (iv).

(iv) If $i>h$, then assign $u_{qs+1}[k+l]=u_{ps}[j-1+l]$ for l=1,...,h+1-j. Otherwise, if $j>h$, then assign $u_{qs+1}[k+l]=u_{rs}[i-1+l]$ for l=1,...,h+1-i. STOP.

Discussion

Algorithm 2.1 forms log$N$+1 sorting levels where on level s=1,...,log$N$+1 it separately sorts $\frac{N}{2^{s-1}}$ arrays of length 2^{s-1}. To form the first level, s=1, Algorithm 2.1 divides array $a[1,\ldots,N]$ into N elementary arrays (Step 1) each of which is trivially sorted. Then it conquers (combines) two sorted arrays at level s into one sorted array at level s+1 by recursively merging them (Step 3). Once the recursion reaches level s=log$N$+1, array $a[1,\ldots,N]$ has been sorted and stored in array $u_{1\log N+1}$.

The number of merges at each level s equals $\frac{N}{2^s}$, while the number of operations of the merge subroutine is $O(2^s)$. Thus, the total complexity of merging at level s is $O(N)$. Since s runs through log$N$+1, Algorithm 2.1 requires $O(N\log N)$ operations. This complexity remains the same when Algorithm 2.1 is modified for N not a power of 2.

2.1.2 BREADTH-FIRST SEARCH

Breadth-first search is a procedure which traverses nodes of a graph in some predetermined order starting from a source node. Formally, *graph* G is a pair (E,A), where E is a set of *nodes* and A is a set of *arcs* that pair-wise connect the nodes. If each arc has a direction associated with it, the graph is called a *directed graph*. Given a graph, if its arcs have values associated with them, the values are called weights and the graph is called a *weighted graph*. In many applications, weights represent distances, therefore, we talk of a path from one node to another. In a directed graph, a *path* is a sequence of nodes such that there is an arc from each node to its *successor*. The *length* of a path in a weighted graph is the sum of the weights (distances) on the path.

Clearly, there can be a great variety of techniques to visit all nodes of a graph. The breadth-first search is the technique which prescribes visiting first the source node (level 0), then all its immediate successors (level 1). Proceeding to a successor of the next level can only be done after visiting all successors of the current one. Algorithm 2.2 below visits all nodes e_i, i=1,...,n of graph G starting from a given source node e_1.

Algorithm 2.2 (Coloring breadth-first search)

Step 1:

Color all nodes from E white. Initialize first-in, first-out queue Q. Visit, color gray and enqueue the source node e_1.

Step 2:

Dequeue the first node in Q. Visit all its white immediate successors, color them gray and enqueue them.

Step 3:

If Q is not empty, then go to Step 2, otherwise STOP.

Discussion

Algorithm 2.2 initially colors all nodes white. Then it visits all white nodes and re-colors them gray to avoid repeat visits to the nodes. It thus requires $O(n)$ time to visit all the nodes.

2.1.3 SHORTEST PATH

The Shortest Path Problem is an optimization problem. An instance of an optimization problem can have more than one candidate solution characterized by a value associated with it. In the case of the Shortest Path Problem, a candidate solution is a path in a graph from one node to another; the value is the length of the path; and the optimal value is the minimum of these lengths.

Let graph G=(E,A) be characterized by an $n \times n$ array W where n is the number of nodes and $W[i,j]$ is the weight of the arc from node e_i to node e_j, $e_i, e_j \in E$. If arc (e_i, e_j) does not exist, $W[i,j]$ can be set at ∞. Algorithm 2.3 below finds the shortest paths from a given source node e_1 to all the other nodes e_i, i=2,...,n of the graph.

Algorithm 2.3 (Dijkstra's algorithm)

Step 1:

Assign k=1, set $F = \{2,\ldots,n\}$, array $u[i] = 1$ and array $length[i] = W[1,i]$ for i=2,...,n.

Step 2:

Assign min= ∞ and i=2.

Step 3:

If $i \in F$ and $0 \le length[i] < min$, then assign $min = length[i]$ and $j = i$.

Step 4:

If $i < n$, then $i = i + 1$ and go to Step 3. Otherwise go to the next step.

Step 5:

Assign i=2.

Step 6:

If $length[j] + W[j,i] < length[i]$, then assign $length[i] = length[j] + W[j,i]$ and $u[i] = j$.

Step 7:

If $i < n$, then $i = i + 1$ and go to Step 6. Otherwise, go to the next step.

Step 8:

Remove j from F. If $k < n - 1$, then assign k=k+1 and go to Step 2. Otherwise, STOP.

Discussion

Algorithm 2.3 initializes a working set, F, of nodes whose shortest paths are to be determined. This set first contains all nodes. Steps 2-7 determine a shortest path to a node from the working set. At Step 8 this node is removed from the working set by assigning a negative value to the path length to this node from the initial one. Step 6 ensures the path is updated each time when a shorter arc is selected. Array $u[i]$ gives the path and array $length[i]$ gives the length of the path.

The complexity of Algorithm 2.3 is due to two loops made up by Steps 3-4 and Steps 6-7 respectively. Each loop is carried out n-1 times. In addition, Steps 2-7 make up an external loop which repeats n-1 times. As a result, the complexity of the algorithm is $O(n^2)$.

2.2 MAXIMUM PRINCIPLE

In optimal control theory, the maximum principle is an instrument which allows analytical investigation of an optimal solution for a large class of dynamic problems with complex constraints. The maximum principle provides the necessary conditions for identifying whether a solution is optimal or not. This facilitates understanding of the system's behavior and construction of exact combinatorial and/or approximate numerical methods. Sub-sections 2.2.1 and 2.2.2 present the canonical formulations of an optimal control problem for fixed and variable planning horizons respectively and corresponding necessary optimality conditions. These conditions comprise the dual formulation of the problems which are derived from the maximum principle. The methodology of integrating maximum principle analysis, combinatorial algorithms and numerical methods is discussed in Section 2.4.

2.2.1 FIXED PLANNING HORIZON

Statement of an Optimal Control Problem

We address a continuous-time dynamic system whose behavior is described by a set of state differential equations

$$\dot{X}_i(t) = f_i(u_1(t),\ldots,u_J(t),t), \quad i=1,\ldots,I, \tag{2.1}$$

where $X_i(t)$ is a state variable, $i=1,\ldots,I$ and $u_j(t)$ is a control variable, $j=1,\ldots,J$.

A number of constraints restricts the behavior of the system. Specifically, the area of admissible controls is bounded

$$g_l(u_1(t),\ldots,u_J(t)) \le 0, \quad l=1,\ldots,L. \tag{2.2}$$

Additionally, boundary constraints may be involved to restrict the admissible area of the initial and final states of the system

$$p_k(X_1(0),\ldots,X_I(0),X_1(T),\ldots,X_I(T)) = 0, \quad k=1,\ldots,K. \tag{2.3}$$

The objective is to minimize an integral measure of the system's behavior along a given planning horizon $[0,T]$

$$\int_0^T \tilde{J}(X_1(t),\ldots,X_I(t),u_1(t),\ldots,u_J(t),t)dt \Rightarrow \min. \tag{2.4}$$

From this point on, we assume that the state variables $X_i(t)$, $t \in [0,T]$ are continuous functions, and that controls $u_j(t)$ are piece-wise continuous functions with a finite number of jumps. Function $\tilde{J}$ is assumed to be convex. All functions defining the system dynamics, constraints and objective, f, g, $\tilde{J}$, p are assumed to be continuous and piece-wise continuously differentiable in arguments X and u. With respect to argument t, the functions are allowed to be discontinuous.

Dual Formulation

According to the maximum principle, the optimal control strategy is achieved by maximizing for each t, the Hamiltonian

$$H(t)=\sum_i \psi_i(t)f_i(X_1(t),...,X_I(t),u_1(t),...,u_J(t),t)-\tilde{J}(X_1(t),...,X_I(t),u_1(t),...,u_J(t),t)\,, \quad (2.5)$$

as a function of controls $u_j(t)$, j=1,...,J, where $\psi_i(t)$, i=1,...,I are continuous co-state variables which satisfy the following dual (co-state) differential equations

$$\dot{\psi}_i(t)=-\frac{\partial H(\psi_1(t),...,\psi_I(t),X_1(t),...,X_I(t),u_1(t),...,u_J(t),t)}{\partial X_i(t)}, \quad i=1,...,I. \quad (2.6)$$

The boundary conditions for equations (2.6) are

$$\psi_i(0)=\sum_k \rho_k \frac{\partial\varphi_k}{\partial X_i(0)}, \quad \psi_i(T)=-\sum_k \rho_k \frac{\partial\varphi_k}{\partial X_i(T)}, \quad i=1,...,I. \quad (2.7)$$

where multipliers ρ_k, k=1,...,K are arbitrary numbers.

Note that the Hamiltonian is not required to be always differentiable with respect to $X_i(t)$. At a point, $X_i^*(t)$, where the derivative does not exist, the co-state equation changes to involve the sub-gradient of the Hamiltonian

$$\dot{\psi}_i(t)\in-\partial H(X_i^*(t))=[\alpha,\beta]\,, \quad (2.8)$$

where

$$\alpha=\lim_{\substack{X_i\to X_i^*\\ X_i<X_i^*}}-\frac{H(X_i(t))-H(X_i^*(t))}{X_i(t)-X_i^*(t)} \text{ and } \beta=\lim_{\substack{X_i\to X_i^*\\ X_i>X_i^*}}-\frac{H(X_i(t))-H(X_i^*(t))}{X_i(t)-X_i^*(t)}. \quad (2.9)$$

Example 2.1

The dual formulation for the problem $1|N|\,u(t)\in[0,\infty]\,|d(t),T|\,L_u,L_x$ stated in Example 1.1 is as follows

$$H(t)=\sum_n \psi_n(t)\big(u_n(t)-d_n(t)\big)-\sum_n \big(c^+_{xn}X^+_n(t)+c^-_{xn}X^-_n(t)+c_{un}u_n(t)\big)\to \max$$

subject to $u_n(t)\geq 0$.

The co-state equation decomposes into three cases determined by (2.6), (2.8) and (2.9) which are due to point $X_n(t)=0$ where the Hamiltonian is not differentiable

$$\dot{\psi}_n(t)=\begin{cases} c^+_{xn}, \text{ if } X_n(t)>0, \\ -c^-_{xn}, \text{ if } X_n(t)<0, \\ a,\ a\in[-c^-_{xn}, c^+_{xn}], \text{ if } X_n(t)=0. \end{cases} \qquad \psi_n(T)=0.$$

■

Dual Formulation for Problems with State Constraints

If problem (2.1)-(2.4) has, in addition, a number of state constraints

$$h_m(X_1(t),\ldots,X_I(t))\leq 0, \quad m=1,\ldots,M, \tag{2.10}$$

the co-state equation changes as

$$d\psi_i(t)=-\frac{\partial H(\psi_1(t),\ldots,\psi_I(t),X_1(t),\ldots,X_I(t),u_1(t),\ldots,u_J(t),t)}{\partial X_i(t)}dt+\sum_m \frac{\partial h_m(X_1(t),\ldots,X_I(t))}{\partial X_i(t)}d\mu_m(t) \tag{2.11}$$

where measure-functions $d\mu_m(t)\geq 0$ are Lagrange multipliers corresponding to the state constraints. The measure-functions must satisfy the complementary slackness

$$\int_0^T h_m(X_1(t),\ldots,X_I(t))d\mu_m(t)=0, \qquad m=1,\ldots,M. \tag{2.12}$$

2.2.2 VARIABLE PLANNING HORIZON

Statement of an Optimal Control Problem

In this section we address a problem with a planning horizon not fixed, that is, T is a decision variable. The problem is characterized by the same constraints (2.1)-(2.3) while the objective is to minimize an integral measure of the system's behavior along a variable planning horizon $[0,T]$ together with a measure of the horizon itself, $\widetilde{T}(T)$,

$$\tilde{I}(T)+\int_0^T \tilde{J}(X_1(t),\ldots,X_I(t),u_1(t),\ldots,u_J(t),t)dt \Rightarrow \min, \tag{2.13}$$

where $\tilde{I}(T)$ is the cost of the planning horizon.

Dual Formulation

Optimality conditions (2.5)-(2.9) remain the same for a problem with a variable planning horizon. However, this variability induces the additional constraint,

$$H(t)=\frac{\partial \tilde{I}(T)}{\partial T}, \quad t\in[0,T]. \tag{2.14}$$

Example 2.2

The dual formulation for the problem $1|N|u(t)\in[0,U], X(T)=X_T \| L_u, L_x, T$ stated in Example 1.5 is as follows

$$H(t)=\sum_n \psi_n(t)u_n(t)-\sum_n \left(c_{xn}^+ X_n(t)+c_{un}u_n(t)\right)\rightarrow \max$$

subject to $\quad 0\le u_n(t)\le U_n, \quad \sum_n \frac{u_n(t)}{U_n}\le 1.$

The co-state equation takes the following form

$$\dot{\psi}_n(t)=c_{xn}^+.$$

According to (2.7), there are no boundary conditions for the co-state equation. The additional condition (2.14), which is due to the variable planning horizon, requires the Hamiltonian to be equal identically to the cost coefficient of the planning horizon

$$H(t)=c_T, \quad t\in[0,T].$$

■

2.3 LINEAR DIFFERENTIAL EQUATIONS

The maximum principle presented in the previous section reduces a dynamic optimization problem to a system of state and co-state differential equations with boundary conditions. Therefore, the possibility of solving differential equations is critical for tractability of the entire optimization problem. Below, we show a general solution to a system of linear ordinary differential equations and two special cases, frequently encountered throughout this book, which have closed-form solutions.

2.3.1 ANALYTICAL SOLUTION

Consider a system of linear differential equations in unknown functions $Y_i(t)$, i=1,...,n, with initial boundary conditions

$$\frac{d}{dt}\begin{vmatrix} Y_1(t) \\ \vdots \\ Y_n(t) \end{vmatrix} = A\begin{vmatrix} Y_1(t) \\ \vdots \\ Y_n(t) \end{vmatrix} + \begin{vmatrix} b_1(t) \\ \vdots \\ b_n(t) \end{vmatrix}, \quad Y_i(0) = Y_{i0}, \quad i=1,\ldots,n, \tag{2.15}$$

where $A = \|a_{ik}\|_1^n$ is $n \times n$ matrix of constant coefficients and $b_i(t)$ are given functions of time.

To solve system (2.15), let us first introduce new variables $Z_i(t)$, i=1,...,n, as follows

$$\begin{vmatrix} Z_1(t) \\ \vdots \\ Z_n(t) \end{vmatrix} = e^{-At}\begin{vmatrix} Y_1(t) \\ \vdots \\ Y_n(t) \end{vmatrix}, \tag{2.16}$$

where $e^{At} = \sum_{i=0}^{\infty} \frac{t^i}{i!} A^i$ is a matrix exponent. By substituting (2.16) into (2.15), we find that

$$\frac{d}{dt}\begin{vmatrix} Z_1(t) \\ \vdots \\ Z_n(t) \end{vmatrix} = e^{-At}\begin{vmatrix} b_1(t) \\ \vdots \\ b_n(t) \end{vmatrix}. \tag{2.17}$$

Thus,

$$\begin{vmatrix} Z_1(t) \\ \vdots \\ Z_n(t) \end{vmatrix} = \begin{vmatrix} c_1 \\ \vdots \\ c_n \end{vmatrix} + \int_0^t e^{-A\tau}\begin{vmatrix} b_1(\tau) \\ \vdots \\ b_n(\tau) \end{vmatrix} d\tau, \tag{2.18}$$

where c_i, i=1,...,n are constants determined by the initial boundary conditions.

By changing the variables back, we obtain

$$\begin{vmatrix} Y_1(t) \\ \vdots \\ Y_n(t) \end{vmatrix} = e^{At} \begin{vmatrix} c_1 \\ \vdots \\ c_n \end{vmatrix} + \int_0^t e^{A(t-\tau)} \begin{vmatrix} b_1(\tau) \\ \vdots \\ b_n(\tau) \end{vmatrix} d\tau .$$

By setting t=0, we find that $c_i = Y_{i0}$, i=1,...,n. It implies that the solution of (2.15) is

$$\begin{vmatrix} Y_1(t) \\ \vdots \\ Y_n(t) \end{vmatrix} = e^{At} \begin{vmatrix} Y_{10} \\ \vdots \\ Y_{n0} \end{vmatrix} + \int_0^t e^{A(t-\tau)} \begin{vmatrix} b_1(\tau) \\ \vdots \\ b_n(\tau) \end{vmatrix} d\tau . \qquad (2.19)$$

In what follows we present three examples of how to exactly solve first- and second order linear differential equations and how to derive a numerical approximation of a solution to an n-order linear differential equation.

Example 2.3
Consider a first-order linear differential equation with a k-order polynomial $b(t)$

$$\dot{Y}(t) = aY(t) + \sum_{i=0}^{k} b_i t^i , \quad Y(0) = Y_0 .$$

According to (2.19), the solution takes the following form:

$$Y(t) = e^{at} Y_0 + \int_0^t e^{a(t-\tau)} \sum_{i=0}^{k} b_i \tau^i d\tau = \left(Y_0 - B_0\right) e^{at} + \sum_{i=0}^{k} B_i t^i ,$$

where constants B_i are recursively calculated as

$$B_k = -\frac{b_k}{a}, \; B_i = \frac{(i+1)B_{i+1} - b_{i+1}}{a}, \; i=0,\dots,k-1.$$

■

Example 2.4
Consider a second-order linear differential equation with a k-order polynomial $b(t)$

$$\ddot{Y}(t) = a_1 Y(t) + a_2 \dot{Y}(t) + \sum_{i=0}^{k} b_i t^i , \quad Y(0) = Y_{10}, \; \dot{Y}(0) = Y_{20} . \qquad (2.20)$$

By denoting $Y_1(t)=Y(t)$ and $Y_2(t)=\dot{Y}(t)$, we reduce (2.20) to a two-dimensional case of (2.15)

$$\frac{d}{dt}\begin{vmatrix}Y_1(t)\\ Y_2(t)\end{vmatrix}=\begin{vmatrix}0 & 1\\ a_1 & a_2\end{vmatrix}\begin{vmatrix}Y_1(t)\\ Y_2(t)\end{vmatrix}+\begin{vmatrix}0\\ \sum_{i=0}^{k} b_i t^i\end{vmatrix},\ Y_i(0)=Y_{i0},\ i=1,2. \tag{2.21}$$

If $a_1 \neq 0$, then from (2.19), we find

$$Y(t)=Y_1(t)=C_1 e^{y_1 t}+C_2 e^{y_2 t}+\sum_{i=0}^{k} B_i t^i, \tag{2.22}$$

where y_1 and y_2 are the roots of the characteristic equation $y^2-a_2 y-a_1=0$, constants B_i are recursively calculated as

$$B_k=-\frac{b_k}{a_1},\ B_{k-1}=-\frac{k a_2 B_k+b_{k-1}}{a_1},$$

$$B_i=\frac{(i+2)(i+1)B_{i+2}-(i+1)a_2 B_{i+1}-b_i}{a_1},\ i=0,\ldots,k\text{-}2, \tag{2.23}$$

and constants C_1 and C_2 satisfy the following system of two linear algebraic equations

$$Y_{10}=C_1+C_2+B_0,\qquad Y_{20}=y_1 C_1+y_2 C_2+B_1. \tag{2.24}$$

If $a_1=0$, then by denoting $Z(t)=Y_2(t)$, we find a solution for $\dot{Y}(t)=Y_2(t)$ as described in Example 2.3. Then, $Y(t)=Y_1(t)$ is immediately obtained by integrating $Y_2(t)$.

■

Example 2.5

Consider a special case of the previous example with $a_2=0$, $a_1>0$ and $b(t)=b$

$$\ddot{Y}(t)=a_1 Y(t)+b,\quad Y(0)=Y_{10},\ \dot{Y}(0)=Y_{20}. \tag{2.25}$$

By substituting $a_2=0$ and setting $k=0$ into (2.22)-(2.24), the solution of Example 2.4 simplifies to

$$Y(t) = C_1 sh(\sqrt{a_1}\, t) + C_2 ch(\sqrt{a_1}\, t) + B, \tag{2.26}$$

where constants $B = -\dfrac{b}{a_1}$, $C_1 = \dfrac{Y_{20} - B}{\sqrt{a_1}}$ and $C_2 = Y_{10} - B$, and functions $sh(x) = \dfrac{e^x - e^{-x}}{2}$ and $ch(x) = \dfrac{e^x + e^{-x}}{2}$.

■

2.3.2 NUMERICAL SOLUTION

Consider a system of non-linear differential equations in unknown functions $Y_i(t)$, i=1,...,n, with initial boundary conditions

$$\dot{Y}_i(t) = f_i(Y_1(t),\ldots,Y_n(t),t),\quad 0 \le t \le T,\quad Y_i(0) = Y_{i0},\quad i=1,\ldots,n, \tag{2.27}$$

where $f_i(Y_1(t),\ldots,Y_n(t),t)$ is a differentiable function in its arguments. Due to the non-linearity of functions f_i, the analytical approach discussed in the previous section cannot generally be applied to solving equations (2.27). Therefore, a numerical approach for determining functions $Y_i(t)$ is considered here. It is based on integrating equation (2.27) from left to right over a time grid which is generated by choosing *mesh points* $t_0, t_1, \ldots, t_K$ (K - a positive integer) within the interval [0,T]. The mesh points, for example, can be equally distributed throughout the interval with distance $\varepsilon = t_{k+1} - t_k = T / K$.

There are various schemes for approximating the solution to non-linear differential equations on a grid. Simple, yet sufficiently accurate schemes go back to Euler and are based on the first two terms of the Taylor series. Indeed, the increment of $Y_i(t)$ between the adjacent mesh points is proportional to the derivative of $Y_i(t)$

$$Y_i(t_{k+1}) = Y_i(t_k) + \dot{Y}_i(t_k)\varepsilon + \ddot{Y}_i(\xi_k)\frac{\varepsilon^2}{2} =$$

$$= Y_i(t_k) + f_i(Y_1(t_k),\ldots,Y_n(t_k),t_k)\varepsilon + \ddot{Y}_i(\xi_k)\frac{\varepsilon^2}{2},\quad t_k < \xi_k < t_{k+1}.$$

If we delete the error (third) term, then the approximate solution $\hat{Y}_i(t_k) \approx Y_i(t_k)$ can be calculated as

$$\hat{Y}_i(t_{k+1}) = \hat{Y}_i(t_k) + f_i(\hat{Y}_1(t_k),\ldots,\hat{Y}_n(t_k),t_k)\varepsilon. \tag{2.28}$$

The last expression presents the Euler method and allows the approximate solution to be built recursively, starting from the initial condition $\hat{Y}_i(0) = Y_{i0}$.

The method assumes that on an interval between subsequent mesh points, the curve $Y_i(t)$ is approximately a straight line with the slope of this line $f_i(Y_1(t_k),\ldots,Y_n(t_k),t_k)$ calculated at the beginning of that interval. Since such an assumption is usually very rough, it causes a low degree of accuracy in the approximation

$$\left|Y_i(t_k) - \hat{Y}_i(t_k)\right| \le kO(\varepsilon^2) \le \frac{T}{\varepsilon}O(\varepsilon^2) = O(T\varepsilon) \quad \text{for } k=1,\ldots,K.$$

The complexity of the described scheme is $O\left(\frac{n^2T}{\varepsilon}\right)$.

In practice, some less crude assumptions are usually made, so that more elements of the Taylor series are accounted for. The modified Euler method improves the accuracy by an order of magnitude $O(T\varepsilon^2)$, while the complexity remains the same, $O\left(\frac{n^2T}{\varepsilon}\right)$. The method estimates the slope twice, at the beginning and at the end of each mesh interval, and then approximates $Y_i(t_k)$ by averaging the two estimates (see Figure 2.1).

Algorithm 2.4 (Modified Euler's scheme)

Step 1:

Assign k=0 and $\hat{Y}_i(t_k) = Y_{i0}$ for i=1,...,n.

Step 2:

Calculate the first estimate $\overline{Y}_i(t_{k+1})$ by the straightforward Euler method

$$\overline{Y}_i(t_{k+1}) = \hat{Y}_i(t_k) + f_i(\hat{Y}_1(t_k),\ldots,\hat{Y}_n(t_k),t_k)\varepsilon. \tag{2.29}$$

Calculate the second estimate $\overline{\overline{Y}}_i(t_{k+1})$ by the slope at the new point

$$\overline{\overline{Y}}_i(t_{k+1}) = \hat{Y}_i(t_k) + f_i(\overline{Y}_1(t_{k+1}),\ldots,\overline{Y}_n(t_{k+1}),t_{k+1})\varepsilon. \tag{2.30}$$

Calculate the approximation of $Y_i(t_{k+1})$ as

$$\hat{Y}_i(t_{k+1}) = \frac{1}{2}(\overline{Y}_i(t_{k+1}) + \overline{\overline{Y}}_i(t_{k+1})). \tag{2.31}$$

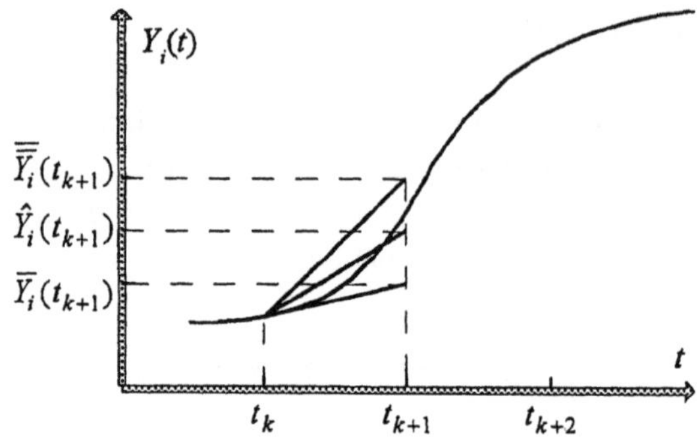

Figure 2.1. Graphical illustration of a step of the modified Euler method.

Step 3:

If $k < K-1$, then $k=k+1$ and go to Step 2.
Otherwise, STOP.

Discussion

To show that the scheme adopted by Algorithm 2.4 captures three terms of the Taylor expansion, thereby having accuracy $O(T\varepsilon^2)$, we substitute (2.29) and (2.30) in (2.31)

$$\hat{Y}_i(t_{k+1}) = \hat{Y}_i(t_k) + \frac{\varepsilon}{2}\left[f_i(\hat{Y}_1(t_k),\dots,\hat{Y}_n(t_k),t_k) + f_i(\overline{Y}_1(t_{k+1}),\dots,\overline{Y}_n(t_{k+1}),t_{k+1})\right] =$$

$$= \hat{Y}_i(t_k) + \frac{\varepsilon}{2}\left[2f_i(\hat{Y}_1(t_k),\dots,\hat{Y}_n(t_k),t_k) + \sum_{j=1}^{n}\frac{\partial f_i}{\partial Y_j}f_i(\hat{Y}_1(t_k),\dots,\hat{Y}_n(t_k),t_k)\varepsilon + \frac{\partial f_i}{\partial t}\varepsilon\right] +$$

$$+O\left(\varepsilon^3\right) = \hat{Y}_i(t_k) + \dot{\hat{Y}}_i(t_k)\varepsilon + \ddot{\hat{Y}}_i(t_k)\frac{\varepsilon^2}{2} + O\left(\varepsilon^3\right).$$

2.4 A CONSTRUCTIVE SOLUTION APPROACH

All scheduling problems defined by the classification described in Section 1.5 contain only linear constraints and either linear or convex performance measures. Therefore the maximum principle presents not only necessary, but also sufficient optimality conditions. In light of this fact, if a solution $(u(t), X(t), \psi(t))$ that we construct is feasible with respect to both primal and dual formulations, then it is optimal.

The constructive solution approach adopted in this book consists of the following steps:

- deriving properties of the optimal solution by analyzing conditions of the maximum principle
- constructing solutions on the basis of the derived properties by solving state and co-state differential equations
- verifying feasibility of the constructed solutions with respect to both primal and dual formulations.

Maximization of the Hamiltonian, subject to the control constraints, often results in partitioning the space of state and co-state variables $(X(t), \psi(t))$ into a number of regions, each of which is associated with certain control rules (regimes). Such rules as well as their possible sequences along the planning horizon constitute the most vital properties which are derived from the maximum principle and significantly reduce the solution space.

Once the properties have been derived, one can either guess a feasible solution because the solution space is reduced to a few points or construct it by a combinatorial search among legitimate sequences of the control rules. No matter whether a solution is readily guessed or iteratively searched by a combinatorial algorithm, the process is stopped if the constructed solution is verified to be feasible with respect to both primal and dual formulations.

In this book we consider only cases for which fast combinatorial algorithms can be developed to obtain either exact optimal solution or its approximation with a given accuracy.

The following example illustrates the constructive approach for a simple problem. In this example, we first determine two optimal production regimes, then guess a solution based on these regimes and prove its feasibility, and thus its optimality. This is accomplished by two different ways widely used throughout the book. One way is to check explicitly the feasibility of the solution by verifying the constraints of the primal and dual problems. The other way is to prove the feasibility of the solution implicitly by checking that no other solution can be feasible.

Example 2.6

Consider the $1|1|u(t)\in[0,\infty]|d(t),T|\,L_x$ problem which is stated as follows

$$J=\int_0^T\left[c_x^+X^+(t)+c_x^-X^-(t)\right]dt\rightarrow \min,$$

$$X^+(t)=\max\{X(t),0\},\ X^-(t)=\max\{-X(t),0\}$$

subject to

$$\dot{X}(t) = u(t) - d(t), \qquad X(0) = 0,$$

$$u(t) \geq 0.$$

The dual formulation of the problem is as follows. The Hamiltonian, H, resulting from the maximum principle is

$$H(t) = -\left[c_x^+ X^+(t) + c_x^- X^-(t)\right] + \psi(t)\left[u(t) - d(t)\right].$$

The co-state variable, $\psi(t)$, satisfies the following co-state (dual) differential equation with the corresponding transversality (boundary) condition

$$\dot{\psi}(t) = \begin{cases} c_x^+, \text{ if } X(t) > 0; \\ -c_x^-, \text{ if } X(t) < 0; \\ a, \ a \in [-c_x^-, c_x^+], \text{ if } X(t) = 0; \end{cases} \qquad \psi(T) = 0.$$

According to the maximum principle, the Hamiltonian is the objective function of the dual problem. The Hamiltonian is maximized at every point of time with respect to the admissible control, $u(t)$. This implies that since the control is unlimited from above, $\frac{\partial H(t)}{\partial u(t)} = \psi(t)$ must be non-positive. Therefore, if $\psi(t) = 0$ holds at an interval of time, then the optimal control $u(t)$ is arbitrary (production regime). Otherwise, $\psi(t) < 0$ and no control should be undertaken at this interval, i.e. $u(t)=0$ (no-production regime).

We now guess a solution based on the production regime only. It is to exactly trace the demand along the entire planning horizon, $u(t) = d(t)$, $0 \leq t \leq T$.

Next, we prove with two ways that the solution is optimal. The first way is the explicit feasibility check. Since the solution consists of only the production regime, $\psi(t) \equiv 0$. By differentiating the last equality, we obtain $\dot{\psi}(t) \equiv 0$.

From the co-state equation, it immediately follows that $\dot{\psi}(t) \equiv 0$ can be met only if $X(t) \equiv 0$. By differentiating the last equality and substituting there the state equation, we finally obtain $u(t) \equiv d(t)$. Thus, solution $u(t) \equiv d(t)$, $\psi(t) \equiv 0$, $X(t) \equiv 0$ satisfies all constraints of the primal and dual problems, i.e. it is feasible and thus optimal.

The second way to prove the feasibility of the solution is to show that the no-production regime cannot be a part of the optimal solution. The proof is by contradiction. Let the production regime $u(t) = d(t)$ be followed by the no-production regime $u(t) = 0$ at interval $[t_1, t_2] \subseteq [0, T]$. Then, from the state and co-state equations, $X(t) < X(t_1) = 0$ and $\psi(t) < \psi(t_1) = 0$ for $t \in (t_1, t_2]$. Consequently, no-production is continued until the end of the planning horizon and $\psi(T) < 0$ that contradicts the co-state boundary condition $\psi(T) = 0$. Let the no-production regime $u(t) = 0$ at interval $[t_1, t_2] \subseteq [0, T]$ be followed by the production regime $u(t) = d(t)$. Then, again from the state and co-state equations, $X(t) > X(t_2) = 0$ and $\psi(t) < \psi(t_2) = 0$ for $t \in (t_1, t_2]$. Consequently, no-production initiates at t=0 with $X(0) > 0$ that contradicts the state boundary condition $X(0) = 0$.

■

SELECTED BIBLIOGRAPHY

Arrow, K.J. and S. Karlin, 1958, "Smooth Production Plans" in *Studies in the Mathematical Theory of Inventory and Production*, Stanford University Press, pp. 70-85.

Blazewicz, J., K.H. Ecker, E. Pesch, G. Schmidt and J. Weglarz, 1996, *Scheduling Computer and Manufacturing Processes*, Springer-Verlag.

Bollen, N.P.B., 1999, "Real Options and Product Life Cycles", *Management Science*, 45(5), pp. 670-684.

Bryson, A.E. and J.Y-C. Ho, 1975, *Applied Optimal Control*, Hemisphere Publishing Corporation, Washington.

Burden, R.L. and J.D. Faires, 1989, *Numerical Analysis*, PWS-KENT, Boston.

Collatz, L., 1986, *Differential Equations: an Introduction with Applications*, Wiley, Chichester.

Dubovitsky, A.Y. and A.A. Milyutin, 1981, "Theory of the Maximum Principle" in *Methods of Theory of Extremum Problems in Economy*, V.L. Levin, ed., Nauka, Moscow, 6-47 (in Russian).

Gershwin, S., 1994, *Manufacturing Systems Engineering*, Prentice Hall, Englewood Cliffs, NJ.

Gohberg, I, P. Lancaster and L. Rodman, 1982, *Matrix Polynomials*, Academic Press, NY.

Graves S.C. *et al.* (Eds.), 1993, *Handbooks in OR and MS*, Elsevier, Amsterdam.

Hartl, R.F., S.P. Sethi and R.G.Vickson, 1995, "A survey of the maximum principles for optimal control problems with state constraints", *SIAM Review*, 37(2), 181-218.

Holt, C.C., F. Modigliani, J.F. Muth, and H.A. Simon, 1960, *Planning Production, Inventories and Work Force*, Prentice-Hall, Englewood Cliffs, NJ.

Maimon, O., E. Khmelnitsky and K. Kogan, 1998, *Optimal Flow Control in Manufacturing Systems: Production Planning and Scheduling*, Kluwer Academic Publishers, Boston.

Milyutin, A.A., N. Osmolovsky, S. Chukanov and A. Ilyutovich, 1993, *Optimal Control in Linear Systems*, Nauka, Moscow (in Russian).

Neapolitan, R. and K. Naimipour, 1996, *Foundations of Algorithms*, Heath, Lexington, MA.

Ostwald, P.F., 1974, *Cost Estimating for Engineering and Management*, Prentice Hall, Englewood Cliffs, NJ.

Pinedo, M., 1995, *Scheduling: Theory, Algorithms and Systems*, Prentice Hall, Englewood Cliffs, NJ.

Sethi, S.P. and Q., Zhang, 1995, *Hierarchical Decision Making in Stochastic Manufacturing Systems*, Birkhauser, Boston, MA.

Sethi, S.P., M.I. Taksar and Q. Zhang, 1992, "Capacity and Production Decisions in Stochastic Manufacturing Systems: an Asymptotic Optimal Hierarchical Approach", *Production and Operations Management*, 1(4), pp. 367-392.

Van Trees H.L., 1968, *Detection, Estimation and Modulation Theory*, Part I, John Wiley&Sons.

Wagner, H.M., and T.M. Whitin, 1958, "A dynamic version of the economic lot size model", *Management Science*, 5, pp. 89-96.

PART 2

ONE-MACHINE, ONE-PRODUCT-TYPE SCHEDULING

3

UNLIMITED CAPACITY PROBLEMS

3.1 DEMAND OVER FINITE PLANNING HORIZON

This section addresses the scheduling problems characterized by demand as a function of time specified over a finite planning horizon. The problems differ by inventory-related constraints that allow or disallow backlogs and by the form of the objective function.

3.1.1 LINEAR PERFORMANCE MEASURE

Backlogs

The first problem to be considered is one-machine, one-product-type scheduling with backlogs to minimize the total linear cost of inventory and production, that is, the $1|1|u(t)\in[0,\infty]|d(t),T|\, L_u, L_x$ problem. The problem is stated as follows

$$J=\int_0^T\left[c_x^+X^+(t)+c_x^-X^-(t)+c_u u(t)\right]dt\to\min, \tag{3.1}$$

$$X^+(t)=\max\{X(t),0\},\ X^-(t)=\max\{-X(t),0\}$$

subject to

$$\dot{X}(t)=u(t)-d(t),\qquad X(0)=0, \tag{3.2}$$

$$u(t)\geq 0. \tag{3.3}$$

Dual Formulation

Problem (3.1)-(3.3) is stated in the canonical form of optimal control, where $u(t)$ is a control variable and $X(t)$ is a state variable whose behavior is described by the state differential equation (3.2). The Hamiltonian, H, resulting from the maximum principle is as follows

$$H(t) = -\left[c_x^+ X^+(t) + c_x^- X^-(t) + c_u u(t)\right] + \psi(t)\left[u(t) - d(t)\right]. \tag{3.4}$$

The co-state variable, $\psi(t)$ in (3.4), measures sensitivity of the performance criterion at each time t to changes in inventory $X(t)$. The co-state variable also satisfies the following co-state differential equation with its transversality (boundary) condition

$$\dot{\psi}(t) = \begin{cases} c_x^+, \text{ if } X(t) > 0; \\ -c_x^-, \text{ if } X(t) < 0; \\ a, \ a \in [-c_x^-, c_x^+], \text{ if } X(t) = 0; \end{cases} \qquad \psi(T) = 0. \tag{3.5}$$

According to the maximum principle, the Hamiltonian is the objective function of the dual problem and is maximized at every point of time with respect to admissible control $u(t)$. This implies that given the control is unlimited from above,

$$\frac{\partial H(t)}{\partial u(t)} = -c_u + \psi(t) \le 0. \tag{3.6}$$

Therefore, if $\psi(t) = c_u$ holds at an interval of time, then any non-negative control effort $u(t)$ can be exercised. Otherwise, $\psi(t) < c_u$, and no control should be undertaken at this interval, i.e. $u(t)=0$.

Algorithm 3.1

Step 1:

Calculate the unit backlog cost over the planning horizon, $c_x^- T$.

If $c_u > c_x^- T$, then solution $u(t)=0$ is optimal over the entire planning horizon, STOP; otherwise go to the next step.

Step 2:

Calculate $t_1 = T - \dfrac{c_u}{c_x^-}$ and set the optimal solution as

$u(t) = d(t)$ for $0 \le t \le t_1$,

$u(t)=0$ for $t_1 < t \le T$; STOP.

Discussion

From Algorithm 3.1, one can observe that there can be only two types of optimal control regimes: no-production $u(t)$=0 and fractional production $u(t)$=$d(t)$. Specifically, unless the unit cost of control effort c_u is higher than the unit backlog cost over the planning horizon $c_x^- T$, the production at a time point must be carried out as much as the demand requires at this point. This production control regime is exercised always from the very beginning of the planning horizon; the less costly it is, the longer the portion of the planning horizon on which the production tracks the demand. It is interesting to point out that there is a certain point of time in the schedule, t_1, beyond which it is not worthwhile to exercise any additional control, since the cumulative effect of the additional backlog will not be very significant. In other words, towards the end of the planning horizon, it is not worthwhile trying to bring production back to track.

The complexity of Algorithm 3.1 is $O(1)$ and its correctness is easily verified by solving the dual formulation in the following lemma.

LEMMA 3.1. (Correctness of Algorithm 3.1) Given problem (3.1)-(3.3), the optimal control effort is determined as follows

- if $c_u \le c_x^- T$, then $u(t) = d(t)$, for $0 \le t \le t_1$, $t_1 = T - \dfrac{c_u}{c_x^-}$ and $u(t)$=0 for $t_1 < t \le T$
- otherwise, $u(t)$=0 for $0 \le t \le T$.

PROOF. By considering only the control dependent term of the Hamiltonian (3.4), we obtain

$$H_u(t) = u(t)\big(\psi(t) - c_u\big). \tag{3.7}$$

Expression (3.7) implies that if $\psi(t) < c_u$ then $u(t)$=0, otherwise condition (3.6) must be satisfied as equality, that is, $\psi(t) = c_u$. By differentiating this equality, we find

$$\dot{\psi}(t) = 0. \tag{3.8}$$

From (3.5) it immediately follows that equation (3.8) can be met only if $X(t)$=0. By differentiating the last equality and substituting the state equation (3.2), we finally obtain

$$u(t) = d(t). \tag{3.9}$$

Note that the optimal value of the control function is found under the conditions $\psi(t) = c_u$ and $X(t)$=0. However, the transversality condition in (3.5) requires $\psi(T) = 0$. This implies that there must be a switching point t_1 before the end of the planning horizon, T, so that the co-state variable ends up at zero, $u(t)$=0 and $X(t)$<0 for $t_1 < t \le T$

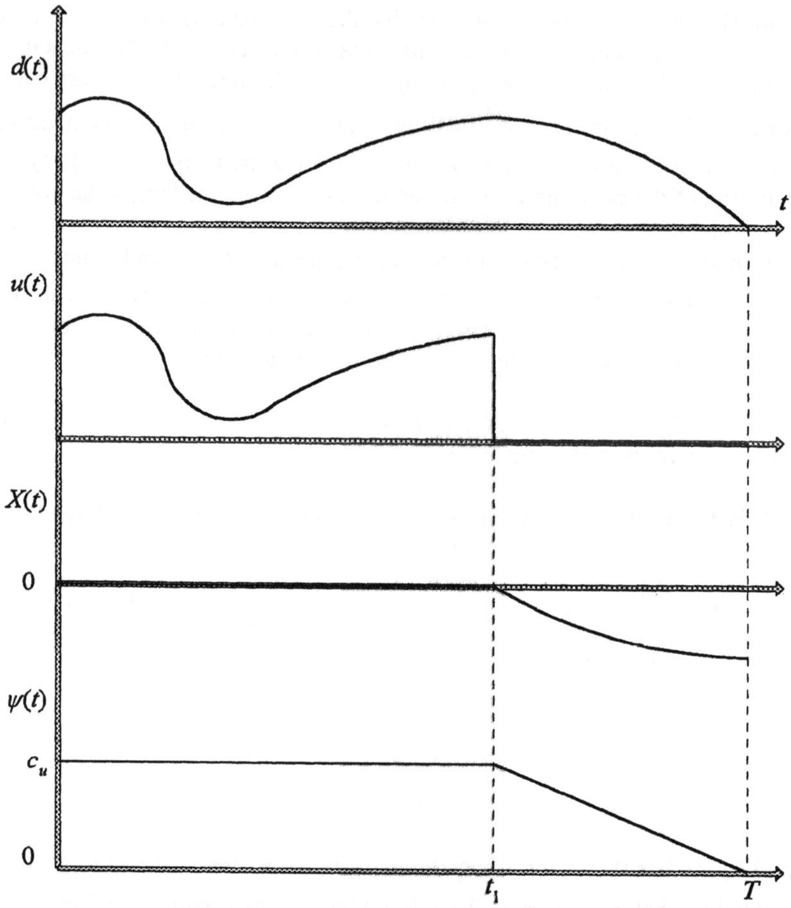

Figure 3.1. Demand $d(t)$, optimal control $u(t)$, state $X(t)$ and co-state $\psi(t)$ variables for the $1|1|u(t)\in[0,\infty]|d(t),T|\,L_u, L_x$ problem.

(see Figure 3.1). By integrating the corresponding condition from (3.5), we thus determine the behavior of the co-state variable

$\psi(t)=c_u$, $0\le t\le t_1$; $\psi(t)=c_x^-(T-t)$, $t_1\le t\le T$. To find the unknown switching time point, we solve the following equation for t_1

$$\psi(t_1)=c_x^-(T-t_1)=c_u,$$

from where $t_1 = T - \frac{c_u}{c_x^-}$. Note that $t_1 \geq 0$ if $c_u \leq c_x^- T$. However, if this is not the case, then $\psi(t) < c_u$, $\dot{\psi}(t) < 0$, $X(t) \leq 0$ and, as stated in the lemma, $u(t)$=0 over the entire planning horizon. Thus, the solution we found is feasible with respect to the state and co-state equations and satisfies condition (3.6).

Finally, according to the maximum principle, a feasible solution of the system of state and co-state equations, which satisfies the optimality condition (3.6) is the optimal solution. Moreover, due to the fact, that the objective function of our problem is convex and the constraints are linear, the problem is unimodal. Thus, the maximum principle presents not only necessary, but also sufficient optimality conditions and the found solution is globally optimal.
■

Special Cases of Zero Costs

First, consider the zero production cost, $c_u = 0$, i.e. the $1|1|u(t) \in [0,\infty]|d(t),T|L_x$ problem. Algorithm 3.1 in this case simplifies. Indeed, no production cost implies that Step 1 of Algorithm 3.1 will never indicate the no-production regime, and Step 2 will always result in $t_1 = T$. As that, the fractional production $u(t) = d(t)$ is optimal along the entire planning horizon.

Secondly, consider the zero inventory cost, i.e. the $1|1|u(t) \in [0,\infty]|d(t),T|L_u, L_x, c_x^+ = 0$ problem. Since the optimal solution is characterized by no inventory (see Lemma 3.1), c_x^+ does not enter any expression of Algorithm 3.1. As a result, the optimal solution will not change and no modification to Algorithm 3.1 is required.

No Backlog

For the case when backlogs are disallowed, i.e. for the $1|1|u(t) \in [0,\infty], X(t) \geq 0|d(t),T|L_u, L_x$ problem, the solution algorithm can be obtained by an easy modification of Algorithm 3.1. Indeed, no backlog means having c_x^- very large, that is, Step 1 of Algorithm 3.1 will never indicate the no-production regime, and Step 2 will always result in $t_1 = T$. This implies that the fractional production $u(t) = d(t)$ is optimal over the entire planning horizon.

3.1.2 NON-LINEAR INVENTORY COST

Next we consider a linear production cost function and non-linear inventory and backlog costs, which will be approximated by a quadratic function. It will be shown below that the non-linearity affects neither the optimal control value nor the fact that there can be at most one switching point at which the control function changes over. However, the non-linearity does affect the timing for the optimal control value, which now depends on change in demand along the planning horizon rather than only production and backlog costs.

Backlogs

The problem to be considered is one-machine, one-product-type scheduling with backlogs to minimize quadratic inventory and linear production costs, that is, the $1|1|u(t)\in[0,\infty]|d(t),T|\,L_u, Q_x$ problem which is stated as follows

$$J=\int_0^T\left[c_x^+X^+(t)^2+c_x^-X^-(t)^2+c_u u(t)\right]dt\rightarrow\min, \tag{3.10}$$

subject to (3.2) and (3.3).

Dual Formulation

The Hamiltonian, H, for problem (3.10), (3.2) and (3.3) is

$$H(t)=-\left[c_x^+X^+(t)^2+c_x^-X^-(t)^2+c_u u(t)\right]+\psi(t)\left[u(t)-d(t)\right]. \tag{3.11}$$

The co-state variable, $\psi(t)$ in (3.11), satisfies the following co-state differential equation with its transversality condition

$$\dot{\psi}(t)=\begin{cases}2c_x^+X(t), & \text{if } X(t)>0;\\ 2c_x^-X(t), & \text{if } X(t)\le 0;\end{cases}\qquad \psi(T)=0. \tag{3.12}$$

According to the maximum principle, if $\psi(t)=c_u$ holds at an interval of time, then any non-negative control effort $u(t)$ can be exercised. Otherwise, $\psi(t)<c_u$, and no control should be undertaken at this interval, i.e. $u(t)$=0.

Algorithm 3.2

Step 1:

Define the backlog cost related to the demand over the planning horizon as a function of t_1

$$f(t_1)=2c_x^-\int_{t_1}^{T}\int_{t_1}^{t}d(\tau)d\tau dt.$$

If equation $f(t_1)=c_u$ does not have a root falling within the planning horizon, $t_1\in[0,T]$, then solution $u(t)$=0 over the entire planning horizon is optimal, STOP; otherwise, calculate t_1 and go to the next step.

Step 2:

Set the optimal solution as
$u(t) = d(t)$ for $0 \le t \le t_1$, and $u(t)=0$ for $t_1 < t \le T$; STOP.

Discussion

Similar to Algorithm 3.1, Algorithm 3.2 operates with only two types of optimal control regimes: no-production and fractional production. Specifically, unless the unit cost of control effort c_u is higher than the backlog cost related to the demand over the planning horizon $2c_x^- \int_0^T \int_0^t d(\tau)d\tau dt$, the production at a time point must be carried out to the extent the demand requires at this point. This control is always exercised from the very beginning of the planning horizon. Again, as shown in Figure 3.1, there is a certain point of time in the schedule, t_1, beyond which it is not worthwhile to exercise any additional control, as the cumulative effect of the additional backlog will not be very significant.

The complexity of Algorithm 3.2 is $O(1)$ only if the root of equation $f(t_1) = c_u$ can be found analytically. This, of course, is not always the case. If demand is an arbitrary complex function of time, only a numerical approximation of the switching point t_1 is possible. However, if demand is a K-level, piece-wise constant function of time

$$d(t) = d_k \text{ for } t \in [t_{k-1}, t_k],\ k=1,...,K,\ t_0 = 0,\ t_K = T,$$

then the exact solution is always available in $O(K)$ time as provided by Subroutine to Algorithm 3.2 presented below. The correctness of Algorithm 3.2 for arbitrary demand is verified by solving the dual formulation in Lemma 3.2.

Subroutine to Algorithm 3.2

Step 1:

Set $f(t_K) = 0$ and $k=K+1$.

Step 2:

(i) Assign $k=k-1$. If $k \le 0$, then STOP; equation $f(t_1) = c_u$ does not have a root $t_1 \in [0, T]$. Otherwise go to (ii).

(ii) Calculate $f(t_{k-1}) = f(t_k) + c_x^- d_k (t_k - t_{k-1})(2T - t_k - t_{k-1})$.

(iii) If $f(t_{k-1}) < c_u$, then go to (ii);

otherwise, set $Discr = T^2 - t_k(2T - t_k) + \dfrac{c_u - f(t_k)}{c_x^- d_k}$ and

$t_1 = T - \sqrt{Discr}$; STOP.

Example 3.1
Consider the $1|1|u(t)\in[0,\infty]|d(t),T|L_u,Q_x$ problem with the following parameters: $T=10$ hr, c_u=9\$/unit, c_x^+=0.002\$/hr.unit², c_x^-=0.01\$/hr.unit² and the two-level, piece-wise constant demand

$$d(t)=\begin{cases}17\text{unit/hr, if } 0\le t\le 5,\\ 23\text{unit/hr, if } 5<t\le 10.\end{cases}$$

Step 1 of Algorithm 3.2 invokes Subroutine which calculates $f(10)$=0, $f(5)$=5.75, $f(0)$=18.5, *Discr*=44.12 and t_1=3.36hr.

Since the obtained value of t_1 is feasible, Step 2 sets the optimal solution as $u(t)=17$ unit/hr for $0\le t\le 3.36$ and $u(t)$=0 for $3.36<t\le 10$.
■

LEMMA 3.2. (Correctness of Algorithm 3.2) Given problem (3.10), (3.2) and (3.3), the optimal control is determined as follows

- if equation $c_u=2c_x^-\int_{t_1}^{T}\int_{t_1}^{t}d(\tau)d\tau dt$ has a root $t_1\in[0,T]$, then

$u(t)=d(t)$, for $0\le t\le t_1$, and $u(t)$=0 for $t_1<t\le T$;

- otherwise $u(t)$=0 for $0\le t\le T$.

PROOF. First note, that introducing the quadratic inventory cost does not change the control-dependent term of the Hamiltonian (3.7). Consequently, the optimality condition proven in Lemma 3.1 remains the same. That is, if $\psi(t)<c_u$, then $u(t)$=0, otherwise equation (3.6) must be satisfied, implying that $\psi(t)=c_u$. As is the case with Lemma 3.1, by differentiating the last equality, we obtain equation (3.8). Equations (3.12) and (3.8) can be simultaneously met only if $X(t)$=0. Thus, again, we have $u(t)=d(t)$, $X(t)$=0 for $0\le t\le t_1$ and $u(t)$=0, $X(t)$<0, for $t_1<t\le T$. Consequently, the single switching point t_1 can be found by integrating the corresponding condition from (3.12)

$$\psi(t)=c_u,\quad 0\le t\le t_1;\quad \psi(t)=-2c_x^-\int_{t}^{T}X(\tau)d\tau,\quad t_1\le t\le T. \tag{3.13}$$

By taking into account that $X(t)=\int_{t_1}^{t}\left(u(\tau)-d(\tau)\right)d\tau$, we finally find the algebraic equation for the unknown switching point, t_1,

$$\psi(t_1)=c_u=2c_x^-\int_{t_1}^{T}\int_{t_1}^{t}d(\tau)d\tau dt. \qquad (3.14)$$

Note that the right-hand side of equation (3.14) is a non-increasing function in t_1. Therefore, if there is no root of (3.14), then $\psi(0)<c_u$, $\dot{\psi}(t)<0$, $X(t)\leq 0$ and, as shown above, the feasible solution for the problem is $u(t)=0$ over the entire planning horizon.
■

Special Cases of Zero Costs

The problem with zero production cost, $1|1|u(t)\in[0,\infty]|d(t),T|Q_x$, is solved by the fractional production regime $u(t)=d(t)$ over the entire planning horizon. This follows from Step 1 of Algorithm 3.2 in which substituting $c_u=0$ results in $t_1=T$ regardless of the demand function. For the problem with zero inventory cost, $1|1|u(t)\in[0,\infty]|d(t),T|L_u,Q_x,c_x^+=0$, no modification to Algorithm 3.2 is required, since no inventory can occur on the optimal solution (see Lemma 3.2).

No Backlog

When backlogs are disallowed, i.e. for the $1|1|u(t)\in[0,\infty],X(t)\geq 0|d(t),T|L_u,Q_x$ problem, the solution algorithm can be obtained in a way similar to the case with the linear performance measure. Indeed, having c_x^- very large, the first step of Algorithm 3.2 will never indicate the no-production regime, i.e. $t_1=T$. This implies that the fractional production $u(t)=d(t)$ is optimal over the entire planning horizon.

3.1.3 NON-LINEAR PRODUCTION COST

In this section we consider the combination of a piece-wise, linear inventory/backlog cost function and a non-linear production cost function. Such cost function implies that additional control effort may become increasingly more expensive. This change complicates significantly the optimal behavior of the system. Unlike the previous cases, where the optimal control was re-switched only once and was characterized by either zero value (no-production regime) or fractional production, $u(t)=d(t)$, this time there can be different non-zero control functions and a large number of switching points.

Backlogs

The problem to be considered is one-machine, one-product-type scheduling with backlogs to minimize a piece-wise linear inventory cost and quadratic production cost, that is, the $1|1|u(t)\in[0,\infty]|d(t),T|Q_u, L_x$ problem which is stated as follows

$$J=\int_0^T\left[c_x^+X^+(t)+c_x^-X^-(t)+c_u u(t)^2\right]dt\to\min, \tag{3.15}$$

subject to (3.2) and (3.3).

Dual Formulation

The Hamiltonian, H, for problem (3.15), (3.2) and (3.3) is

$$H(t)=-\left[c_x^+X^+(t)+c_x^-X^-(t)+c_u u(t)^2\right]+\psi(t)\left[u(t)-d(t)\right]. \tag{3.16}$$

The co-state variable, $\psi(t)$ in (3.16), satisfies the co-state differential equation (3.5).

According to the maximum principle, the Hamiltonian is maximized at every point of time with respect to admissible controls $u(t)$. This implies that if the control effort function $u(t)$ that satisfies

$$\frac{\partial H}{\partial u(t)}=-2c_u u(t)+\psi(t)=0 \tag{3.17}$$

is non-negative, then it is optimal. Otherwise, no control should be undertaken at point t, i.e. $u(t)=0$.

Algorithm 3.3

Step 1:

If the average demand as a fuction of time, $\frac{1}{t}\int_0^t d(\tau)d\tau$ is greater than $\frac{c_x^-}{4c_u}(2T-t)$ for all $t \in [0,T]$, then set the optimal solution as

$u(t) = \frac{c_x^-}{2c_u}(T-t)$ over the entire planning horizon, STOP;

otherwise go to the next step.

Step 2:

Define the following function of t_1

$$f(t_1) = \frac{1}{2c_u}\left(c_x^-(T-t_1)t_1 - c_x^+\frac{t_1^2}{2}\right) - \int_0^{t_1} d(t)dt;$$

If the smallest positive root of this function is such that $0 < t_1 \le \frac{c_x^- T}{c_x^- + c_x^+}$ and $\frac{1}{t-t_1}\int_{t_1}^t d(\tau)d\tau \ge \frac{c_x^-}{4c_u}\left(2T-t-t_1\right)$ for all $t \in (t_1, T]$, then set the optimal solution as

$$u(t) = \frac{1}{2c_u}\left(c_x^-(T-t_1) - c_x^+(t_1-t)\right) \text{ for } 0 \le t \le t_1,$$

$$u(t) = \frac{c_x^-}{2c_u}(T-t) \text{ for } t_1 \le t \le T; \quad \text{STOP.}$$

Otherwise, go to the next step.

Step 3:

If the demand function $d(t)$ is continuous, almost everywhere differentiable, $-c_x^- \le 2c_u\dot{d}(t) \le c_x^+$ for all $t \in [0,T]$ and $d(0) \le \frac{c_x^- T}{2c_u}$, then find $t_1 \in [0,T]$ as the largest root of the equation $c_x^-(T-t_1) = 2c_u d(t_1)$, and set the optimal solution as

$$u(t) = d(t) \text{ for } 0 \le t \le t_1,$$

$$u(t) = \frac{c_x^-}{2c_u}(T-t) \quad \text{for } t_1 \le t \le T; \quad \text{STOP.}$$

Example 3.2

Consider the $1|1|u(t) \in [0,\infty]|d(t),T|Q_u, L_x$ problem with the following parameters: $T = 10$ hr, c_u=0.1\$hr/unit^2, c_x^+=0.2\$/hr.unit, c_x^-=1\$/hr.unit and linearly increasing demand, $d(t) = (20+4t)$ unit/hr.

Step 1 of Algorithm 3.3 compares two functions $\frac{1}{t}\int_0^t d(\tau)d\tau = 20+2t$ and $\frac{c_x^-}{4c_u}(2T-t) = 50-2.5t$ and concludes that the first function is not greater than the second one throughout the entire planning horizon. Therefore, the algorithm proceeds to Step 2.

At Step 2, the smallest positive root of function $f(t_1) = -7.5t_1^2 + 30t_1$ is calculated as $t_1 = 4$ hr. Then it checks $0 < t_1 \le \frac{c_x^- T}{c_x^- + c_x^+}$=8.33hr and compares two functions $\frac{1}{t-t_1}\int_{t_1}^t d(\tau)d\tau = 2t+28$ and $\frac{c_x^-}{4c_u}(2T-t-t_1) = 40-2.5t$. It concludes that the first function exceeds the second one everywhere at interval $t \in [4,10]$. Therefore, the optimal solution is set as

$u(t) = (26+t)$ unit/hr for $0 \le t \le 4$,

$u(t) = (50-5t)$ unit/hr for $4 < t \le 10$.

■

Example 3.3

Consider the $1|1|u(t) \in [0,\infty]|d(t),T|Q_u, L_x$ problem with parameters from Example 3.2 but where demand is linearly decreasing, $d(t) = (40-3t)$ unit/hr. Steps 1 and 2 of Algorithm 3.3 do not find the optimal control and the algorithm proceeds to Step 3.

Step 3 checks the following conditions:

demand is everywhere differentiable,

$d(0)$=40, $d(0) \le \frac{c_x^- T}{2c_u} = 50$,

$2c_u\dot{d}(t) = -0.6$, $-1 \le 2c_u\dot{d}(t) \le 0.2$, for all $t \in [0,T]$.

Since these conditions hold, the largest root of the equation $10-t_1=0.2(40-3t_1)$ is found as $t_1=5$ hr. The optimal solution is set as
$u(t)=(40-3t)$ unit/hr for $0\le t\le 5$,
$u(t)=(50-5t)$ unit/hr for $5\le t\le 10$.
■

Definition 3.1. Graph $G=(E,A)$, where E is a set of numbered nodes and A is a set of arcs, is called *L-weighted* if each arc $arc_{nm}\in A$ linking nodes $\tau_n\in E$ and $\tau_m\in E$ is characterized by L non-negative numbers $a_l(arc_{nm})$, $l=1,\dots,L$.

Algorithm 3.4

Step 1:

Set accuracy ξ required for computing the optimal solution. Select time intervals $\tau_k=[t^k_{left},t^k_{right}]$ at which $-c^-_x\le 2c_u\dot{d}(t)\le c^+_x$, $t^k_{right}<t^{k+1}_{left}$, and number them beginning from k=2. In addition, define τ_1 and τ_K as the intervals which are degenerated to a point: $\tau_1=\{0\}$ and $\tau_K=\{T\}$, where K is the total number of obtained intervals. Calculate accuracy of integration $\varepsilon=\dfrac{c_u}{KT^2\left(\max\{c^+_x,c^-_x\}\right)^2}\xi$.

Step 2:

Form a 3-weighted directed graph, $G=(E,A)$, with nodes $E=\{\tau_k\}$ as time intervals τ_k. For each node τ_k, assign $K\times K$ matrix of weights, SD^k=0. Define arcs $arc_{nm}=(\tau_n,\tau_m)\in A$ and calculate their weights, $a(arc_{nm})$, $b(arc_{nm})$ and $c(arc_{nm})$ by linking nodes $\tau_n=[t^n_{left},t^n_{right}]$ and $\tau_m=[t^m_{left},t^m_{right}]$, for all $n<m$, n=1,2,...,K-1 as follows:
if $n\ne 1$ and $m\ne K$ then do Subroutine 1, otherwise
if $n=1$ and $m\ne K$ then do Subroutine 2, otherwise
if $n=1$ and $m=K$ then do Subroutine 3, otherwise
if $n\ne 1$ and $m=K$ then do Subroutine 4.

If all pairs of nodes have been checked in this step, go to Step 3, otherwise, link a pair of nodes which has not yet been checked.

Step 3:

Find the shortest feasible path from node τ_1 to node τ_K in graph G by the following sequence-dependent, shortest path procedure

(i) Define the solved and unsolved sets of nodes, $S=\varnothing$ and $\bar{S}=E$, respectively. Assign distance $dist_1=0$ and $dist_k=\infty$, k=2,...,K. Color all arcs of graph G white.

(ii) Select an unsolved node (nodes) $k\in\bar{S}$ with the minimum distance, i.e. $dist_k = \min_{\substack{i\in S, j\in\bar{S}, m\in\bar{S} \\ SD_{im}^{j}>0}} \{dist_j + c(arc_{jm}) + SD_{im}^{j}\}$. Denote i^* as the node (nodes) $i\in S$ for which the minimum is achieved. Color arc (arcs) arc_{i^*k} black. Withdraw node k from $\bar{S}$ and put it into S.

(iii) For each arc_{kj}, $j\in\bar{S}$, such that $SD_{i^*j}^{k}$>0, update the distance as follows

$$\text{if } dist_j > dist_k + c(arc_{kj}) + SD_{i^*j}^{k} \text{ then } dist_j = dist_k + c(arc_{kj}) + SD_{i^*j}^{k}.$$

(iv) If S=E, then trace back the shortest feasible path (paths) from the last node to the first one by passing from a current black arc to any of its black predecessors (in the direction opposite to the direction of the arcs) until it enters node 1; STOP. Otherwise go to (ii).

Subroutine 1 to Algorithm 3.4

(i) Set time points $t_x = t_{left}^{n}$, $t_y = t_{right}^{n}$, $t_{out} = t_{left}^{n}$.

(ii) Integrate twice, for $\dot{\psi}(t_{out}) = c_x^+$ and $\dot{\psi}(t_{out}) = -c_x^-$, the following system of differential equations

$$\dot{X}(t) = \frac{\psi(t)}{2c_u} - d(t), \quad X(t_{out}) = 0,$$

$$\dot{\psi}(t) = \begin{cases} c_x^+, \text{ if } X(t) > 0, \\ -c_x^-, \text{ if } X(t) < 0, \end{cases} \qquad \psi(t_{out}) = 2c_u d(t_{out})$$

over the interval $[t_{out}, t_{right}^{m}]$.

If $X(t)$<0 for all $t\in\tau_m$, then no arc links nodes τ_n and τ_m; STOP.

Otherwise, set $t_{out} = t_{right}^{n}$ and integrate the above system twice, for $\dot{\psi}(t_{out}) = c_x^+$ and $\dot{\psi}(t_{out}) = -c_x^-$.

If $X(t)$>0 for all $t\in\tau_m$, then no arc links nodes τ_n and τ_m; STOP.

Otherwise go to (iii)

(iii) Integrate the system from (ii) twice, for $\dot{\psi}(t_{out}) = c_x^+$ and $\dot{\psi}(t_{out}) = -c_x^-$ with $t_{out} = 0.5(t_x + t_y)$ and set time point $t_{in} \in \tau_m$ such that $X(t_{in}) = 0$.

If $\psi(t_{in}) > 2c_u d(t_{in})$, then $t_x = t_{out}$ and go to (iv);
if $\psi(t_{in}) < 2c_u d(t_{in})$, then $t_y = t_{out}$ and go to (iv).

(iv) If $t_y - t_x \le \varepsilon$, then set $arc_{nm} = (\tau_n, \tau_m) \in A$ and its three weights

$a(arc_{nm}) = t_{out}$, $b(arc_{nm}) = t_{in}$ and

$$c(arc_{nm}) = \int_{t_{out}}^{t_{in}} c_x^+ X^+(t) + c_x^- X^-(t) + c_u u(t)^2\, dt \text{ ; go to (v).}$$

Otherwise go to (iii).

(v) If there exist arcs arc_{jn} such that $\left| b(arc_{jn}) - a(arc_{nm}) \right| < \varepsilon$, then set $a(arc_{nm}) = \max_j b(arc_{jn})$. For each arc arc_{jn} entering node n, update weights

$$SD^n_{jm} = \begin{cases} \displaystyle\int_{b(arc_{jn})}^{a(arc_{nm})} c_u d(t)^2\, dt, & \text{if } a(arc_{nm}) \ge b(arc_{jn}); \\ -1, & \text{otherwise.} \end{cases} \quad \text{STOP.}$$

Otherwise, STOP.

Subroutine 2 to Algorithm 3.4

(i) Set $\psi_x = 0$, $\psi_y = c_x^- T$.

(ii) Integrate twice, for $\dot{\psi}(0) = c_x^+$ and $\dot{\psi}(0) = -c_x^-$, the following system of differential equations

$$\dot{X}(t) = \frac{\psi(t)}{2c_u} - d(t), \quad X(0) = 0,$$

$$\dot{\psi}(t) = \begin{cases} c_x^+, & \text{if } X(t) > 0, \\ -c_x^-, & \text{if } X(t) < 0, \end{cases} \qquad \psi(0) = \psi_x$$

over the interval $[0, t^m_{right}]$.

If $X(t) < 0$ for all $t \in \tau_m$, then no arc links nodes τ_1 and τ_m; STOP.

Otherwise, integrate the above system twice, for $\dot{\psi}(0)=c_x^+$ and $\dot{\psi}(0)=-c_x^-$ with $\psi(0)=\psi_y$.
If $X(t)>0$ for all $t\in\tau_m$, then no arc links nodes τ_1 and τ_m; STOP.
Otherwise go to (iii).

(iii) Integrate the system from (ii) twice, for $\dot{\psi}(0)=c_x^+$ and $\dot{\psi}(0)=-c_x^-$ with $\psi(0)=0.5(\psi_x+\psi_y)$ and set time point $t_{in}\in\tau_m$ such that $X(t_{in})=0$.
If $\psi(t_{in})>2c_u d(t_{in})$, then $\psi_x=\psi(0)$ and go to (iv);
if $\psi(t_{in})<2c_u d(t_{in})$, then $\psi_y=\psi(0)$ and go to (iv).

(iv) If $\psi_y-\psi_x\le\varepsilon$, then set $arc_{1m}=(\tau_1,\tau_m)\in A$ and its three weights $a(arc_{1m})=0$, $b(arc_{1m})=t_{in}$ and

$$c(arc_{1m})=\int_0^{t_{in}} c_x^+X^+(t)+c_x^-X^-(t)+c_u u(t)^2\,dt\text{; STOP.}$$

Otherwise go to (iii).

Subroutine 3 to Algorithm 3.4

(i) Set $\psi_x=0$, $\psi_y=c_x^-T$.

(ii) Integrate twice, for $\dot{\psi}(0)=c_x^+$ and $\dot{\psi}(0)=-c_x^-$, the following system of differential equations

$$\dot{X}(t)=\frac{\psi(t)}{2c_u}-d(t),\quad X(0)=0,$$

$$\dot{\psi}(t)=\begin{cases}c_x^+, & \text{if } X(t)>0,\\ -c_x^-, & \text{if } X(t)<0,\end{cases}\qquad \psi(0)=\psi_x$$

over the interval $[0,T]$.
If $\psi(T)<0$, then no arc links nodes τ_1 and τ_K; STOP.
Otherwise, integrate the above system twice, for $\dot{\psi}(0)=c_x^+$ and $\dot{\psi}(0)=-c_x^-$ with $\psi(0)=\psi_y$.
If $\psi(T)>0$, then no arc links nodes τ_1 and τ_K; STOP.
Otherwise go to (iii).

(iii) Integrate the system from (ii) twice, for $\dot{\psi}(0) = c_x^+$ and $\dot{\psi}(0) = -c_x^-$ with $\psi(0) = 0.5(\psi_x + \psi_y)$.

If $\psi(T) > 0$, then $\psi_x = \psi(0)$ and go to (iv);

if $\psi(T) < 0$, then $\psi_y = \psi(0)$ and go to (iv).

(iv) If $\psi_y - \psi_x \le \varepsilon$, then set $arc_{1K} = (\tau_1, \tau_K) \in A$ and its three weights $a(arc_{1K}) = 0$, $b(arc_{1K}) = T$ and

$$c(arc_{1K}) = \int_0^T c_x^+ X^+(t) + c_x^- X^-(t) + c_u u(t)^2 \, dt \text{; STOP.}$$

Otherwise go to (iii).

Subroutine 4 to Algorithm 3.4

(i) Set time points $t_x = t_{left}^n$, $t_y = t_{right}^n$, $t_{out} = t_{left}^n$.

(ii) Integrate twice, for $\dot{\psi}(t_{out}) = c_x^+$ and $\dot{\psi}(t_{out}) = -c_x^-$, the following system of differential equations

$$\dot{X}(t) = \frac{\psi(t)}{2c_u} - d(t), \quad X(t_{out}) = 0,$$

$$\dot{\psi}(t) = \begin{cases} c_x^+, & \text{if } X(t) > 0, \\ -c_x^-, & \text{if } X(t) < 0, \end{cases} \qquad \psi(t_{out}) = 2c_u d(t_{out})$$

over the interval $[t_{out}, T]$.

If $\psi(T) < 0$, then no arc links nodes τ_n and τ_K; STOP.

Otherwise, set $t_{out} = t_{right}^n$ and integrate the above system twice, for $\dot{\psi}(t_{out}) = c_x^+$ and $\dot{\psi}(t_{out}) = -c_x^-$.

If $\psi(T) > 0$, then no arc links nodes τ_n and τ_K; STOP.

Otherwise go to (iii).

(iii) Integrate the system from (ii) twice, for $\dot{\psi}(t_{out}) = c_x^+$ and $\dot{\psi}(t_{out}) = -c_x^-$ with $t_{out} = 0.5(t_x + t_y)$.

If $\psi(T) > 0$, then $t_x = t_{out}$ and go to (iv);

if $\psi(T) < 0$, then $t_y = t_{out}$ and go to (iv).

(iv) If $t_y - t_x \le \varepsilon$, then set $arc_{nK} = (\tau_n, \tau_K) \in A$ and its three weights $a(arc_{nK}) = t_{out}$, $b(arc_{nK}) = T$ and

$$c(arc_{nK}) = \int_{t_{out}}^{T} c_x^+ X^+(t) + c_x^- X^-(t) + c_u u(t)^2 \, dt \text{; go to (v).}$$

Otherwise go to (iii).

(v) If there exist arcs arc_{jn} such that $|b(arc_{jn}) - a(arc_{nK})| < \varepsilon$, then set $a(arc_{nK}) = \max_j b(arc_{jn})$.

For each arc arc_{jn} entering node n, update weights

$$SD_{jK}^n = \begin{cases} \int_{b(arc_{jn})}^{a(arc_{nK})} c_u d(t)^2 \, dt, & \text{if } a(arc_{nK}) \ge b(arc_{jn}); \\ -1, & \text{otherwise.} \end{cases} \quad \text{STOP.}$$

Example 3.4

Consider the $1|1|u(t) \in [0,\infty]|d(t),T|Q_u, L_x$ problem with the following parameters: $T = 10$ hr, c_u=0.05\$hr/unit², c_x^+=0.2\$/hr.unit, c_x^-=1\$/hr.unit and quadratic demand, $d(t) = 38 + 3(t-5)^2$ unit/hr.

At Step 1 we set accuracy of the objective at ξ=\$150 and select a single time interval $\tau_2 = [3.33, 5.33]$ at which $-c_x^- \le 2c_u \dot{d}(t) \le c_x^+$, in addition to $\tau_1 = \{0\}$ and $\tau_K = \{T\}$. This results in K=3 and $\varepsilon = 0.025$. Step 2 of Algorithm 3.4 forms a 3-weighted graph of three nodes with arcs set by Subroutines 1-4 as shown in Figure 3.2.

The weights of the arcs are calculated as follows:

$$a(arc_{12}) = 0 \text{ hr},\ b(arc_{12}) = 5.0 \text{ hr},\ c(arc_{12}) = \$986,$$

$$a(arc_{23}) = 5.01 \text{ hr},\ b(arc_{23}) = 10.0 \text{ hr},\ c(arc_{23}) = \$425,$$

$$a(arc_{13}) = 0 \text{ hr},\ b(arc_{13}) = 10.0 \text{ hr},\ c(arc_{13}) = \$1540.$$

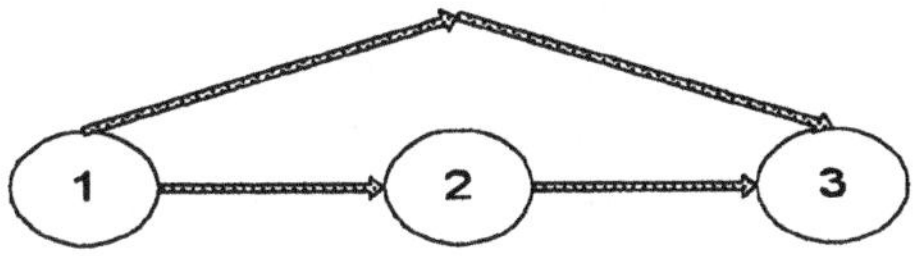

Figure 3.2. Graph G for Example 3.4.

All elements of matrices SD^k, k=1,2,3 which form sequence-dependent weights, equal zero except for $SD_{13}^2 = \$0.8$.

Step 3 compares two feasible paths: $P_1 = (arc_{12}, arc_{23})$ and $P_2 = (arc_{13})$. The shortest path is found to be P_1 with distance (objective value) equal to 986+0.8+425=\$1411.8. The distance of the other possible path is \$1540. The optimal control corresponding to the arcs of P_1 is

$$u(t) = 88 - 10t,\ 0 \le t < 5.0,$$
$$u(t) = 38 + 3(t-5)^2,\ 5.00 \le t < 5.01,$$
$$u(t) = 38 + 2(t - 5.01),\ 5.01 \le t < 6.0,$$
$$u(t) = 99.98 - 10t,\ 6.0 \le t \le 10.0.$$

■

Definition 3.2. Path P= $(arc_{n_1n_2}, arc_{n_2n_3}, \dots, arc_{n_{p-1}n_p})$ in a 2-weighted directed graph G with weights $a(arc_{nm})$ and $b(arc_{nm})$ is called *feasible*, if it satisfies the following conditions:

(1) it starts at node τ_1 $(n_1 = 1)$ of graph G and ends in a node $\tau_p \in E$;

(2) $a(arc_{n_jn_{j+1}}) \ge b(arc_{n_{j-1}n_j})$ for all j=2,...,p-1.

Algorithm 3.5

Step 1:

If demand is K-interval, single-mode, piece-wise constant, then denote K-2 time intervals $\tau_k = [t_{k-2}, t_{k-1}]$, k=2,...,K-1, $t_0 = 0$, $t_{K-2} = T$ at which the demand is constant. In addition, define τ_1 and τ_K as the intervals which are degenerated to a point: $\tau_1 = \{0\}$ and $\tau_K = \{T\}$.

Step 2:

Form a 2-weighted directed graph, $G = (E, A)$, with nodes $E = \{\tau_k\}$ as time intervals τ_k. Define arcs $arc_{nm} = (\tau_n, \tau_m) \in A$ and calculate

their weights, $a(arc_{nm})$ and $b(arc_{nm})$ by linking nodes τ_n and τ_m for all $n<m$, $n=1,2,\ldots,K-1$ as follows:
if $n \neq 1$ and $m \neq K$ then do Subroutine 1, otherwise
if $n = 1$ and $m \neq K$ then do Subroutine 2, otherwise
if $n = 1$ and $m = K$ then do Subroutine 3, otherwise
if $n \neq 1$ and $m = K$ then do Subroutine 4.

If all pairs of nodes have been checked in this step, go to Step 3, otherwise, link a pair of nodes which has not yet been checked.

Step 3:

Find a feasible path from node τ_1 to node τ_K in graph G by coloring its arcs:

(i) Initially, all arcs start out white and may later become colored. Set $count=1$ and paint arcs arc_{1m}, $m=2,\ldots,K$, gray. (We assume here that arc_{1K} is absent, otherwise the problem is solved trivially). All gray arcs are stored in a first-in, first-out queue Q.

(ii) Set $count=count+1$, $2 \leq count \leq |A|$ and check each gray arc v (taking each in turn at the head of the queue Q). Examine all white arcs, w, that are successors to v, and, if $a(w) \geq b(v)$, then arc w is colored gray (otherwise the latter arc remains white). After all successor arcs to v have been examined, this gray arc is painted either black or red. Namely, arc v becomes black if it is found to have at least one successor; otherwise it becomes red. All new gray arcs are inserted in turn at the end of the queue Q. The painted (black or red) v is removed from Q. Proceed with checking until a gray arc $v = arc_{nK}$ is discovered. This arc is painted black.

(iii) Move backward from node K to node 1, each time passing from a current black arc to any of its predecessors (in the opposite direction to the arcs) until it enters node 1. Thus, the desired feasible path from 1 to K is constructed; STOP.

Subroutine 1 to Algorithm 3.5

(i) Assign $k=n+1$ and calculate unknown time points

$$t_{out} = t_x + \frac{2c_u(d_n - d_m) - c_x^- y}{c_x^+},$$

$$t_x = \frac{1}{d_m - d_k}\left(\frac{c_x^-}{4c_u}y^2 - d_k t_{k-1} + d_m t_{m-2} - \sum_{j=k+1}^{j=m-1} d_j(t_{j-1} - t_{j-2})\right) \quad \text{and}$$

$t_{in} = t_x + y$,

where y is a positive root of the following algebraic quadratic equation

$$\frac{c_x^-}{4c_u}\left(1 - \frac{c_x^-(d_m - d_k)}{c_x^+(d_k - d_n)}\right)y^2 - \frac{c_x^-(d_m - d_k)}{c_x^+(d_k - d_n)}y - \frac{c_u(d_m - d_k)}{c_x^+(d_k - d_n)}(d_m - d_n)^2$$

$$-d_k t_{k-1} + d_m t_{m-2} - \sum_{j=k+1}^{j=m-1} d_j(t_{j-1} - t_{j-2}) +$$

$$\frac{d_m - d_k}{d_k - d_n}\, d_n t_{n-1} - d_k t_{k-2} + \sum_{j=n+1}^{j=k-1} d_j(t_{j-1} - t_{j-2}) = 0.$$

(ii) If $t_x \in \tau_k$, $t_{out} \in \tau_n$ and $t_{in} \in \tau_m$, then set $arc_{nm} = (\tau_n, \tau_m) \in A$ directed from node τ_n to node τ_m and assign its weights $a(arc_{nm}) = t_{out}$, $b(arc_{nm}) = t_{in}$. Set the optimal control $u(t) = d_m + \frac{c_x^-}{2c_u}y + \frac{c_x^+}{2c_u}(t - t_x)$ for $t_x - \frac{c_x^- y - 2c_u(d_n - d_m)}{c_x^+} \le t \le t_x$,

$u(t) = d_m + \frac{c_x^-}{2c_u}(t_x + y) - \frac{c_x^-}{2c_u}t$ for $t_x \le t \le t_x + y$; STOP. Otherwise go to (iii).

(iii) If $k<m-1$, then assign $k=k+1$ and go to (i). Otherwise go to (iv).

(iv) Calculate time points

$$t_{out} = \frac{d_m}{d_m - d_n}y - \frac{2c_u}{c_x^+}d_m \quad \text{and} \quad t_{in} = \frac{d_m}{d_m - d_n}y - \frac{2c_u}{c_x^+}d_n, \quad \text{where}$$

$$y = \frac{1}{d_m}\left(\frac{c_u}{c_x^+}(d_m - d_n)^2 + \frac{2c_u}{c_x^+}(d_m - d_n)d_n - d_n t_{n-1} + d_m t_{m-2} - \right.$$

$$\left. - \sum_{j=n+1}^{j=m-1} d_j(t_{j-1} - t_{j-2})\right).$$

If $t_{out} \in \tau_n$, $t_{in} \in \tau_m$ and $d_m > d_n$, then set $arc_{nm} = (\tau_n, \tau_m) \in A$ directed from node τ_n to node τ_m and assign its weights $a(arc_{nm}) = t_{out}$, $b(arc_{nm}) = t_{in}$. Set the optimal control

$u(t) = d_n + \frac{c_x^+}{2c_u}(t - t_{out})$ for $t_{out} \le t \le t_{in}$; STOP. Otherwise go to (v).

(v) Calculate time points

$t_{out} = \frac{d_m}{d_m - d_n} y + \frac{2c_u}{c_x^-} d_m$, $t_{in} = \frac{d_m}{d_m - d_n} y + \frac{2c_u}{c_x^-} d_n$, where

$$y = \frac{1}{d_m}\left(-\frac{c_u}{c_x^-}(d_m - d_n)^2 - \frac{2c_u}{c_x^-}(d_m - d_n)d_n - d_n t_{n-1} + d_m t_{m-2} - \right.$$

$$\left. - \sum_{j=n+1}^{j=m-1} d_j (t_{j-1} - t_{j-2}) \right).$$

If $t_{out} \in \tau_n$, $t_{in} \in \tau_m$ and $d_m < d_n$, then set $arc_{nm} = (\tau_n, \tau_m) \in A$ directed from node τ_n to node τ_m and assign its weights $a(arc_{nm}) = t_{out}$, $b(arc_{nm}) = t_{in}$. Set the optimal control

$u(t) = d_n - \frac{c_x^-}{2c_u}(t - t_{out})$ for $t_{out} \le t \le t_{in}$.

Otherwise, no arc links nodes τ_n and τ_m; STOP.

Discussion

The three algorithms presented in this section solve the $1|1|u(t) \in [0, \infty] |d(t), T| Q_u, L_x$ problem for different forms of the demand function. Algorithms 3.3 and 3.5 provide an exact optimal solution in strictly polynomial time, while Algorithm 3.4 approximates the optimal solution with a given accuracy in pseudopolynomial time.

Specifically, Algorithm 3.3 finds an exact optimal solution for three special cases. Unlike Algorithms 3.1 and 3.2, the first two cases are characterized by only one type of optimal control regime which is production proportional to the co-state variable. In such a case, the fractional production regime $u(t)=d(t)$ never occurs and the control function has multiple switching points. That is to say, the optimal control is presented by a number of linear functions of time and time points where these functions are changed over. The third case is similar to those solved by Algorithms 3.1 and 3.2, i.e. there is a single switching point between the fractional production $u(t)=d(t)$ and the production proportional to the co-state variable.

The first solvable case emerges when the average demand by a time point t, $\frac{1}{t}\int_0^t d(\tau)d\tau$, exceeds $\frac{c_x^-}{4c_u}(2T-t)$ at this time point, as proven in the following lemma. Note that the greater the backlog cost and the lower the production cost, the greater the average demand is needed for the case to occur.

LEMMA 3.3. (Correctness of Algorithm 3.3 Step 1) Given problem (3.15), (3.2) and (3.3), if $\frac{1}{t}\int_0^t d(\tau)d\tau \geq \frac{c_x^-}{4c_u}(2T-t)$ for $0 \leq t \leq T$, then $u(t) = \frac{c_x^-}{2c_u}(T-t)$.

PROOF. Let the optimal solution to the problem be such that

$$\psi(t) > 0 \text{ for } 0 \leq t < T \text{ and } X(t) \leq 0 \text{ for } 0 \leq t \leq T,$$

as illustrated in Figure 3.3.

Then, according to (3.5) and (3.17), we have

$$\psi(t) = 2c_u u(t) = c_x^-(T-t)$$

and thus

$$u(t) = \frac{c_x^-}{2c_u}(T-t). \tag{3.18}$$

To prove the condition stated in this lemma, we determine when the control effort function as stated in (3.18) results in a feasible behavior of the state variable. We achieve this by integrating its equation (3.2) and requiring that $X(t) \leq 0$

$$X(t) = X(0) + \int_0^t \left(\frac{c_x^-(T-\tau)}{2c_u} - d(t) \right) d\tau = \frac{c_x^-}{4c_u}\left(2Tt - t^2\right) - \int_0^t d(t)d\tau \leq 0.$$

■

The second solvable case emerges when there is a single switching point such that the average demand, up to this point and after this point, is balanced by production, inventory and backlog cost related expressions, $\frac{1}{t_1}\int_0^{t_1} d(t)dt = \frac{c_x^-}{2c_u}(T-t_1) - \frac{c_x^+}{4c_u}t_1$ and

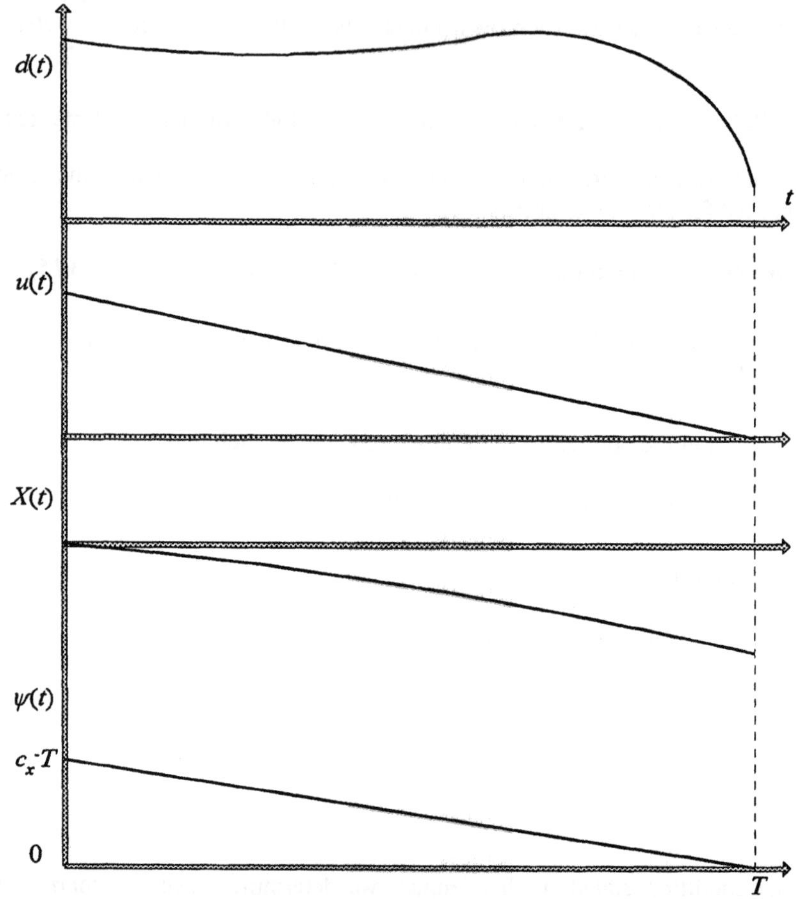

Figure 3.3. Demand *d*(*t*), optimal control *u*(*t*), state *X*(*t*) and co-state $\psi(t)$ variables for the first solvable case of the $1|1|u(t)\in[0,\infty]|d(t),T|\, Q_u, L_x$ problem.

$\frac{1}{t-t_1}\int_{t_1}^{t} d(\tau)d\tau \geq \frac{c_x^-}{4c_u}\left(2T-t-t_1\right)$. In addition, the unit backlog cost related to the time interval after the switching point must be greater than the unit inventory cost related to the time interval preceding the switching point, $c_x^-(T-t_1)\geq c_x^+ t_1$ as proved in the following lemma.

LEMMA 3.4. (Correctness of Algorithm 3.3 Step 2) Given problem (3.15), (3.2) and (3.3), if the smallest positive root t_1 of the function

$$f(t_1) = \frac{1}{2c_u}\left(c_x^-(T-t_1)t_1 - c_x^+\frac{t_1^2}{2}\right) - \int_0^{t_1} d(t)dt$$

is such that $0 < t_1 \le \dfrac{c_x^- T}{c_x^- + c_x^+}$ and $\dfrac{1}{t-t_1}\displaystyle\int_{t_1}^{t} d(\tau)d\tau \ge \dfrac{c_x^-}{4c_u}\left(2T - t - t_1\right)$ for all $t \in [t_1, T]$,

then the optimal solution is

$u(t) = \dfrac{1}{2c_u}\left(c_x^-(T-t_1) - c_x^+(t_1 - t)\right)$ for $0 \le t \le t_1$, $\quad u(t) = \dfrac{c_x^-}{2c_u}(T-t)$ for $t_1 \le t \le T$.

PROOF. Let the optimal solution to the problem be such that

$X(t) \ge 0$ for $0 \le t \le t_1$, $X(t) \le 0$ for $t_1 \le t \le T$, $\psi(t) > 0$ for $0 \le t < T$,

as shown in Figure 3.4.

Based on the described behavior of $X(t)$, we integrate the co-state equation (3.5) as follows

$$\psi(t) = \psi(0) + c_x^+ t,\ \ 0 \le t \le t_1, \tag{3.19}$$

$$\psi(t) = \psi(0) + c_x^+ t_1 - c_x^-(t - t_1),\ \ t_1 \le t \le T, \tag{3.20}$$

$$\psi(T) = \psi(0) + c_x^+ t_1 - c_x^-(T - t_1) = 0. \tag{3.21}$$

One can observe from (3.19)-(3.21) that if $\psi(0) > 0$, then $\psi(t) > 0$ for $0 \le t < T$. According to (3.21), $\psi(0) \ge 0$ if $c_x^-(T - t_1) \ge c_x^+ t_1$, as stated in the lemma. Next, given $\psi(t)$ along the entire planning horizon, the optimal control is defined by substituting the corresponding equations for $\psi(t)$, i.e. (3.19), (3.20) and for $\psi(0)$, i.e. (3.21) in (3.17)

$$u(t) = \frac{\psi(t)}{2c_u} = \frac{1}{2c_u}\left(c_x^-(T-t_1) - c_x^+(t_1 - t)\right) \text{ for } 0 \le t \le t_1, \tag{3.22}$$

$$u(t) = \frac{\psi(t)}{2c_u} = \frac{c_x^-}{2c_u}(T-t) \text{ for } t_1 \le t \le T. \tag{3.23}$$

Finally, to find the switching point, we integrate (3.2) with control (3.22) taking into account that at this point the inventory is zero

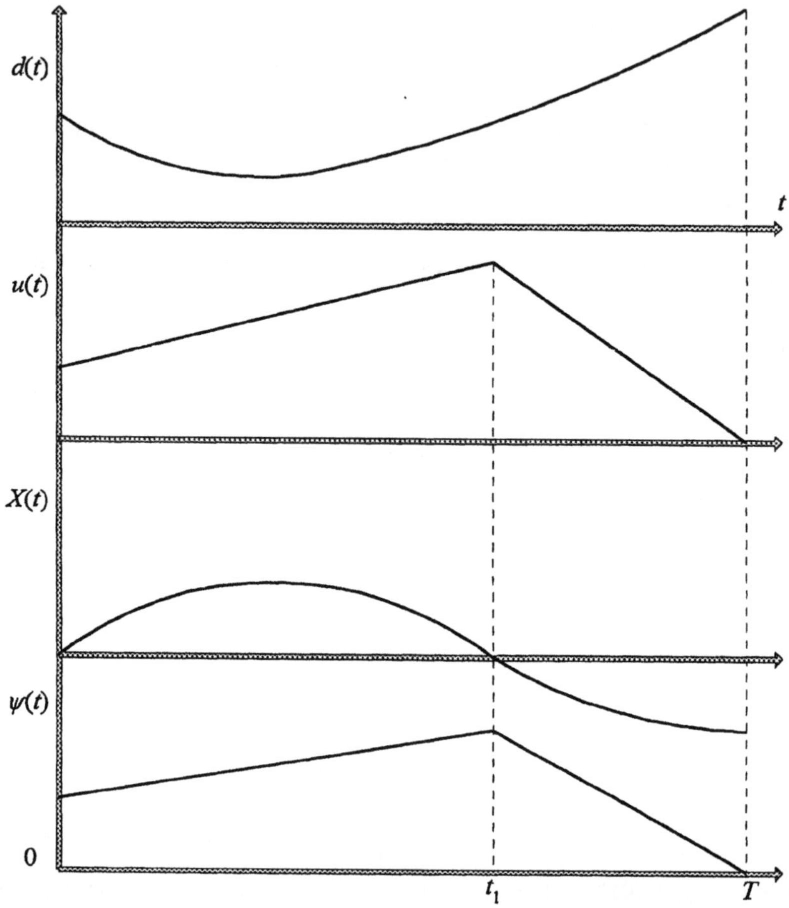

Figure 3.4. Demand $d(t)$, optimal control $u(t)$, state $X(t)$ and co-state $\psi(t)$ variables for the second solvable case of the $1|1|u(t)\in[0,\infty]|d(t),T|Q_u, L_x$ problem.

$$X(t_1) = -\int_0^{t_1} d(t)dt + \int_0^{t_1} \frac{1}{2c}\left(c_x^+(t_1 - t) - c_x^-(T - t_1)\right)dt = -\int_0^{t_1} d(t)dt - \frac{1}{2c_u}\left(c_x^+\frac{t_1^2}{2} - c_x^-(T - t_1)t_1\right) = 0 .$$

Thus, a non-negative root of the last equation is the switching point, t_1, for which we have been searching. However, the found solution is correct if $X(t) < 0$ after the switching point, as stated in the lemma.

■

The last case solved by Algorithm 3.3 emerges when the demand function is non-lumpy in the sense that its rate of change is bounded by the ratio of the inventory/backlog cost to the production cost

$$-\frac{c^-}{2c_u} \le \dot{d}(t) \le \frac{c^+}{2c_u}.$$

In this case, before a single switching point it is optimal to maintain zero inventory, i.e. $u(t)=d(t)$. After this point, some backlog becomes profitable as proven in the following lemma.

LEMMA 3.5. (Correctness of Algorithm 3.3 Step 3) Given problem (3.15), (3.2) and (3.3), if the demand function $d(t)$ is continuous, almost everywhere differentiable, non-lumpy $-c^- \le 2c_u \dot{d}(t) \le c^+$ for all $t \in [0,T]$ and $d(0) \le c_x^- T$, then the optimal solution is $u(t) = d(t)$ for $0 \le t \le t_1$, $u(t) = \frac{c_x^-}{2c_u}(T-t)$ for $t_1 \le t \le T$, where switching point t_1 is the root of equation $c^-(T-t_1) = 2c_u d(t_1)$.

PROOF. Let the optimal solution to the problem be such that

$$X(t) = 0 \text{ for } 0 \le t \le t_1,\ X(t) < 0 \text{ for } t_1 < t \le T,\ \psi(t) > 0 \text{ for } 0 \le t < T,$$

as illustrated in Figure 3.5.

Since $X(t)$ is constant at the time interval before the switching point, $u(t)=d(t)$ must hold at this interval. Therefore, according to (3.17), $\psi(t) = 2c_u d(t)$ and due to the co-state differential equation (3.5), the last expression can hold only if $-c^- \le 2c_u \dot{d}(t) \le c^+$. To find the co-state variable behavior at the time interval after the switching point, we integrate equation (3.5) as follows

$$\psi(t) = \psi(t_1) - c_x^-(t-t_1),\quad t_1 \le t \le T, \tag{3.24}$$

By substituting $\psi(T) = 0$ and $\psi(t_1) = 2c_u d(t_1)$ in (3.24), we obtain the equation for the switching point stated in the lemma. The control after the switching point immediately follows from (3.17) and (3.24).

■

All three cases are solved by Algorithm 3.3 in $O(1)$ time if the root of the corresponding equation for the switching point can be found analytically. As has been discussed in previous sections, this is not always the case. However, if demand is a K-level, piece-wise constant function of time, then the exact solution is available in $O(K)$ time.

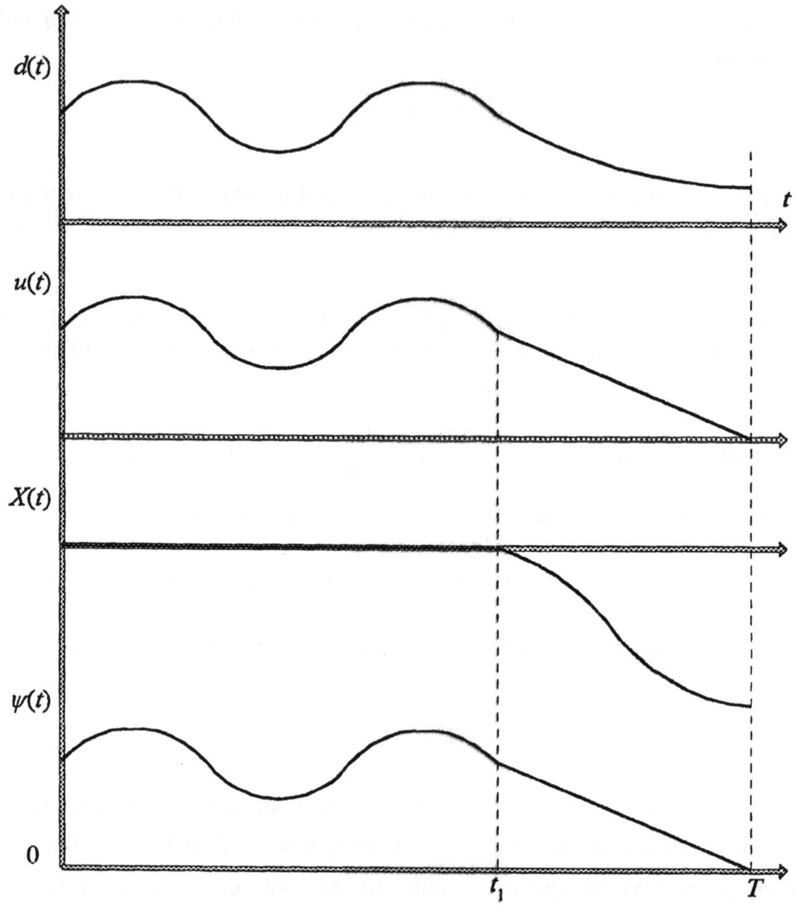

Figure 3.5. Demand $d(t)$, optimal control $u(t)$, state $X(t)$ and co-state $\psi(t)$ variables for the third solvable case of the $1|1|u(t)\in[0,\infty]|d(t),T|Q_u, L_x$ problem.

Since the maximum principle does not provide enough information about the optimal solution for the $1|1|u(t)\in[0,\infty]|d(t),T|Q_u, L_x$ problem with demand as an arbitrary function of time, Algorithm 3.4 has been presented to approximate the optimal solution with an *a priori* given accuracy.

The following lemma states that even if the demand function is arbitrary, that is, conditions of Lemmas 3.3-3.5 are not satisfied, then the general form of the optimal solution has a definite structure characterized by controls providing zero inventories only at time intervals where demand is non-lumpy, i.e. $-c^- \le 2c_u\dot{d}(t) \le c^+$.

LEMMA 3.6. (Correctness of Algorithm 3.4 Step 1) Given problem (3.15), (3.2) and (3.3), if $-c^- \le 2c_u\dot{d}(t) \le c^+$ at K intervals of time τ_k, $k = 1,...,K$, then the optimal solution contains at most K time intervals of zero inventory, $\chi_k = \{t \in \tau_k | X(t) = 0\}$.

PROOF. The proof immediately follows from Lemma 3.5 and the fact that at an interval of time at which $-c^- \le 2c_u\dot{d}(t) \le c^+$ does not hold, $X(t)$ cannot be zero at any sub-interval of this interval. The detailed proof of this fact is left as an exercise for the reader.

■

Algorithm 3.4 makes use of condition $-c^- \le 2c_u\dot{d}(t) \le c^+$ for non-lumpy demands. Since this condition is generally not satisfied along the entire planning horizon, the main idea of the algorithm, drawn from Lemma 3.6, is to locate intervals τ_k, $k = 1,...,K$ at which this condition is met. The intervals are linked so that the solution of the dual problem becomes feasible. This is accomplished by constructing a graph with nodes representing these intervals and arcs as the links between pairs of nodes. Thus, the continuous-time scheduling problem is reduced to a combinatorial problem of finding a feasible path in this graph. Since the links in the graph are calculated numerically, the algorithm may identify a number of feasible paths, each of which approximately satisfies the maximum principle (with a given accuracy). Therefore, at the final step of the algorithm, the best path is determined by a generalization of the well-known, shortest path procedure, where the path distance is associated with the objective (3.15).

The link between a couple of nodes τ_n and τ_m is calculated numerically by integrating the system of state and co-state differential equations. Figure 3.6 illustrates a typical evolution of control, state and co-state variables when the demand is non-lumpy over two time intervals τ_2 and τ_3.

In addition, intervals τ_n must be linked with the beginning and end of the planning horizon to satisfy the boundary conditions $X(0)=0$, $\psi(T) = 0$. Therefore, there are four types of links: between two fractional production regimes; between the beginning of the planning horizon and a fractional regime; between a fractional regime and the end of the planning horizon; and between the beginning and the end of the planning horizon. Each type of link is equivalent to finding a root of a specially constructed slack function as defined below. Thus, there are four slack functions which are proved to be monotone (see Lemma 3.7 below). Due to the monotonicity, the dichotomous search is applied in Algorithm 3.4 to efficiently find the root. The dichotomous search is initially applied to

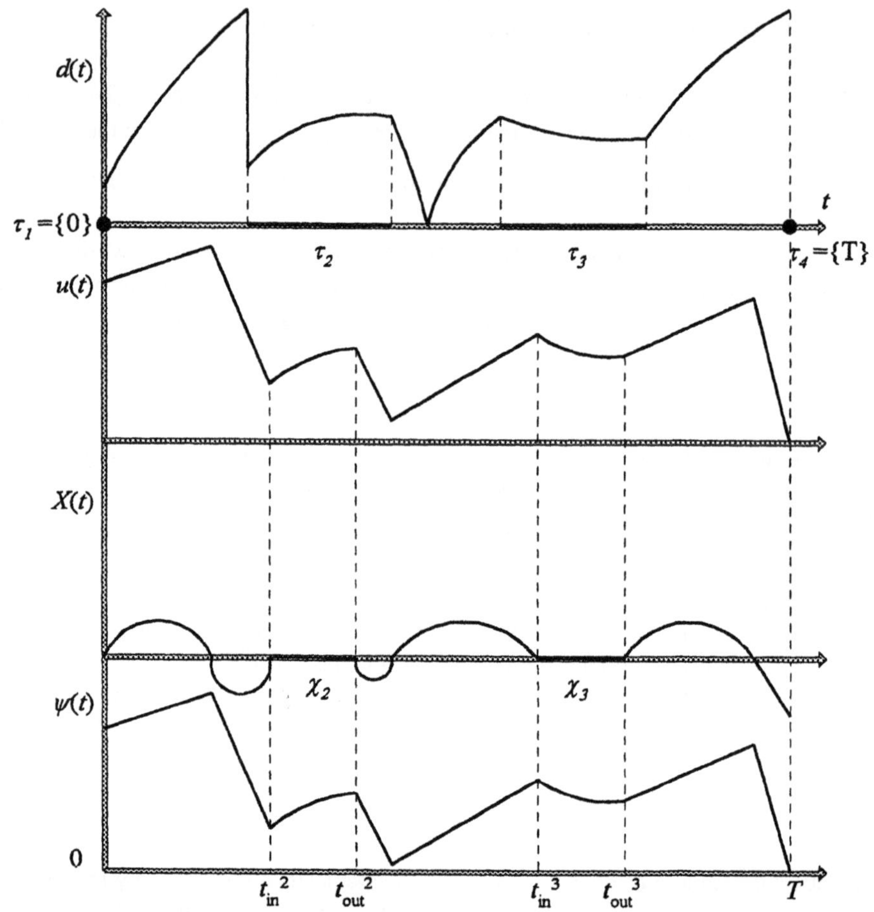

Figure 3.6. Demand $d(t)$, optimal control $u(t)$, state $X(t)$ and co-state $\psi(t)$ variables for a case of the $1|1|u(t)\in[0,\infty]|d(t),T|Q_u, L_x$ problem with arbitrary demand.

the entire interval τ_n and shortens the interval iteratively until the root is located with given accuracy ε. This procedure thus requires at most $\log\frac{T}{\varepsilon}$ iterations.

The definitions of the slack functions are as follows

Definition 3.3. Given a pair of nodes τ_n and τ_m, $n,m\neq 1,K$, F_1 is a function which maps $t_{out}\to\psi(t_{in})-2c_u d(t_{in})$, where $t_{out}\in\tau_n$, $t_{in}\in\tau_m$ and $X(t_{in})=0$.

Definition 3.4. Given a pair of nodes τ_1 and τ_m, $m \neq 1, K$, F_2 is a function which maps $\psi(0) \to \psi(t_{in}) - 2c_u d(t_{in})$, where $t_{in} \in \tau_m$ and $X(t_{in}) = 0$.

Definition 3.5. Given a pair of nodes τ_1 and τ_K, F_3 is a function which maps $\psi(0) \to \psi(T)$.

Definition 3.6. Given a pair of nodes τ_n and τ_K, $n \neq 1, K$, F_4 is a function which maps $t_{out} \to \psi(T)$, where $t_{out} \in \tau_n$.

Subroutines 1-4, invoked in Algorithm 3.4 Step 2, locate the roots of slack functions F_1 - F_4 respectively. The monotonicity of the slack functions is proved in the following lemma.

LEMMA 3.7. Slack functions F_1, F_2, F_3 and F_4 are non-decreasing in their arguments.

PROOF. Since the proof for all slack functions is very similar, we consider here only function F_3. Let $(X_1(t), \psi_1(t))$ and $(X_2(t), \psi_2(t))$ be two solutions of the system of the state and co-state differential equations (see Step 2 of Algorithm 3.4). Without loss of generality, assume also that $\psi_1(0) > \psi_2(0)$.

Consider an interval of time $[0, \hat{t}]$ at which $\psi_1(t) > \psi_2(t)$. From (3.17), it follows that $u(t)$ is a non-decreasing function of $\psi(t)$, i.e. $u_1(t) \geq u_2(t)$, $t \in [0, \hat{t}]$. From equation (3.2), it immediately follows that $\dot{X}_1(t) \geq \dot{X}_2(t)$ at the same time interval. Because $X_1(0) = X_2(0)$ and $\dot{X}_1(t) \geq \dot{X}_2(t)$, then $X_1(t) \geq X_2(t)$. Consequently, from co-state equation (3.5), we have $\dot{\psi}_1(t) \geq \dot{\psi}_2(t)$. Because $\psi_1(0) > \psi_2(0)$ and $\dot{\psi}_1(t) \geq \dot{\psi}_2(t)$, then $\psi_1(\hat{t}) - \psi_2(\hat{t}) \geq \psi_1(0) - \psi_2(0)$. It has been proved that if $\psi_1(0) > \psi_2(0)$, then $X_1(\hat{t}) \geq X_2(\hat{t})$ and $\psi_1(\hat{t}) - \psi_2(\hat{t}) > 0$. Therefore, the proof can be continued in the same way over subsequent time intervals, each of which is no shorter than $[0, \hat{t}]$ until the entire planning horizon is covered. Thus, $\psi_1(T) > \psi_2(T)$, which means that F_3 is a non-decreasing function.
■

The following lemma sets bounds for the co-state variable which are utilized in the dichotomous search performed in Subroutines 2 and 3.

LEMMA 3.8. Given problem (3.15), (3.2) and (3.3), the co-state variable $\psi(t)$ is bounded as $\psi(t) \in [0, \hat{\psi}]$ for $t \in [0, T]$, where $\hat{\psi} = c_x^- T$.

PROOF. The upper bound of the co-state variable is found by assuming zero production rate along the entire planning horizon $u(t)=0$ for $t \in [0,T]$. In this case, $X(t) \le 0$ and from the co-state differential equation $\psi(t)$ equals $c_x^-(T-t)$. Therefore, the maximum of $\psi(t)$, denoted in the lemma by $\hat{\psi}$, is achieved at t=0.

The proof of the lower bound for the co-state variable is by contradiction. Contrary to the lemma, let $\psi(t)<0$ at an interval of time $t \in [t_1,t_2]$. If at a sub-interval of this interval $X(t) \le 0$, then, according to the co-state equation (3.5), $\psi(t)$ does not increase along the sub-interval. Therefore, $\psi(t)$ remains negative and the control effort is always zero on this interval (see (3.17)). Under zero production rate, the buffer level $X(t)$ will not become positive forcing $\psi(t)$ to decrease until the end of the planning horizon. This contradicts the terminal co-state constraint $\psi(T)=0$.

Likewise, assuming that $X(t)>0$ at an interval where $\psi(t)<0$, we conclude that $X(t)$ does not increase anywhere before this interval. This contradicts the initial state constraint $X(0)=0$.
■

The correctness of Step 2 of Algorithm 3.4 immediately follows from Lemmas 3.7 and 3.8. The last step of the algorithm uses a sequence-dependent, shortest path procedure. The procedure is based on Dijkstra's algorithm (see Section 2.1.3) where calculation of the path distances is modified to incorporate matrix SD^n with sequence-dependent weights which correspond to every node n. Specifically, matrix SD^n has three types of entries: positive, zero and minus one. Entry $SD^n_{ij}=-1$ codes that both arc_{in} and arc_{nj} exist but cannot be included into a feasible path, since $b(arc_{in})>a(arc_{nj})$. Entry $SD^n_{ij}=0$ implies that at least one arc arc_{in} or arc_{nj}, does not exist. Otherwise, if both arcs exist and $b(arc_{in}) \le a(arc_{nj})$, the entry codes the numerical value of the increment of the objective function at time interval $[b(arc_{in}),a(arc_{nj})]$.

THEOREM 3.1. Problem (3.15), (3.2) and (3.3) is solvable with accuracy ξ in $O\left(\max\{K^4,\frac{K^3R}{\xi}\log\frac{KR}{\xi}\}\right)$ time, where K is the number of time intervals at which $-c^- \le 2c_u\dot{d}(t) \le c^+$ and $R=\frac{T^3\left(\max\{c_x^+,c_x^-\}\right)^2}{c_u}$.

PROOF. First note that Algorithm 3.4 links time intervals where $-c^- \le 2c_u\dot{d}(t) \le c^+$ so that the primal and dual formulations of the problem are satisfied with a given accuracy ε. The algorithm thus constructs a solution structured in accordance with Lemmas 3.6-3.8 and approximates the global optimum since problem (3.15), (3.2) and (3.3) has convex objective and linear constraints.

To estimate the algorithm's complexity, let us consider it step by step. At Step 1, the algorithm locates K time intervals at which $-c^- \le 2c_u \dot{d}(t) \le c^+$. This takes $O\left(\frac{T}{\varepsilon}\right)$ time. At Step 2, the algorithm repeatedly invokes the four subroutines which solve the system of two differential equations with accuracy ε. The equations are integrated in procedure (ii) of the Subroutines in $O\left(\frac{T}{\varepsilon}\right)$ time. The initial conditions for the differential equations are determined in procedures (iii) and (iv) by dichotomous search in $O\left(\log\frac{T}{\varepsilon}\right)$ time. In addition, procedure (v) of Subroutines 1 and 4 ensures that the computational accuracy of arc weights a and b is comparable to ε and takes at most $|A| = \frac{K(K+1)}{2}$ time, where $|A|$ is the maximum number of the arcs in graph G. Since Step 2 is repeated at most $|A|$ times, it totally needs

$$O\left(\max\{K^4, K^2\frac{T}{\varepsilon}\log\frac{T}{\varepsilon}\}\right) \tag{3.25}$$

time.

Step 3 of Algorithm 3.4 uses a sequence-dependent, shortest path method. Procedure (ii), which minimizes the path distances, is the most complex procedure at this step and requires $O(K^3)$ time. Thus, the algorithm's total complexity is determined by Step 2.

Finally, we estimate the relationship between the computational accuracy of the objective ξ and the accuracy of the dichotomous search ε. By varying the objective function (3.15) we determine that

$$\xi \le T\max\{c_x^+, c_x^-\}\max_t|\Delta X(t)| + 2Tc_u \max_t u(t)\max_t|\Delta u(t)|. \tag{3.26}$$

The value of $\max_t u(t)$ can be estimated from equation (3.17) as $\frac{1}{2c_u}\max_t \psi(t)$, which in turn is estimated from co-state differential equation (3.5) by $\frac{T}{2c_u}\max\{c_x^+, c_x^-\}$. The value of $\max_t|\Delta u(t)|$ is also estimated from (3.17) as $\frac{1}{2c_u}\max_t|\Delta\psi(t)|$, where the error of the co-state variable in the dichotomous search is $\max_t|\Delta\psi(t)| = \varepsilon K \max\{c_x^+, c_x^-\}$ for each link at Step 2 of the algorithm. The estimate of $\max_t|\Delta\psi(t)|$ is due to the fact that every isolated point t^*, at which $X(t^*)=0$ (the maximum number of such points is K), induces an increment in the error of the co-state variable by at most $\varepsilon \max\{c_x^+, c_x^-\}$ as illustrated in Figure 3.7. The value of $\max_t|\Delta X(t)|$ is just $T\max_t|\Delta u(t)| = \frac{\varepsilon K \max\{c_x^+, c_x^-\}}{2c_u}T$. By substituting the found estimates into (3.26) , we obtain maximum error ξ as a function of ε

$$\xi \le \frac{KT^2\left(\max\{c_x^+, c_x^-\}\right)^2}{c_u}\varepsilon . \tag{3.27}$$

To obtain the complexity and accuracy of Algorithm 3.4, we express ε from (3.27), substitute it into (3.25) and introduce parameter R to simplify the presentation.
■

According to Theorem 3.1, Algorithm 3.4 is of pseudopolynomial complexity which is due to multipliers T and ε entering (3.25). However, if demand is single-mode, piece-wise constant, the algorithm becomes strictly polynomial. This is because the approximate integration carried out by Subroutines 1-4 is reduced to an exact solution of the system of two algebraic equations in two unknowns. This case is treated by Algorithm 3.5. Modifications required for the analytical integration of the state and co-state differential equations are presented only for Subroutine 1 to Algorithm 3.5. Modifications to the other three subroutines are left as an exercise for the reader.

In what follows we prove the correctness and complexity of Algorithm 3.5.

LEMMA 3.9. If demand is single-mode, then the dynamics of $\psi(t)$ is also single-mode, i.e. there exists a time point $t_\psi \in [0,T]$ such that $\psi(t)$ neither decreases before t_ψ nor increases after it.

PROOF. Let function $\psi(t)$ have more than one local maximum, and $t_1^{\max}$ and $t_2^{\max}$ be the points of adjacent local maximums. Let $t^{\min}$ be the point of a local minimum between $t_1^{\max}$ and $t_2^{\max}$ (see Figure 3.8). With respect to the co-state equation (3.5), $X(t)$ must be zero at all three points $t_1^{\max}$, $t^{\min}$ and $t_2^{\max}$. By considering the interval $[t_1^{\max}, t^{\min}]$, we conclude that

- $d(t_1^{\max}+0) > u(t_1^{\max}+0)$ (since $X(t)$ decreases right of point $t_1^{\max}$)
- $u(t_1^{\max}+0) \ge u(t^{\min}-0)$ (since $\psi(t)$ decreases at interval $(t_1^{\max}, t^{\min})$

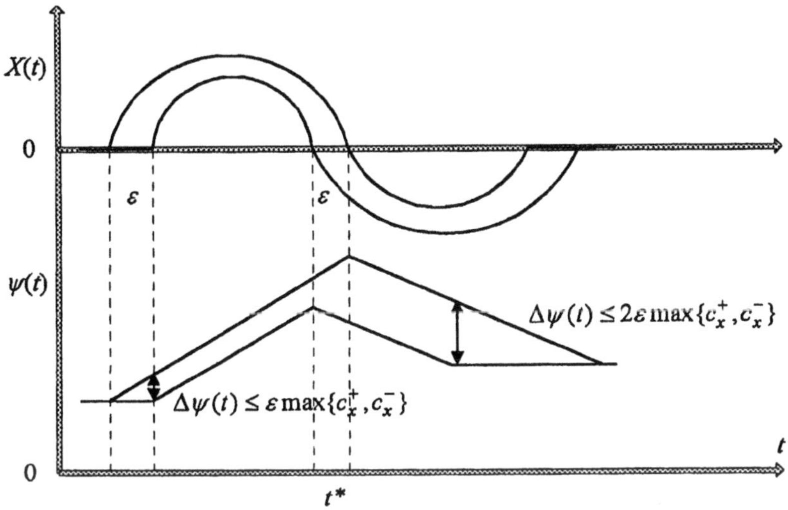

Figure 3.7. Increment in the co-state variable due to the error of dichotomous search of Algorithm 3.4.

- $u(t^{\min}-0) > d(t^{\min}-0)$ (since $X(t)$ increases left of point $t^{\min}$).

Therefore, the demand $d(t)$ must decrease on the interval $[t_1^{\max}, t^{\min}]$. Similarly, considering the interval $[t^{\min}, t_2^{\max}]$, we conclude that the demand $d(t)$ must increase on this interval. Such a behavior contradicts the single-mode dynamics of the demand assumed in the lemma.

■

As illustrated in Figure 3.8, the single-mode of $\psi(t)$ implies that there can be at most one point of time where $X(t)$ equals zero. This fact is employed in the following lemma to analytically determine the optimal control between any two intervals of fractional production when the demand function is single-mode and K-interval,, piece-wise constant

$$d(t) = d_k, \text{ for } t \in [t_{k-1}, t_k), \quad k = 1,\dots,K, \quad t_0 = 0, \ t_K = T.$$

LEMMA 3.10. (Rule for linking nodes in graph G) Given demand is K-interval, single-mode, piece-wise constant, two adjacent fractional regimes $u(t)=d(t)$ at sub-intervals χ_n and χ_m of intervals τ_n and τ_m respectively, if there exists k, such that $n<k<m$ and

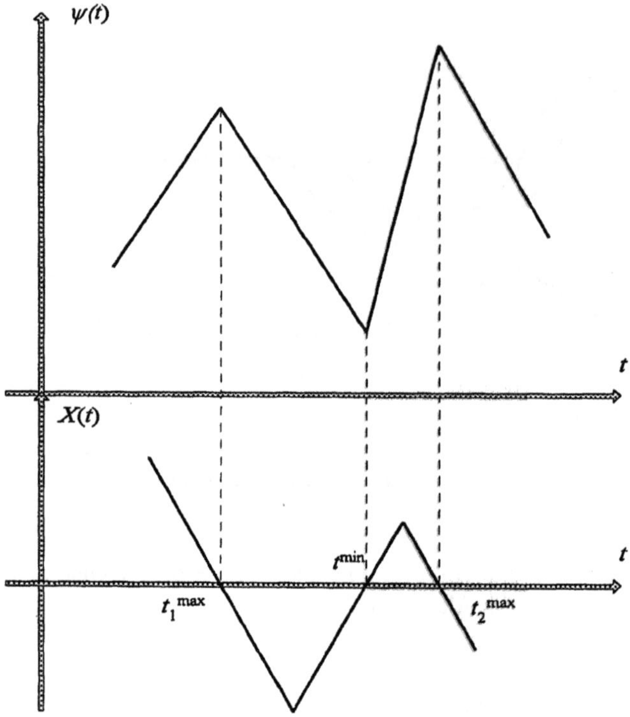

Figure 3.8. Multiple-mode dynamics of the co-state variable $\psi(t)$ and corresponding behavior of the state variable $X(t)$.

$$t_{n-2} \le t_x + \frac{2c_u(d_n - d_m) - c_x^- y}{c_x^+} \le t_{n-1},\ \ t_{m-2} \le t_x + y \le t_{m-1},\ \ t_{k-2} \le t_x \le t_{k-1},$$

where $t_x = \dfrac{1}{d_m - d_k}\left(\dfrac{c_x^-}{4c_u} y^2 - d_k t_{k-1} + d_m t_{m-2} - \sum_{j=k+1}^{j=m-1} d_j (t_{j-1} - t_{j-2})\right)$,

and y is a positive root of the following algebraic quadratic equation

$$\frac{c_x^-}{4c_u}\left(1 - \frac{c_x^-(d_m - d_k)}{c_x^+(d_k - d_n)}\right) y^2 - \frac{c_x^-(d_m - d_k)}{c_x^+(d_k - d_n)} y - \frac{c_u(d_m - d_k)}{c_x^+(d_k - d_n)}(d_m - d_n)^2 - d_k t_{k-1} + d_m t_{m-2} -$$

$$- \sum_{j=k+1}^{j=m-1} d_j (t_{j-1} - t_{j-2}) + \frac{d_m - d_k}{d_k - d_n} d_n t_{n-1} - d_k t_{k-2} + \sum_{j=n+1}^{j=k-1} d_j (t_{j-1} - t_{j-2}) = 0,$$

then the optimal control between these fractional production regimes is

$$u(t) = d_m + \frac{c_x^-}{2c_u} y + \frac{c_x^+}{2c_u}(t - t_x) \text{ for } t_x - \frac{c_x^- y - 2c_u(d_n - d_m)}{c_x^+} \le t \le t_x,$$

$$u(t) = d_m + \frac{c_x^-}{2c_u}(t_x + y) - \frac{c_x^-}{2c_u} t \text{ for } t_x \le t \le t_x + y.$$

Otherwise

$$u(t) = d_n + \frac{c_x^+}{2c_u}(t - t_{out}) \text{ for } t_{out} \le t \le t_{in} \quad \text{if } d_m > d_n,$$

where $t_{out} = \frac{d_m}{d_m - d_n} y - \frac{2c_u}{c_x^+} d_m, \quad t_{in} = \frac{d_m}{d_m - d_n} y - \frac{2c_u}{c_x^+} d_n,$

$$y = \frac{1}{d_m}\left(\frac{c_u}{c_x^+}(d_m - d_n)^2 + \frac{2c_u}{c_x^+}(d_m - d_n)d_n - d_n t_{n-1} + d_m t_{m-2} - \sum_{j=n+1}^{j=m-1} d_j (t_{j-1} - t_{j-2}) \right);$$

$$u(t) = d_n - \frac{c_x^-}{2c_u}(t - t_{out}) \text{ for } t_{out} \le t \le t_{in} \quad \text{if } d_m < d_n,$$

where

$$t_{out} = \frac{d_m}{d_m - d_n} y + \frac{2c_u}{c_x^-} d_m, \quad t_{in} = \frac{d_m}{d_m - d_n} y + \frac{2c_u}{c_x^-} d_n,$$

$$y = \frac{1}{d_m}\left(-\frac{c_u}{c_x^-}(d_m - d_n)^2 - \frac{2c_u}{c_x^-}(d_m - d_n)d_n - d_n t_{n-1} + d_m t_{m-2} - \sum_{j=n+1}^{j=m-1} d_j (t_{j-1} - t_{j-2}) \right).$$

PROOF. According to Lemma 3.9, there exists at most one time point t_x between intervals τ_n and τ_m, such that $X(t_x)$ equals zero. Let us first assume that the point does exist and belongs to interval $\tau_k = [t_{k-2}, t_{k-1}]$, i.e. $X(t) > 0$ for $t_{n-1} < t_x$ and $X(t) < 0$ for $t_x < t_{m-2}$. Then according to (3.17) and (3.5)

$$X(t_x) = X(t_{in}) = X(t_{out}) = 0, \quad \psi(t_{in}) - \psi(t_{out}) = 2c_u(d_m - d_n), \tag{3.28}$$

where t_{out} is the terminal point of the first fractional regime, $t_{out} \in \tau_n$ and t_{in} is the initial point of the other fractional regime, $t_{in} \in \tau_m$.

By integrating the system of the state and co-state differential equations at interval $[t_{out}, t_{in}]$ with three boundary conditions (3.28), we find three corresponding algebraic equations in three unknowns t_x, t_{out} and t_{in}

$$d_n(t_{n-1}-t_{out})+d_k(t_x-t_{k-2})+\sum_{j=n+1}^{j=k-1} d_j(t_{j-1}-t_{j-2})=d_n(t_x-t_{out})+\frac{c_x^+}{4c_u}(t_x-t_{out})^2,$$

$$d_k(t_{k-1}-t_x)+d_m(t_{in}-t_{m-2})+\sum_{j=k+1}^{j=m-1} d_j(t_{j-1}-t_{j-2})=d_m(t_{in}-t_x)+\frac{c_x^-}{4c_u}(t_{in}-t_x)^2,$$

$$2c_u d_m=2c_u d_n+c_x^+(t_x-t_{out})-c_x^-(t_{in}-t_x).$$

By denoting $y=t_{in}-t_x$ in the last system, we find the quadratic equation in y as well as the expression for t_x stated in the lemma. Thus, t_{in} and t_{out} are calculated as

$$t_{in}=t_x+y \quad \text{and} \quad t_{out}=t_x+\frac{2c_u(d_n-d_m)-c_x^- y}{c_x^+},$$

that proves the first condition of the lemma. Given t_x, t_{in} and t_{out}, the control is immediately obtained from (3.17) and (3.5).

The second and third conditions of the lemma are special cases of the first one and are similarly obtained by assuming that the point t_x such that $X(t_x)=0$ does not exist.
■

Lemma 3.10 has proved the correctness of Steps 2 of Algorithm 3.5. The next lemma proves the correctness of the last step of Algorithm 3.5. This step presents a breadth-first search, which, unlike the standard breadth-first search (see Algorithm 2.2), paints the arcs of graph *G* rather than its nodes.

LEMMA 3.11. (Correctness of Algorithm 3.5 Step 3) Suppose that Step 3 of Algorithm 3.5 runs on a given graph *G*. Then there is a feasible path starting at node 1 and terminating at node *K* that consists of black arcs.

PROOF. From the description of Step 3, a white arc remains uncolored if, and only if, it has no gray predecessors. This means that such an arc is unachievable from node 1 and cannot be a part of any feasible path from node 1 to node *K*. Next, if a gray arc *v* becomes red in the course of coloring, then it has no successor satisfying the required condition $a(succ(v)) \geq b(v)$. Thus the arc cannot be a part of a feasible path from 1 to *K* as well.

The main part of the proof is for black arcs. We wish to prove that for any black arc, v, there is a feasible path, starting at node 1 and terminating at the arc v, that consists of black arcs. The proof proceeds by induction on L, the number of arcs making up the feasible path from 1 to the v. Clearly, for L=1, the claim trivially holds. As an inductive hypothesis, we assume that the claim holds for the path of length L-1 (L>1) starting at node 1 and terminating at a predecessor, *pred*(v). But, according to the choice of v, v is black. This implies that: (1) there is a predecessor arc, *pred*(v), to arc v such that $a(v) \geq b(pred(v))$, and (2) there is a successor *succ*(v) such that $a(succ(v)) \geq b(v)$. Then, according to the definition of a feasible path, a path of length L leading from node 1 to arc v through arc *pred*(v) is feasible as well.
■

THEOREM 3.2. If demand is a K-interval, single-mode, piece-wise constant function of time, then problem (3.15), (3.2) and (3.3) is solvable in $O(K^3)$ time.

PROOF. At Step 1, Algorithm 3.5 exactly identifies K intervals where the fractional production may occur. At Step 2, the algorithm invokes one of the four subroutines which calculates the analytical solution in $O(K)$ time. Since Step 2 is repeated at most $|A|$ times, it needs a total of $O(K^3)$ time.

The last step of the algorithm includes three procedures. The first one requires $O(K)$ time. After this procedure starts to work, no colored arc is whitened, and thus no white arc is colored gray twice. This ensures that each arc is inserted into queue Q at most only once and hence removed from Q at most only once. Similarly, no gray arc is painted black or red twice. These operations of enqueuing and dequeuing require $O(1)$ time, so that the total time devoted to queue operations is $O(|A|)$. Further, to estimate the worst-case complexity of procedure (ii) of this step, notice that: (1) each of $|A|$-2 steps of procedure (ii) examines one gray arc and scans all its white successors; (2) each arc in graph G has no more than K-2 successors. Therefore, at most, $O(K|A|)$ time is spent in total scanning all successors of all gray arcs. Procedure (iii), the backward search for the feasible path from node 1 to node K through black arcs, requires only $O(K)$ time. By recalling the magnitude of set A, we conclude that the total running time of Step 3 is $O(K^3)$.
■

Special Cases of Zero Costs

The problem with no production cost, $1|1|u(t) \in [0,\infty]\,|d(t),T|\,L_x$, has been solved in Section 3.1.1. The problem with no inventory cost, $1|1|u(t) \in [0,\infty]\,|d(t),T|\,Q_u, L_x, c_x^+ = 0$, is solved by Algorithms 3.3-3.5 without significant simplifications. However, subroutines to Algorithm 3.5 become simpler so that only procedure (v) has to be carried out.

No Backlog

When backlogs are disallowed, i.e. for the $1|1|u(t)\in[0,\infty], X(t)\geq 0\,|d(t),T|\,Q_u, L_x$ problem, the solution algorithm can be obtained by minor modifications to Algorithms 3.3-3.5 . Namely, since the first two solvable cases proven in Lemmas 3.3 and 3.4 are realized only if negative inventories are allowed, they become unfeasible due to the no-backlog constraint. Assuming c_x^- is very large, the third and the only solvable case (Step 3 of Algorithm 3.3) become simpler because the optimal control has a single function, which is $u(t)=d(t)$ and the condition for this regime to occur is $2c_u\dot{d}(t)\leq c^+$. Prior to presenting a modification to Algorithm 3.4, we introduce the dual formulation for the no-backlog case.

Dual Formulation

Now problem (3.15), (3.2) and (3.3) has the additional constraint $X(t)\geq 0$ which affects neither Hamiltonian (3.16) nor control equation (3.17), but changes the co-state differential equation

$$d\psi(t)=c_x^+dt-d\mu(t),\;\; \psi(T+0)=0, \tag{3.29}$$

where Lagrange multiplier $d\mu(t)\geq 0$ corresponds to the no-backlog constraint and satisfies the complementary slackness condition

$$\int_0^T X(t)d\mu(t)=0. \tag{3.30}$$

Algorithm 3.6

Step 1:

If $2c_u\dot{d}(t)\leq c^+$ along the entire planning horizon, then set the optimal solution as $u(t)=d(t)$ for $0\leq t\leq T$, STOP; otherwise go to the next step.

Step 2:

Set accuracy ξ required for computing the optimal solution. Calculate accuracy of integration $\varepsilon=\dfrac{c_u}{3T^2c_x^{+2}}\xi$. Select time intervals $\tau_k=[t_{left}^k, t_{right}^k]$ at which $2c_u\dot{d}(t)\leq c^+$, $t_{right}^k<t_{left}^{k+1}$ and number them beginning from k=2. In addition, define τ_1 and τ_K as the intervals which are degenerated to a point: $\tau_1=\{0\}$ and $\tau_K=\{T\}$, where K is the total number of obtained intervals.

Step 3 and Step 4 are identical to Steps 2 and 3 of Algorithm 3.4.

Subroutine 1 to Algorithm 3.6

(i) Set time points $t_x = t^n_{left}$, $t_y = t^n_{right}$, $t_{out} = t^n_{left}$.

(ii) Integrate the following system of differential equations

$$\dot{X}(t) = \frac{\psi(t)}{2c_u} - d(t), \quad X(t_{out}) = 0,$$

$$\dot{\psi}(t) = c^+_x, \qquad \psi(t_{out}) = 2c_u d(t_{out}),$$

over the interval $[t_{out}, t^m_{right}]$.

If $X(t) < 0$ for all $t \in \tau_m$, then no arc links nodes τ_n and τ_m; STOP.

Otherwise, set $t_{out} = t^n_{right}$ and integrate the above system of differential equations.

If $X(t)>0$ for all $t \in \tau_m$, then no arc links nodes τ_n and τ_m; STOP.

Otherwise go to (iii)

(iii) Integrate the system from (ii) with $t_{out} = 0.5(t_x + t_y)$ and set time point $t_{in} \in \tau_m$ such that $X(t_{in}) = 0$.

If $\psi(t_{in}) > 2c_u d(t_{in})$, then $t_x = t_{out}$ and go to (iv);

if $\psi(t_{in}) < 2c_u d(t_{in})$, then $t_y = t_{out}$ and go to (iv).

(iv) If $t_y - t_x \le \varepsilon$, then set $arc_{nm} = (\tau_n, \tau_m) \in A$ and its three weights

$a(arc_{nm}) = t_{out}$, $b(arc_{nm}) = t_{in}$ and

$c(arc_{nm}) = \int_{t_{out}}^{t_{in}} c^+_x X(t) + c_u u(t)^2 \, dt$; go to (v). Otherwise go to (iii).

(v) If there exist arcs arc_{jn} such that $\left|b(arc_{jn}) - a(arc_{nm})\right| < \varepsilon$, then set $a(arc_{nm}) = \max_j b(arc_{jn})$. For each arc arc_{jn} entering node n, update weights

$$SD^n_{jm} = \begin{cases} \displaystyle\int_{b(arc_{jn})}^{a(arc_{nm})} c_u d(t)^2 \, dt, & \text{if } a(arc_{nm}) \ge b(arc_{jn}); \\ -1, & \text{otherwise.} \end{cases} \quad \text{STOP.}$$

Otherwise, STOP.

Discussion

Step 1 presents the closed form of the optimal control, $u(t)=d(t)$, along the entire planning horizon when $2c_u\dot{d}(t) \le c^+$. This solution is due to $X(t)=0$, $\psi(t)=2c_u d(t)$, $\dot{\mu}(t)=c_x^+ - 2c_u\dot{d}(t)$ for $0 \le t \le T$ and jump $d\mu(T)=-d\psi(T)=2c_u d(T)$.

The other steps of Algorithm 3.6 approximate the globally optimal solution with the accuracy and complexity of Algorithm 3.4 when c_x^- is eliminated. At Step 2 of Algorithm 3.6, the condition for selecting time intervals τ_k is simplified, while the other two steps are identical to those of Algorithm 3.4. However, Step 3 invokes four subroutines which are now slightly changed. Namely, the co-state differential equation integrated iteratively in the subroutines is modified as presented in procedure (ii) of Subroutine 1 to Algorithm 3.6. Subroutines 2-4 to Algorithm 3.6 are omitted since they are modified accordingly. Note, that with respect to (3.30), $d\mu(t)=0$ at time intervals of positive inventories. However, if inventory drops down to zero, jumps $d\mu(t)>0$ at points $t=t_{out}$ and $t=t_{in}$ are formally possible. The following lemma proves that there are no jumps at t_{out} and t_{in} and thus the system of the state and co-state equations solved at Step 3 is correct.

LEMMA 3.12. (Correctness of Algorithm 3.6) Given two time intervals $[t_1,t_{out}]$ and $[t_{in},t_2]$ at which $X(t)=0$ and $\psi(t)=2c_u d(t)$ (the fractional regime), the following condition holds, $d\mu(t_{in})=d\mu(t_{out})=0$.

PROOF. The proof is by contradiction. Suppose $d\mu(t_{out})>0$, that is the co-state variable jumps down. Then, in a right neighborhood of t_{out}, $\psi(t)<2c_u d(t)$. Thus, $u(t)<d(t)$ and $X(t)<0$ after t_{out}. This contradicts the no-backlog constraint. Next, suppose $d\mu(t_{in})>0$, then, in a left neighborhood of t_{in}, $\psi(t)>2c_u d(t)$. Thus, $u(t)>d(t)$ and $X(t)<0$ before t_{in}.

■

3.1.4 NON-LINEAR INVENTORY AND PRODUCTION COSTS

Both costs entering the objective are now non-linear. This further complicates the optimal behavior of the system. Only special cases, in which demand is a polynomial function in time, are found to be analytically solvable.

Backlogs

The problem to be considered is one-machine, one-product-type scheduling with backlogs to minimize the quadratic inventory and production costs. This is the $1|1|u(t) \in [0,\infty]|d(t),T|Q_u, Q_x$ problem which is stated as follows

$$J = \int_0^T \left[c_x^+ X^+(t)^2 + c_x^- X^-(t)^2 + c_u u(t)^2\right] dt \to \min, \tag{3.31}$$

subject to (3.2) and (3.3).

Dual Formulation

The Hamiltonian, H, resulting from the maximum principle is

$$H(t) = -\left[c_x^+ X^+(t)^2 + c_x^- X^-(t)^2 + c_u u(t)^2\right] + \psi(t)\left[u(t) - d(t)\right], \tag{3.32}$$

where co-state variable $\psi(t)$ satisfies co-state differential equation (3.12). Maximization of the Hamiltonian as a function of $u(t)$ results, as in the previous section, in

$$u(t) = \begin{cases} \dfrac{\psi(t)}{2c_u}, & \text{if } \psi(t) \geq 0; \\ 0, & \text{if } \psi(t) < 0. \end{cases} \tag{3.33}$$

Algorithm 3.7

Step 1:

If the demand function is a polynomial of order P, $d(t) = \sum_{p=0}^{p=P} a_p t^p$,

and $c_x^+ = c_x^-$, then calculate $y = \sqrt{\dfrac{c_x^+}{c_u}}$, $b_P = 2c_u a_P$, $b_{P-1} = 2c_u a_{P-1}$;

$b_p = 2c_u a_p + \dfrac{(p+1)(p+2)}{y^2} b_{p+2}$ for p=0,...,P-2;

$$A = -\frac{b_1}{y} \text{ and } B = \frac{\frac{b_1}{y} sh(yT) - \sum_{p=0}^{p=P} b_p T^p}{ch(yT)}.$$

Set the optimal solution as

$$u(t) = \frac{1}{2c_u}\left(Ash(yt) + Bch(yt) + \sum_{p=0}^{p=P} b_p t^p \right), \ 0 \le t \le T; \text{ STOP.}$$

Otherwise go to the next step.

Step 2:

(i) Set accuracy ξ required for computing the optimal objective value. Calculate upper co-state estimate $\hat{\psi} = 2c_x^- \int_0^T \int_0^t d(\tau) d\tau dt$, parameter $g = \sqrt{\frac{\max\{c_x^+, c_x^-\}}{c_u}}$, accuracy of the dichotomous search $\hat{\varepsilon} = c_u \left(T\hat{\psi}(1 + T^2 g^2) ch(gT) \right)^{-1} \xi$ and accuracy of integration, $\varepsilon = \left(\max\left\{ g^2 T\hat{\psi}, 2c_x^- D(T) \right\} \right)^{-1} \hat{\varepsilon}$. Assign $\psi^{\max} = \psi^{\min} = \hat{\psi}$.

(ii) Integrate the following system of ordinary differential equations

$$\dot{\psi}(t) = \begin{cases} 2c_x^+ X(t), & \text{if } X(t) > 0; \\ 2c_x^- X(t), & \text{if } X(t) \le 0; \end{cases}$$

$$\dot{X}(t) = \frac{\psi(t)}{2c_u} - d(t), \qquad (3.34)$$

over the interval $[0,T]$ with initial conditions $X(0) = 0$ and $\psi(0) = 0.5(\psi^{\min} + \psi^{\max})$.

(iii) If $\psi(T) \ge 0$, then assign $\psi^{\max} = \psi(0)$;
otherwise, assign $\psi^{\min} = \psi(0)$.

(iv) If $\psi^{\max} - \psi^{\min} > \hat{\varepsilon}$, then go to step (ii); otherwise STOP.

Example 3.5
Consider the $1|1|u(t) \in [0,\infty]|d(t),T|Q_u, Q_x$ problem with the following parameters: $T = 10$ hr, c_u=10\$hr/unit^2, c_x^+=0.1\$/hr.unit^2, c_x^-=0.1\$/hr.unit^2 and second-order (P=2) polynomial demand $d(t) = 10 + 0.1t^2$ unit/hr.

Since $c_x^+ = c_x^-$, Step 1 of Algorithm 3.7 calculates γ=0.1, b_2=2, b_1=0, b_0=600, A=0 and B= -800/ch(1). Then it sets the optimal solution as

$$u(t) = 30 + 0.1t^2 - 40\frac{ch(0.1t)}{ch(1)} \text{ unit/hr for } 0 \le t \le 10.$$

■

Discussion

Step 1 of Algorithm 3.7 provides a closed analytical form of the optimal control for the demand function expressed as a polynomial when the inventory cost is equal to the backlog cost. The complexity of this step is $O(P)$ and is due to calculation of coefficients b_p. If, however, the conditions of Step 1 are not met, the exact optimal solution cannot be found. Step 2 of the algorithm effectively approximates the solution for a given accuracy when demand is arbitrary and no *a priori* conditions are imposed on the cost coefficients.

LEMMA 3.13. Given problem (3.31), (3.2) and (3.3), the co-state variable $\psi(t)$ is bounded as $\psi(t) \in [0, \hat{\psi}]$ for $t \in [0,T]$, where $\hat{\psi} = 2c_x^- \int_0^T \int_0^t d(\tau)d\tau dt$.

PROOF. The upper bound of the co-state variable is found by assuming zero production rate along the entire planning horizon, $u(t) = 0$ for $t \in [0,T]$. In this case, integration of the state and co-state differential equations results in $X(t) = -\int_0^t d(\tau)d\tau$ and $\psi(t) = 2c_x^- \int_t^T \int_0^s d(\tau)d\tau ds$ respectively. Since demand is a non-negative function of time, the maximum of $\psi(t)$ is achieved at t=0 that is denoted in the lemma by $\hat{\psi}$.

The proof of the lower bound is identical to that of Lemma 3.8.
■

LEMMA 3.14. (Correctness of Algorithm 3.7 Step 1) Given problem (3.31), (3.2) and (3.3), if the demand function is a polynomial of order P, $d(t) = \sum_{p=0}^{p=P} a_p t^p$, and $c_x^+ = c_x^-$, then the optimal solution is $u(t) = \frac{1}{2c_u}\left(Ash(yt) + Bch(yt) + \sum_{p=0}^{p=P} b_p t^p\right)$,

where $y = \sqrt{\frac{c_x^+}{c_u}}$, $b_P = 2c_u a_P$, $b_{P-1} = 2c_u a_{P-1}$; $b_p = 2c_u a_p + \frac{(p+1)(p+2)}{y^2} b_{p+2}$ for

p=0,...,P-2; $A = -\frac{b_1}{y}$ and $B = \frac{\frac{b_1}{y} sh(yT) - \sum_{p=0}^{p=P} b_p T^p}{ch(yT)}$.

PROOF. By using $c_x^+ = c_x^-$, the co-state differential equation (3.12) simplifies to

$$\dot{\psi}(t) = 2c_x^+ X(t). \tag{3.35}$$

By differentiating (3.35) and substituting (3.2), we obtain

$$\ddot{\psi}(t) = 2c_x^+ (u(t) - d(t)). \tag{3.36}$$

Next, by taking into account Lemma 3.13 and (3.33), we find

$$\ddot{\psi}(t) = \frac{c_x^+}{c_u} \psi(t) - 2c_x^+ d(t).$$

Thus, the conditions of the lemma allow the system of state and co-state differential equations (3.34) to be presented as a single, second-order ordinary differential equation

$$\ddot{\psi}(t) = \frac{c_x^+}{c_u} \psi(t) - 2c_x^+ \sum_{p=0}^{p=P} a_p t^p, \qquad \dot{\psi}(0) = 0, \qquad \psi(T) = 0. \tag{3.37}$$

The solution of (3.37) is (see Section 2.3.1)

$$\psi(t) = Ash(yt) + Bch(yt) + \sum_{p=0}^{p=P} b_p t^p, \tag{3.38}$$

where $y = \sqrt{\frac{c_x^+}{c_u}}$. Coefficients b_p are calculated by substituting (3.38) into (3.37) and equalizing polynomial terms of the same order. Finally, coefficients A and B are found

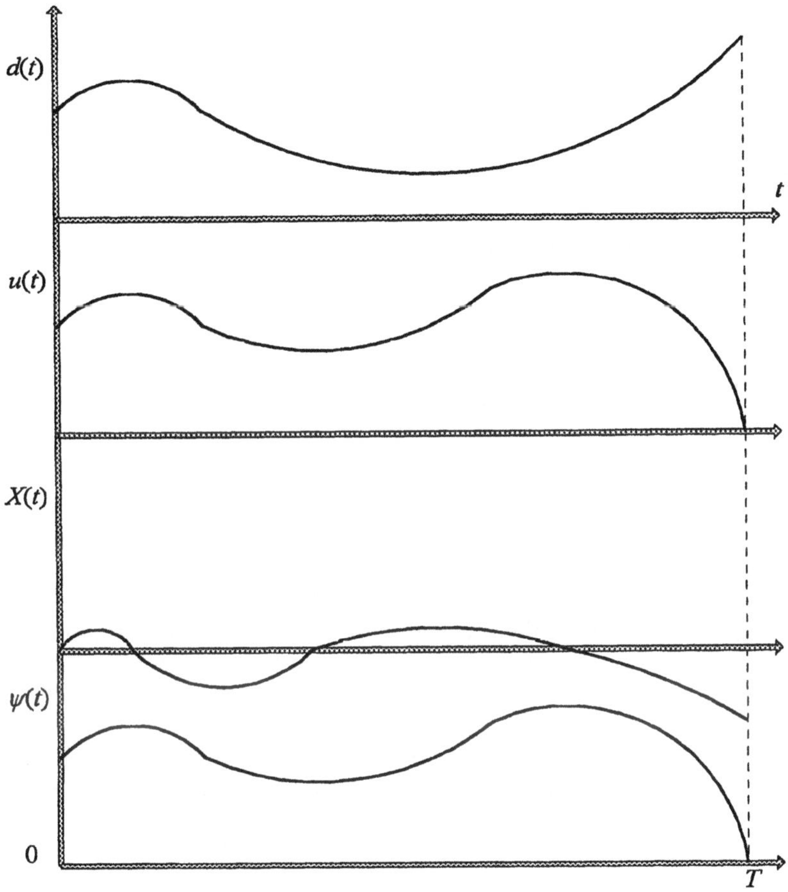

Figure 3.9. Demand $d(t)$, optimal control $u(t)$, state $X(t)$ and co-state $\psi(t)$ variables for the $1|1|u(t)\in[0,\infty]|d(t),T|Q_u,Q_x$ problem.

from the boundary conditions (3.37) substituted in (3.38). The optimal behavior of the state, co-state and control variables is illustrated in Figure 3.9.

■

If the conditions of Algorithm 3.7 Step 1 are not satisfied, Step 2 then approximates the optimal solution. This step uses the dichotomous search for locating $\psi(0)$ - the missing initial condition of the system of state and co-state differential equations (3.34). Via integration of this system, the algorithm obtains a value of $\psi(T)$ for each $\psi(0)$. Since the required value of $\psi(T)$ is zero, the dichotomous search determines a root of a slack function which maps $\psi(0) \to \psi(T)$.

Definition 3.7. Given system (3.34), F_5 is a function which maps $\psi(0) \to \psi(T)$.

LEMMA 3.15. Slack function F_5 is non-decreasing.

PROOF. The proof is similar to that of Lemma 3.7 and, therefore, is omitted.
■

THEOREM 3.3. Problem (3.31), (3.2) and (3.3), is solvable with accuracy ξ in $O\left(\frac{T\max\{g^2T\hat{\psi},2c_x^- D(T)\}}{\hat{\psi}}\frac{\hat{R}}{\xi}\log\frac{\hat{R}}{\xi}\right)$ time, where $\hat{R}=\frac{T\hat{\psi}^2}{c_u}ch(gT)\left(1+g^2T^2\right)$ and $g=\sqrt{\frac{\max\{c_x^+,c_x^-\}}{c_u}}$.

PROOF. First note that Step 2 of Algorithm 3.7 locates $\psi(0)$ such that the primal and dual formulations of the problem are satisfied with a given accuracy. The algorithm thus approximates the global optimum since the primal problem has convex objective and linear constraints.

To estimate the algorithm's complexity, we compare the complexity of the procedures which Step 2 consists of. Calculation of $\hat{\psi}$ with accuracy ε, by procedure (i) as well as integration of (3.34) with the same accuracy by procedure (ii), requires $O\left(\frac{T}{\varepsilon}\right)$ time. Comparisons executed by procedures (iii) and (iv) are of $O(1)$ complexity. Due to the monotonicity of the slack function (see Lemma 3.15), the dichotomous search repeats procedures (ii), (iii) and (iv) $O\left(\log\frac{\hat{\psi}}{\hat{\varepsilon}}\right)$ times. Thus, the total complexity of Algorithm 3.7 Step 2 is

$$O\left(\frac{T}{\varepsilon}\log\frac{\hat{\psi}}{\hat{\varepsilon}}\right). \tag{3.39}$$

Finally, we estimate the relationship between computational accuracy of the objective ξ, accuracy of integration ε and accuracy of the dichotomous search $\hat{\varepsilon}$. By varying the objective function (3.31), we determine

$$\xi \le 2T \max\{c_x^+ \max_t X^+(t), c_x^- \max_t X^-(t)\} \max_t |\Delta X(t)| + 2Tc_u \max_t u(t) \max_t |\Delta u(t)| . \quad (3.40)$$

The value of $\max_t u(t)$ can be estimated from (3.33) as $\frac{1}{2c_u} \max_t \psi(t) = \frac{\hat{\psi}}{2c_u}$. The value of $\max_t |\Delta u(t)|$ is also estimated from (3.33) as $\frac{1}{2c_u} \max_t |\Delta\psi(t)|$. The maximum inventory and backlog are estimated from the state equation as $\max_t X^+(t) = T \max_t u(t) = \frac{T\hat{\psi}}{2c_u}$ and $\max_t X^-(t) = \int_0^T d(t)dt = D(T)$. To estimate the error of the co-state variable, we take into account that accuracy of the dichotomous search $\hat{\varepsilon}$ is related to integration accuracy ε as

$$\hat{\varepsilon} = \max_t |\dot{\psi}(t)| \varepsilon = 2 \max\{c_x^+, c_x^-\} \max_t |X(t)| \varepsilon = \max\{g^2 T\hat{\psi}, 2c_x^- D(T)\} \varepsilon . \quad (3.41)$$

The error of the co-state variable $\Delta\psi(t)$ in the dichotomous search satisfies the following differential equation

$$\Delta\ddot{\psi}(t) = g^2 \Delta\psi(t), \quad \Delta\psi(0) = 2\hat{\varepsilon} \ , \quad \Delta\dot{\psi}(0) = 0 .$$

By solving the last equation, we find $\Delta\psi(t) = 2ch(gt)\hat{\varepsilon}$. Thus, $\max_t \Delta\psi(t) = 2ch(gT)\hat{\varepsilon}$. The upper estimates for $\max_t |\Delta u(t)|$ and $\max_t |\Delta X(t)|$ are $\max_t |\Delta u(t)| = \frac{1}{2c_u} \max_t |\Delta\psi(t)|$ and $\max_t |\Delta X(t)| = \frac{T}{2c_u} \max_t |\Delta\psi(t)|$ respectively. By substituting the found estimates into (3.40) , we bound ξ by a function of $\hat{\varepsilon}$

$$\xi \le \frac{\hat{\psi}T}{c_u} ch(gT)\left(1 + g^2T^2\right)\hat{\varepsilon} . \quad (3.42)$$

To elaborate on the final estimates of complexity and accuracy for Algorithm 3.7 Step 2, we express $\hat{\varepsilon}$ and ε from (3.41) and (3.42), substitute them into (3.39) and introduce parameter $\hat{R}$ to simplify the presentation.
■

Special Cases of Zero Costs

The problem with no production cost, $1|1|u(t) \in [0,\infty]|d(t),T|Q_x$, has been solved in Section 3.1.2. The problem with no inventory cost,

$1|1|u(t) \in [0,\infty]\,|d(t),T|\,Q_u, Q_x, c_x^+ = 0$, is solved by Step 2 of Algorithm 3.7 with no modification, while Step 1 of the same algorithm is not applicable since it requires $c_x^+ = c_x^-$.

No Backlog

To solve the $1|1|u(t) \in [0,\infty], X(t) \geq 0\,|d(t),T|\,Q_u, Q_x$ problem, we suggest a modification to approximation Algorithm 3.4 which copes with arbitrary demands. The case of a non-increasing demand function is solved here exactly.

Dual Formulation

Now problem (3.31), (3.2) and (3.3) has the additional constraint $X(t) \geq 0$ which affects neither Hamiltonian (3.32) nor control equation (3.33), but changes the co-state differential equation

$$d\psi(t) = c_x^+ X(t)dt - d\mu(t), \quad \psi(T+0) = 0, \tag{3.43}$$

where Lagrange multiplier $d\mu(t) \geq 0$ corresponds to the no-backlog constraint and satisfies the complementary slackness condition (3.30).

Algorithm 3.8

Step 1:

If demand is a non-increasing function along the entire planning horizon, then set the optimal solution as

$u(t)=d(t)$ for $0 \leq t \leq T$, STOP.

Otherwise if demand is a continuous and differentiable function of time, then go to the next step.

Step 2:

Set accuracy ξ required for computing the optimal solution. Calculate $A = \max_t d(t)$, $B = \max_t |\dot{d}(t)|$ and accuracy of integration $\varepsilon = \left(2TABKc_u ch(yT)(1+y^2T^2)\right)^{-1}\xi$. Select time intervals $\tau_k = [t_{left}^k, t_{right}^k]$ at each of which the demand function decreases, $t_{right}^k < t_{left}^{k+1}$, and number them beginning from k=2. In addition, define τ_1 and τ_K as the intervals which are degenerated to a point: $\tau_1 = \{0\}$ and $\tau_K = \{T\}$, where K is the total number of obtained intervals.

Step 3 and Step 4 are identical to Steps 2 and 3 of Algorithm 3.4.

Subroutine 1 to Algorithm 3.8

(i) Set time points $t_x = t^n_{left}$, $t_y = t^n_{right}$, $t_{out} = t^n_{left}$.

(ii) Integrate the following system of differential equations:

$$\dot{X}(t) = \frac{\psi(t)}{2c_u} - d(t), \quad X(t_{out}) = 0,$$

$$\dot{\psi}(t) = 2c_x^+ X(t), \qquad \psi(t_{out}) = 2c_u d(t_{out}),$$

over the interval $[t_{out}, t^m_{right}]$.

If $X(t) < 0$ for any $t < t^m_{left}$, then no arc links nodes τ_n and τ_m; STOP.

Otherwise, set $t_{out} = t^n_{right}$ and integrate the above system.

If $X(t) > 0$ for all $t \in \tau_m$, then no arc links nodes τ_n and τ_m; STOP.

Otherwise go to (iii)

(iii) Integrate the system from (ii) with $t_{out} = 0.5(t_x + t_y)$;

If $X(t)$<0 for any $t < t^m_{left}$, then $t_y = t_{out}$ and go to (iv),

otherwise, set time point $t_{in} \in \tau_m$ such that $X(t_{in}) = 0$.

If $\psi(t_{in}) > 2c_u d(t_{in})$, then $t_x = t_{out}$ and go to (iv);

if $\psi(t_{in}) < 2c_u d(t_{in})$, then $t_y = t_{out}$ and go to (iv).

(iv) If $t_y - t_x \leq \varepsilon$, then set $arc_{nm} = (\tau_n, \tau_m) \in A$, assign its three weights $a(arc_{nm}) = t_{out}$, $b(arc_{nm}) = t_{in}$ and

$$c(arc_{nm}) = \int_{t_{out}}^{t_{in}} c_x^+ X^+(t)^2 + c_x^- X^-(t)^2 + c_u u(t)^2 \, dt \text{ and go to (v).}$$

Otherwise go to (iii).

(v) If there exist arcs arc_{jn} such that $|b(arc_{jn}) - a(arc_{nm})| < \varepsilon$, then set $a(arc_{nm}) = \max_j b(arc_{jn})$. For each arc arc_{jn} entering node *n*, update weights

$$SD^n_{jm} = \begin{cases} \int_{b(arc_{jn})}^{a(arc_{nm})} c_u d(t)^2 \, dt, & \text{if } a(arc_{nm}) \geq b(arc_{jn}); \\ -1, & \text{otherwise.} \end{cases} \quad \text{STOP.}$$

Otherwise, STOP.

Discussion

Step 1 of Algorithm 3.8 presents a closed form of the optimal control, $u(t)$=$d(t)$, when demand is a non-decreasing function along the entire planning horizon. This control function implies that the solution $X(t)=0$, $\psi(t)=2c_u d(t)$, $\dot{\mu}(t)=-2c_u \dot{d}(t)$ for $0 \le t \le T$ and jump $d\mu(T)=-d\psi(T)=2c_u d(T)$ are feasible.

The other steps of Algorithm 3.8 are very similar to the steps of Algorithm 3.4 and therefore approximate the globally optimal solution if demand is a continuous and differentiable function of time. At Step 2, the condition of selecting time intervals τ_k, where the fractional production can occur, is changed. The other two steps are identical to those of Algorithm 3.4. However, four subroutines invoked by Step 3 are now slightly changed. Namely, verification of the no-backlog condition is added and the co-state differential equation in the subroutines is modified as presented in Subroutine 1. Subroutines 2-4 to Algorithm 3.8 are omitted since they are modified accordingly.

Note, that according to (3.30), $d\mu(t)=0$ at time intervals of positive inventory. Although jumps of the measure function, $d\mu(t)>0$, at points $t=t_{out}$ and $t=t_{in}$ are formally possible, their values at these points are zero, which is proved similarly to the proof of Lemma 3.12.

The new condition for selecting intervals of possible fractional production $u(t)$=$d(t)$ is proved in the following lemma. The optimal solution is illustrated in Figure 3.10.

LEMMA 3.16. (Correctness of Algorithm 3.8 Steps 1 and 2) Given problem (3.31), (3.2) and (3.3) with backlog disallowed, if demand decreases at K intervals of time τ_k, $k=1,\dots,K$, then the optimal solution contains at most K time intervals of zero inventory, $\chi_k=\{t|X(t)=0\}\subseteq\tau_k$.

PROOF. The proof is by contradiction. Let the fractional production $u(t)$=$d(t)$ and $X(t)$=0 occur at an interval of time where demand increases. Then, according to (3.33), $\psi(t)$ must increase. However, from (3.43), $\psi(t)$ cannot increase, since $X(t)$=0 and $d\mu(t)\ge 0$ at this interval.

■

Prior to declaring complexity and accuracy of Algorithm 3.8 in Theorem 3.4, we need to estimate the upper and lower bounds of the co-state variable.

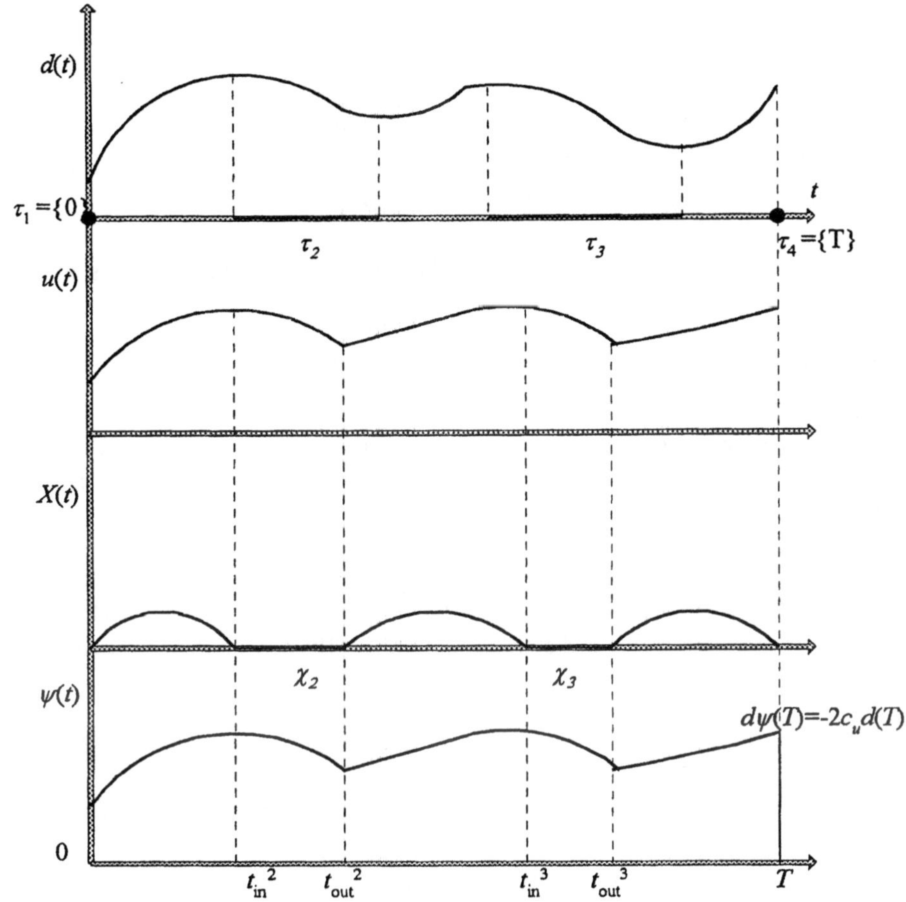

Figure 3.10. Demand $d(t)$, optimal control $u(t)$, state $X(t)$ and co-state $\psi(t)$ variables for the $1|1|u(t)\in[0,\infty], X(t)\geq 0|d(t),T|Q_u, Q_x$ problem.

LEMMA 3.17. Given problem (3.31), (3.2) and (3.3) with backlog disallowed, co-state variable $\psi(t)$ is bounded as $\psi(t)\in[0,2c_u \max_{t'} d(t')]$ for $t\in[0,T]$.

PROOF. From (3.33) we have $\psi(t)=2c_u d(t)$ at intervals of fractional production, $u(t)$=$d(t)$. At intervals between the fractional production, $\psi(t)$ increases to reach the value of $2c_u d(t)$ at the subsequent fractional production interval. Therefore, the upper bound of the co-state variable is determined by $2c_u \max_t d(t)$. The proof of the lower bound is identical to that of Lemma 3.8.

■

THEOREM 3.4. Given that demand is a continuous and differentiable function of time, bounded as $d(t) \le A$ and $|\dot{d}(t)| \le B$ for $t \in [0,T]$, the problem (3.31), (3.2) and (3.3) with backlog disallowed is solvable with accuracy ξ in $O\left(\max\{K^4, \frac{K^3 R}{\xi}\log\frac{KR}{\xi}\}\right)$ time, where K is the number of time intervals at which demand decreases, $R = 2ABT^2 c_u ch(yT)\left(1+y^2T^2\right)$ and $y = \sqrt{\frac{c_x^+}{c_u}}$.

PROOF. Complexity estimate (3.25) is evidently valid for Algorithm 3.8, since only Step 1, which needs $O(K)$ operations, has been changed in comparison with Algorithm 3.4.

However, this complexity estimate involves a new accuracy. To take this into account, we evaluate the relationship between required accuracy of the objective ξ and computational accuracy of the dichotomous search ε. By varying the objective function (3.31), we obtain (3.40) and estimate its terms.

The value of $\max_t u(t)$ can be estimated from equation (3.33) as $\frac{1}{2c_u}\max_t \psi(t) \le \max_t d(t) \le A$. Thus, the upper estimate of the maximum value for $X(t)$ is TA. The value of $\max_t|\Delta u(t)|$ is also estimated from (3.33) as $\frac{1}{2c_u}\max_t|\Delta\psi(t)|$. From (3.36) the error of the co-state variable $\Delta\psi(t)$ in the dichotomous search satisfies

$$\Delta\ddot{\psi}(t) = \frac{c_x^+}{c_u}\Delta\psi(t) \tag{3.44}$$

for each link starting at point t_{out} (see Step 2 of the algorithm). The boundary conditions for this second-order differential equation are $\Delta\psi(t_{out}) = 2c_u\dot{d}(t_{out})\varepsilon$ and $\Delta\dot{\psi}(t_{out}) = 0$. By solving equation (3.44) with these boundary conditions, we obtain $\Delta\psi(t) = 2c_u\dot{d}(t_{out})\varepsilon ch(y(t-t_{out}))$. By taking into account that there can be at most K links along the planning horizon T, $\max_t|\Delta\psi(t)| \le 2c_u BK\varepsilon ch(yT)$ as shown in Figure

3.11. The value of $\max_t |\Delta X(t)|$ is just $T \max_t |\Delta u(t)| = TBK\varepsilon ch(yT)$. By substituting the found estimates into (3.40), we bound maximum error ξ as a function of ε

$$\xi \leq 2TABKc_u ch(yT)\left(1 + y^2 T^2\right)\varepsilon. \tag{3.45}$$

To find the complexity and accuracy of Algorithm 3.8, we express ε from (3.45), substitute it into (3.25) and introduce parameter R to simplify the presentation.
■

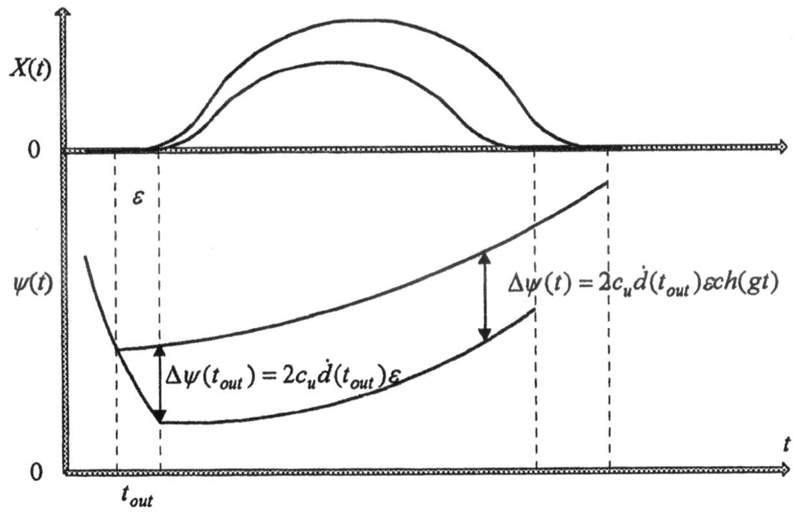

Figure 3.11. Increment in the error of the co-state variable due to the dichotomous search of Algorithm 3.8.

3.2 DUE DATE

This section addresses scheduling problems characterized by a special form of demand. This demand is defined by a pre-specified date which is commonly referred to as the due date, t_d. If backlogs are disallowed, then the end of the planning horizon is set at the due date, $T = t_d$. However, if backlogs are allowed and the demand is not completed before the due date, the production may continue after the due date until the end of the planning horizon. In such cases, the planning horizon plays the role of the *deadline*.

3.2.1 LINEAR PERFORMANCE MEASURE

Backlogs

The first problem to be considered is one-machine, one-product-type scheduling with backlogs to minimize the total linear cost of inventory and production. This is the $1|1|u(t)\in[0,\infty]|D,T|\, L_u, L_x$ problem which is stated as follows

$$J = \int_0^T \left[c_x^+ Z^+(t) + c_x^- Z^-(t) + c_u u(t)\right] dt \to \min, \tag{3.46}$$

$$Z^+(t) = \max\{0, X(t) - D(t)\}, Z^-(t) = \max\{0, D(t) - X(t)\},$$

subject to

$$\dot{X}(t) = u(t), \quad X(0) = 0, \tag{3.47}$$

$$u(t) \ge 0, \tag{3.48}$$

where

$$D(t) = \begin{cases} D, & \text{if } t \ge t_d; \\ 0, & \text{otherwise}; \end{cases} \tag{3.49}$$

and D is the amount of the product type to be produced by the due date t_d.

Dual Formulation

The Hamiltonian, H, resulting form the maximum principle is

$$H(t) = -\left[c_x^+ Z^+(t) + c_x^- Z^-(t) + c_u u(t)\right] + \psi(t) u(t). \tag{3.50}$$

The co-state variable, $\psi(t)$ in (3.50), satisfies the following co-state differential equation with its transversality condition

$$\dot{\psi}(t) = \begin{cases} c_x^+, \text{ if } X(t) > D(t); \\ -c_x^-, \text{ if } X(t) < D(t); \\ a, \ a \in [-c_x^-, c_x^+], \text{ if } X(t) = D(t); \end{cases} \qquad \psi(T) = 0. \tag{3.51}$$

According to the maximum principle, the Hamiltonian is the objective function of the dual problem and is maximized at every point of time with respect to the admissible controls $u(t)$. This implies inequality (3.6). That is, if $\psi(t) = c_u$ holds at an interval of time, then the optimal control $u(t)$ is non-negative. Otherwise, $\psi(t) < c_u$ and no control should be undertaken at this interval, i.e. $u(t)$=0.

Algorithm 3.9

Step 1:

Calculate the sum of the due date and the ratio of the production and backlog cost coefficients $\hat{T} = t_d + \dfrac{c_u}{c_x^-}$.

If $\hat{T} > T$ then set the optimal solution as
$u(t)$=0 over the entire planning horizon, STOP;
otherwise go to the next step.

Step 2:

Set the optimal solution as
$u(t) = 0$ for $0 \le t < t_d$ and $t_d < t \le T$,

$u(t_d) = \infty$, $\int_{t_d-0}^{t_d+0} u(t)dt = D$. STOP.

Discussion

Although the optimal solution that Algorithm 3.9 Step 2 finds in $O(1)$ time seems to be impractical, it states an important scheduling rule which can be straightforwardly interpreted in manufacturing. Namely, if a company is able to mobilize a very significant capacity, then no matter how expensive this will be, all the capacity must be utilized to start and carry out the production as close as possible to the due date. However, if the length of the planning horizon (deadline) is set too close to the required due date, $\hat{T} > T$, then it is not worthwhile to carry out the production at all. Capacity should not then be mobilized as determined by Step 1 of Algorithm 3.9.

One can observe that the dual formulation of the problem is almost identical to formulation (3.4)-(3.6) presented for the same problem but with general demands. This, of course, is not surprising. Although the due date is modeled differently, it is a special case of the general demand. Thus, condition (3.9) from Lemma 3.1 is valid, requiring no production (which is $X(t)=D(t)=0$) until the due date; instant satisfaction of the demand at the due date; and no-production thereafter. However, as with Lemma 3.1, there are cases where no-production at all is more profitable. The following lemma formalizes this property.

LEMMA 3.18. (Correctness of Algorithm 3.9) Given problem (3.46)-(3.49), the optimal control is determined as follows

- if $t_d + \frac{c_u}{c_x^-} \le T$ then $u(t)=0$, for $0 \le t < t_d$ and $t_d < t \le T$; $u(t_d) = \infty$, $\int_{t_d-0}^{t_d+0} u(t)dt = D$;
- otherwise $u(t)=0$ for $0 \le t \le T$.

PROOF. By considering the control dependent term of the Hamiltonian (3.50), we obtain

$$H_u(t) = u(t)[\psi(t) - c_u]. \tag{3.52}$$

Maximization of (3.52) implies that if $\psi(t) < c_u$ then $u(t)=0$, otherwise equation (3.6) must be satisfied, that is, $\psi(t) = c_u$. By differentiating this equality, we obtain

$$\dot{\psi}(t) = 0. \tag{3.53}$$

From (3.51), it immediately follows that equation (3.53) can be met only if $X(t)=D(t)$. However, the transversality condition in (3.51) requires $\psi(T) = 0$. This implies that there must be a switching point, t_1, before the end of the planning horizon, T, so that the co-state variable ends up at zero, $u(t)=0$ and $X(t)=D(t)$, $t_1 < t \le T$. Due to the special demand form (3.49) and equation (3.53), the optimal control must be set at zero before and after the due date. It must be also infinitely large to complete the overall demand at the due date, that is $\int_{t_d-0}^{t_d+0} u(t)dt = D$. Thus, the only feasible solution which satisfies the optimality condition (3.6) is $t_1 = t_d$. By integrating the corresponding condition from (3.51) and considering the fastest possible rate of the co-state variable decrease c_x^-, we find

$$\psi(t) = c_u,\ 0 \le t \le t_d;\ \psi(t) = c_x^-(\hat{T} - t),\ t_d \le t \le \hat{T}.$$

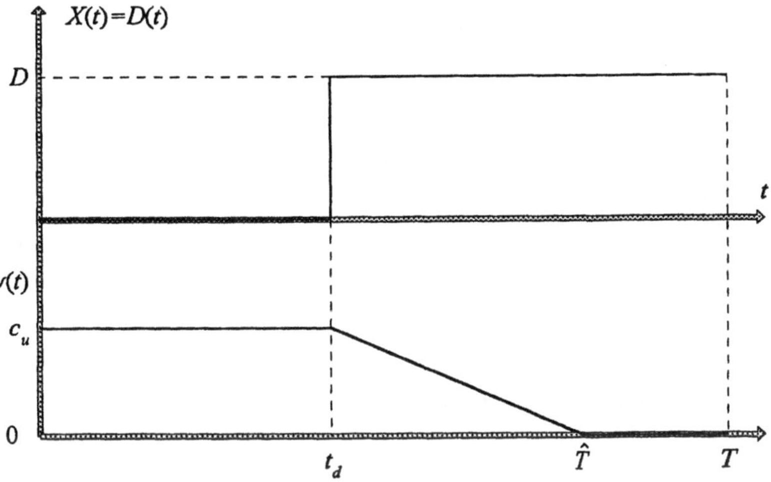

Figure 3.12. Demand $D(t)$, optimal state $X(t)$ and co-state $\psi(t)$ variables for the $1|1|u(t)\in[0,\infty]|D,T|\,L_u, L_x$ problem.

Note, although $\psi(t)$ becomes equal to zero, this does not necessarily take place at the end of the planning horizon, i.e. $\hat{T}\le T$. This is due to the fact that $X(t)$=$D(t)$ through the end of the planning horizon. Once $\psi(\hat{T})=0$, the co-state variable can remain zero for $\hat{T}\le t\le T$. This optimal solution is illustrated in Figure 3.12.

To find the condition, when it is profitable to produce the product, we simply verify

$$\psi(t_d)=c_x^-(\hat{T}-t_d)=c_u,\ \ \hat{T}\le T,$$

from where condition $t_d+\dfrac{c_u}{c_x^-}\le T$ stated in the lemma is obtained.

■

Special Cases of Zero Costs

First, consider the zero production cost, i.e. the $1|1|u(t)\in[0,\infty]|D,T|\,L_x$ problem. The solution algorithm can be obtained by an easy modification to Algorithm 3.9. Indeed, zero production cost implies that Step 1 of Algorithm 3.9 will never indicate the no-production regime over the entire planning horizon. The second step will always result in completing the demand at the due date.

Considering the zero inventory cost $1|1|u(t)\in[0,\infty]|D,T|L_u,L_x,c_x^+=0$ problem, we note that c_x^+ does not enter any expression of Algorithm 3.9. Thus, the optimal solution does not change and no modification to Algorithm 3.9 is required.

No Backlog

For the case when backlogs are disallowed, i.e. for the $1|1|u(t)\in[0,\infty],X(t)\geq 0|D,T|L_u,L_x$ problem, the solution algorithm can be obtained by an easy modification to Algorithm 3.9. Indeed, having c_x^- very large, the first step of Algorithm 3.9 will never indicate the no-production regime, and the second step will always determine the only optimal solution.

3.2.2 NON-LINEAR INVENTORY COST

It will be shown below that introducing the non-linearity of inventory costs does not influence the optimal control regimes. However, this non-linearity does affect the optimal control value and may cause backlogs, i.e. incomplete demand, by the end of the planning horizon. When backlog is disallowed, the optimal solution is very similar to that derived for the linear cost cases.

Backlogs

The problem to be considered is one-machine, one-product-type scheduling with backlogs to minimize the quadratic inventory cost and linear production cost, that is, the $1|1|u(t)\in[0,\infty]|D,T|L_u,Q_x$ problem which is stated as follows

$$J=\int_0^T\left[c_x^+Z^+(t)^2+c_x^-Z^-(t)^2+c_u u(t)\right]dt\rightarrow\min, \tag{3.54}$$

subject to (3.47) and (3.48).

Dual Formulation

The Hamiltonian, H, resulting from the maximum principle is

$$H(t)=-\left[c_x^+Z^+(t)^2+c_x^-Z^-(t)^2+c_u u(t)\right]+\psi(t)u(t). \tag{3.55}$$

The co-state variable, $\psi(t)$ in (3.55), satisfies the following co-state differential equation with its transversality condition

$$\dot{\psi}(t)=\begin{cases}2c_x^+(X(t)-D(t)), \text{ if } X(t)>D(t);\\ 2c_x^-(X(t)-D(t)), \text{ if } X(t)\leq D(t);\end{cases}\qquad \psi(T)=0. \tag{3.56}$$

According to the maximum principle, the Hamiltonian is maximized at every point of time with respect to the admissible controls $u(t)$. This implies that the control function $u(t)$ must satisfy (3.6).

Algorithm 3.10

Step 1:

If the difference between the length of the planning horizon and the due date is such that $T-t_d<\frac{c_u}{2c_x^- D}$ then set the optimal solution as $u(t)=0$ over the entire planning horizon, STOP; otherwise go to the next step.

Step 2:

Set the optimal solution as

$u(t)=0$ for $0\le t<t_d$ and $t_d<t\le T$,

$$u(t_d)=\infty,\quad \int_{t_d-0}^{t_d+0} u(t)dt = D-\frac{c_u}{2c_x^-(T-t_d)}.$$

STOP.

Discussion

Similar to Algorithm 3.9, Algorithm 3.10 is of $O(1)$ complexity and operates with only two types of optimal control regimes: no-production and fractional production. Specifically, if the length of the planning horizon (deadline) is not close to the required due date, $T-t_d\ge\frac{c_u}{2c_x^- D}$, then the production is worthwhile. Since the only point where control is exercised is at the due date, the control effort at the due date is infinite. The integral of such an effort, however, is bounded, as proven in the following lemma.

LEMMA 3.19. (Correctness of Algorithm 3.10) Given problem (3.54), (3.47) and (3.48), the optimal control is

- if $T-t_d\ge\frac{c_u}{2c_x^- D}$ then $u(t)=0$, for $0\le t<t_d$ and for $t_d<t\le T$,

$$u(t_d)=\infty,\quad \int_{t_d-0}^{t_d+0} u(t)dt = D-\frac{c_u}{2c_x^-(T-t_d)};$$

- otherwise $u(t)=0$ for $0\le t\le T$.

PROOF. First note, that with the introduction of the quadratic inventory cost, the control dependent term of the Hamiltonian (3.55) does not change. Consequently, the optimality condition proven in Lemma 3.18 remains unchanged. That is, if $\psi(t)<c_u$ then $u(t)=0$, otherwise equation (3.6) must be satisfied, implying that $\psi(t)=c_u$. By

differentiating this equality, we again obtain equation (3.53) as with Lemma 3.18. Equations (3.56) and (3.53) can be simultaneously met only if $X(t)=D(t)$. Thus, again we have $u(t)=0$, $X(t)=0$ for $0 \le t < t_d$ and $u(t)=0$, $X(t)-D(t)=Y<0$, for $t_d < t \le T$. The value of Y is found from the co-state transversality condition

$$\psi(T) = c_u - 2c_x^-(T - t_d)Y = 0.$$

From the last expression, the constant inventory level at interval $t_d < t \le T$ is

$$Y = -\frac{c_u}{2c_x^-(T - t_d)}. \tag{3.57}$$

Then the optimal control at t_d satisfies the following equation

$$Y = \int_{t_d-0}^{t_d+0} u(t)dt - D.$$

The lemma is immediately proved by substituting (3.57) into the last equation. The optimal solution which has been found is illustrated in Figure 3.13. Finally, the condition $T - t_d \ge \frac{c_u}{2c_x^- D}$ follows from the requirement $\int_{t_d-0}^{t_d+0} u(t)dt = Y + D \ge 0$. Otherwise, if $T - t_d < \frac{c_u}{2c_x^- D}$, the feasible solution for the stated primal and dual problems is $u(t)=0$, $X(t)=0$ and $\psi(t) = 0$ over the entire planning horizon.

■

Special Cases of Zero Costs

First, consider the zero production cost, i.e. the $1|1|u(t) \in [0,\infty]|D,T|Q_x$ problem. The solution algorithm can be obtained by an easy modification to Algorithm 3.10. Indeed, the zero production cost implies that Step 1 of Algorithm 3.10 will never indicate the no-production regime over the entire planning horizon, while the second step will always result in completing the demand at the due date.

The problem with zero inventory cost, i.e. the $1|1|u(t) \in [0,\infty]|D,T|L_u, Q_x, c_x^+ = 0$ problem is solved by Algorithm 3.10 with no modification.

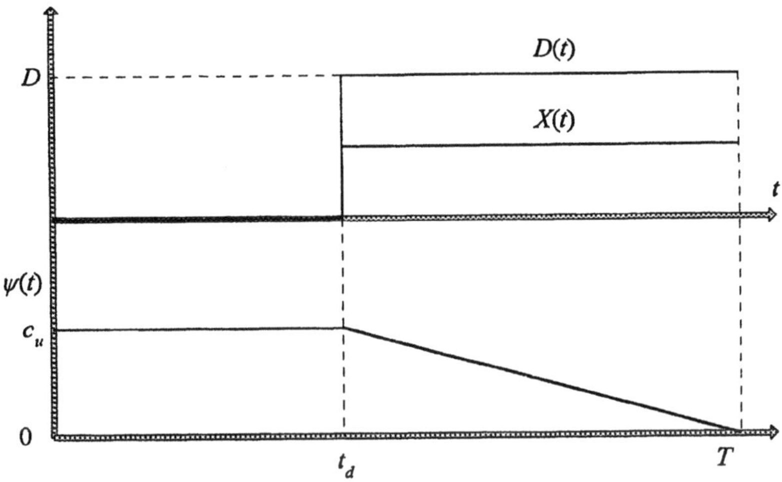

Figure 3.13. Demand $D(t)$, optimal state $X(t)$ and co-state $\psi(t)$ variables for the $1|1|u(t)\in[0,\infty]|D,T|\,L_u,Q_x$ problem.

No Backlog

When backlogs are disallowed, i.e. for the $1|1|u(t)\in[0,\infty],X(t)\geq 0|D,T|\,L_u,Q_x$ problem, the solution algorithm can be obtained by an easy modification to Algorithm 3.10. Indeed, having c_x^- very large, the first step of Algorithm 3.10 will never indicate the no-production regime. This implies that the fractional production $X(t)=D(t)$ is optimal along the entire planning horizon and that the demand must be completed by the due date, i.e. $\int_{t_d-0}^{t_d+0} u(t)dt = D$.

3.2.3 NON-LINEAR PRODUCTION COST

In this section we consider a piece-wise linear function of inventory and backlog costs combined with a non-linear production cost function. This change influences the optimal behavior of the system. Unlike Sections 3.2.1 and 3.2.2, where the zero optimal control was re-switched at most once to an infinite value at the due date, here, there can be two switching points which identify a sequence of regimes: no-production, fractional production with finite control and again no-production.

Backlogs
The problem to be considered is one-machine, one-product-type scheduling with backlogs to minimize the piece-wise inventory cost and quadratic production cost, that is, the $1|1|\,u(t)\in[0,\infty]\,|D,T|\,Q_u, L_x$ problem which is stated as follows

$$J=\int_0^T\left[c_x^+Z^+(t)+c_x^-Z^-(t)+c_u u(t)^2\right]dt\rightarrow \min, \tag{3.58}$$

subject to (3.47) and (3.48).

Dual Formulation
The Hamiltonian, H, resulting from the maximum principle is

$$H(t)=-\left[c_x^+Z^+(t)+c_x^-Z^-(t)+c_u u(t)^2\right]+\psi(t)u(t). \tag{3.59}$$

The co-state variable, $\psi(t)$ in (3.59), satisfies the co-state differential equation (3.51).

According to the maximum principle, if the control function $u(t)$ that satisfies

$$\frac{\partial H}{\partial u(t)}=-2c_u u(t)+\psi(t)=0 \tag{3.60}$$

is non-negative, then it is optimal. Otherwise, no control should be undertaken at point t, i.e. $u(t)$=0.

Algorithm 3.11
Step 1:

Calculate $2\sqrt{\dfrac{c_x^-c_uD}{c_x^+\left(c_x^-+c_x^+\right)}}$ to check whether production starts from the very beginning of the planning horizon.

If $t_d\geq 2\sqrt{\dfrac{c_x^-c_uD}{c_x^+\left(c_x^-+c_x^+\right)}}$, then

calculate the time required to complete the demand

$t_d+2\sqrt{\dfrac{c_x^+c_uD}{c_x^-\left(c_x^-+c_x^+\right)}}$ and proceed to the next step,

otherwise go to Step 3.

Step 2:

If $t_d + 2\sqrt{\dfrac{c_x^+ c_u D}{c_x^-\left(c_x^- + c_x^+\right)}} \le T$, then

calculate the time points where the production starts and ends up

$$t_1 = t_d - 2\sqrt{\frac{c_x^- c_u D}{c_x^+\left(c_x^- + c_x^+\right)}} \text{ and } t_2 = t_d + 2\sqrt{\frac{c_x^+ c_u D}{c_x^-\left(c_x^- + c_x^+\right)}}.$$

Set the optimal solution as

$u(t) = \dfrac{c_x^+}{2c_u}(t - t_1),\ t_1 \le t \le t_d$ and

$u(t) = \dfrac{1}{2c_u}\left(c_x^+(t_d - t_1) - c_x^-(t - t_d)\right),\ t_d \le t \le t_2,$

$u(t) = 0,\ 0 \le t \le t_1$ and $t_2 \le t \le T$; STOP.

Otherwise, if $t_d + 2\sqrt{\dfrac{c_x^+ c_u D}{c_x^-\left(c_x^- + c_x^+\right)}} > T$, then calculate the time point at which the production starts, $t_1 = t_d - \dfrac{c_x^-}{c_x^+}(T - t_d)$. If $t_1 \ge 0$, then set the optimal solution as

$u(t) = \dfrac{c_x^+}{2c_u}(t - t_1),\ t_1 \le t \le t_d$, $u(t) = \dfrac{c_x^-}{2c_u}(T - t),\ t_d \le t \le T$, and

$u(t) = 0,\ 0 \le t \le t_1$, STOP.

Otherwise, go to the next step.

Step 3:

If $\sqrt{\dfrac{4c_u D}{c_x^-} + \dfrac{c_x^+ + c_x^-}{c_x^-} t_d^2} \le T$, then calculate the time point at which the production ends up, $t_2 = \sqrt{\dfrac{4c_u D}{c_x^-} + \dfrac{c_x^+ + c_x^-}{c_x^-} t_d^2}$,

and set the optimal solution as

$u(t) = \dfrac{c_x^-(t_2 - t_d) - c_x^+(t_d - t)}{2c_u},\ 0 \le t \le t_d$,

$u(t) = \dfrac{1}{2c_u} c_x^-(t_2 - t),\ t_d \le t \le t_2$ and $u(t) = 0$, $t_2 \le t \le T$, STOP.

Otherwise, set the optimal solution as

$$u(t)=\frac{c_x^-(T-t_d)-c_x^+(t_d-t)}{2c_u},\ 0\le t\le t_d \text{ and}$$

$$u(t)=\frac{1}{2c_u}c_x^-(T-t),\ t_d\le t\le T,\ \text{STOP.}$$

Example 3.6
Consider the $1|1|u(t)\in[0,\infty]|D,T|Q_u,L_x$ problem with the following parameters: $T=10$ hr, c_u=1\$hr/unit^2, c_x^+=0.2\$/hr.unit, c_x^-=1\$/hr.unit and demand $t_d=8$ hr and D=1.5unit.

At Step 1, Algorithm 3.11 calculates $2\sqrt{\dfrac{c_x^- c_u D}{c_x^+\left(c_x^-+c_x^+\right)}}$=5hr. Since $t_d>5$, Step 2 results in $t_1=t_d-5=3$ hr and $t_2=t_d+1=9$ hr. Then it sets the optimal solution as
$u(t)=-0.3+0.1t,\ 3\le t\le 8$ and $u(t)=4.5-0.5t,\ 8\le t\le 9$,
$u(t)=0,\ 0\le t\le 3$ and $9\le t\le 10$.
■

Discussion
According to Algorithm 3.11, problem (3.58), (3.47) and (3.48) is solvable in $O(1)$ and the solution contains two types of optimal production regimes: no-production and fractional production. Specifically, since the no-production regime may or may not appear before or after the fractional production regime, there exist four types of optimal solutions. Each is identified at the corresponding steps of Algorithm 3.11. The main distinction between the types of optimal solutions constructed by the algorithm is that the demand is fully completed, or not, and the production starts, or does not start, from the beginning of the planning horizon.

LEMMA 3.20. (Correctness of Algorithm 3.11) Given problem (3.58), (3.47) and (3.48),

- if $t_d\ge 2\sqrt{\dfrac{c_x^- c_u D}{c_x^+\left(c_x^-+c_x^+\right)}}$ and $t_d\le T-2\sqrt{\dfrac{c_x^+ c_u D}{c_x^-\left(c_x^-+c_x^+\right)}}$, then

$$u(t)=\frac{c_x^+}{2c_u}(t-t_1),\ t_1\le t\le t_d \text{ and } u(t)=\frac{1}{2c_u}\left(c_x^+(t_d-t_1)-c_x^-(t-t_d)\right),\ t_d\le t\le t_2,$$

$u(t)=0,\ 0\le t\le t_1$ and $t_2\le t\le T$,

where $t_1 = t_d - 2\sqrt{\dfrac{c_x^- c_u D}{c_x^+\left(c_x^- + c_x^+\right)}}$ and $t_2 = t_d + 2\sqrt{\dfrac{c_x^+ c_u D}{c_x^-\left(c_x^- + c_x^+\right)}}$;

- if $t_d \ge 2\sqrt{\dfrac{c_x^- c_u D}{c_x^+\left(c_x^- + c_x^+\right)}}$ and $t_d > T - 2\sqrt{\dfrac{c_x^+ c_u D}{c_x^-\left(c_x^- + c_x^+\right)}}$, then

  $u(t) = \dfrac{c_x^+}{2c_u}(t - t_1),\ t_1 \le t \le t_d$, $u(t) = \dfrac{1}{2c_u} c_x^- (T - t),\ t_d \le t \le T$ and

  $u(t) = 0,\ 0 \le t \le t_1$, where $t_1 = t_d - \dfrac{c_x^-}{c_x^+}(T - t_d)$.

- if $t_d < 2\sqrt{\dfrac{c_x^- c_u D}{c_x^+\left(c_x^- + c_x^+\right)}}$ and $\sqrt{\dfrac{4c_u D}{c_x^-} + \dfrac{c_x^+ + c_x^-}{c_x^-} t_d^2} \le T$, then

  $u(t) = \dfrac{c_x^-(t_2 - t_d) - c_x^+(t_d - t)}{2c_u},\ 0 \le t \le t_d$ and $u(t) = \dfrac{1}{2c_u} c_x^- (t_2 - t),\ t_d \le t \le t_2$,

  $u(t) = 0,\ t_2 \le t \le T$, where $t_2 = \sqrt{\dfrac{4c_u D}{c_x^-} + \dfrac{c_x^+ + c_x^-}{c_x^-} t_d^2}$;

- if $t_d < 2\sqrt{\dfrac{c_x^- c_u D}{c_x^+\left(c_x^- + c_x^+\right)}}$ and $\sqrt{\dfrac{4c_u D}{c_x^-} + \dfrac{c_x^+ + c_x^-}{c_x^-} t_d^2} > T$, then

  $u(t) = \dfrac{c_x^-(T - t_d) - c_x^+(t_d - t)}{2c_u},\ 0 \le t \le t_d$ and $u(t) = \dfrac{1}{2c_u} c_x^- (T - t),\ t_d \le t \le T$.

PROOF. Let us construct a feasible behavior of the co-state variable that satisfies the maximum principle. Four types of such co-state solutions can be identified as feasible.

The first type is characterized by production which starts from some point $t_1 > 0$, that is $\psi(t) = 0,\ 0 \le t \le t_1$, $\psi(t) > 0$, $X(t) > D(t)$, $u(t) > 0$ for $t > t_1$. According to (3.51) and (3.60), we have $\psi(t) = 2c_u u(t)$ and $\dot{\psi}(t) = c_x^+$. In such a case, the demand is not completely met by the due date, $X(t_d) < D(t_d)$, so that after the due date $\dot{\psi}(t) = -c_x^-$, $t > t_d$. The production finally stops at t_2 when the demand is completed, that is $X(t)$=$D(t)$ and $\psi(t) = 0,\ t_2 \le t \le T$. The described solution is illustrated in Figure 3.14.

By integrating the state and co-state equations, we have the following system for this type of solution

$$\psi(t) = c_x^+(t - t_1),\ t_1 \le t \le t_d,$$

$$\psi(t) = c_x^+(t_d - t_1) - c_x^-(t - t_d),\ t_d \le t \le t_2,$$

$$X(t_d) = \int_{t_1}^{t_d} \frac{c_x^+(t - t_1)}{2c_u} dt = \frac{c_x^+}{2c_u}\frac{(t_d - t_1)^2}{2},$$

$$X(t_2) = X(t_d) + \int_{t_d}^{t_2} \frac{c_x^+(t_d - t_1) - c_x^-(t - t_d)}{2c_u} dt = X(t_d) + \frac{1}{2c_u}\left(c_x^+(t_d - t_1)(t_2 - t_d) - c_x^- \frac{(t_2 - t_d)^2}{2}\right).$$

By taking into account that $\psi(t_2) = 0$ and $X(t_2) = D$, the presented system results in the following two algebraic equations in two unknowns, t_1 and t_2

$$c_x^+(t_d - t_1) = c_x^-(t_2 - t_d), \tag{3.61}$$

$$D = \frac{1}{2c_u}\left(c_x^+ \frac{(t_d - t_1)^2}{2} + c_x^+(t_d - t_1)(t_2 - t_d) - c_x^- \frac{(t_2 - t_d)^2}{2}\right), \tag{3.62}$$

from where

$$t_1 = t_d - 2\sqrt{\frac{c_x^- c_u D}{c_x^+\left(c_x^- + c_x^+\right)}} \quad \text{and} \quad t_2 = t_d + 2\sqrt{\frac{c_x^+ c_u D}{c_x^-\left(c_x^- + c_x^+\right)}}. \tag{3.63}$$

To identify three other types of optimal solutions, we next analyze the conditions when the found form of the optimal solution is not feasible. If $t_2 > T$, that implies the demand D will not be completed by the end of the planning horizon and therefore equation (3.62) no longer holds. To determine the optimal solution for this second type of optimal solution, we solve equation (3.61) by replacing t_2 with T. The other possibility for the solution to be non-feasible is when t_1 becomes negative, that is

$$t_d < 2\sqrt{\frac{c_x^- c_u D}{c_x^+\left(c_x^- + c_x^+\right)}}. \tag{3.64}$$

This implies that $\psi(0) > 0$ and that production starts from the very beginning of the planning horizon, unlike the case with the first two types of solutions characterized by $\psi(0) = 0$. This third type of the optimal solution is characterized by the following equations

$$\psi(t) = \psi(0) + c_x^+ t, \quad 0 \le t \le t_d,$$

$$\psi(t) = \psi(0) + c_x^+ t_d - c_x^- (t - t_d), \quad t_d \le t \le t_2,$$

$$X(t_d) = \int_0^{t_d} \frac{\psi(0) + c_x^+ t}{2c_u} dt = \frac{1}{2c_u}\left(\psi(0) t_d + \frac{c_x^+ t_d{}^2}{2}\right),$$

$$X(t_2) = X(t_d) + \int_{t_d}^{t_2} \frac{\psi(0) + c_x^+ t_d - c_x^-(t - t_d)}{2c_u} dt = X(t_d) + \frac{1}{2c_u}\left((\psi(0) + c_x^+ t_d)(t_2 - t_d) - c_x^- \frac{(t_2 - t_d)^2}{2}\right).$$

Although there is now only one unknown switching point, the initial value of the co-state variable is unknown. By taking into account that $\psi(t_2) = 0$ and $X(t_2) = D$, the last system results in the following two algebraic equations in two unknowns $\psi(0)$ and t_2

$$\psi(0) + c_x^+ t_d - c_x^- (t_2 - t_d) = 0 \tag{3.65}$$

$$D = \frac{1}{2c_u}\left(\psi(0) t_2 + c_x^+ \frac{t_d{}^2}{2} + c_x^+ t_d (t_2 - t_d) - c_x^- \frac{(t_2 - t_d)^2}{2}\right) \tag{3.66}$$

from where

$$\psi(0) = c_x^- (t_2 - t_d) - c_x^+ t_d \quad \text{and} \quad t_2 = \sqrt{\frac{4c_u D}{c_x^-} + \frac{c_x^+ + c_x^-}{c_x^-} t_d^2}\,. \tag{3.67}$$

Note, we assumed $\psi(0) > 0$ to find a solution of the third type. By using the found equation for $\psi(0)$, one can readily verify condition (3.64) which identifies the case when a solution of the first type is not feasible. Finally, since from (3.67) t_2 is always greater than t_d, the solution is feasible. However, if $t_2 > T$, the demand D will not be completed by the end of the planning horizon and therefore equation (3.66) no longer holds. To determine the optimal solution of the fourth type, we simply solve equation (3.65) for $\psi(0)$ by replacing t_2 with T. The found behavior of the co-state variable is then used to set the optimal control value $u(t) = \dfrac{\psi(t)}{2c_u}$ as stated in the lemma.

■

Special Cases of Zero Costs

The problem with zero production cost, i.e. the $1|1|u(t)\in[0,\infty]|D,T|L_x$ problem, has been solved in Section 3.2.1. Looking at the zero inventory cost problem, i.e. the $1|1|u(t)\in[0,\infty]|D,T|Q_u, L_x, c_x^+=0$ problem, one can observe that Steps 1 and 2 of Algorithm 3.11 are no longer needed and the optimal solution is always set at Step 3. Thus, for the zero inventory cost, Algorithm 3.11 reduces to Step 3 only.

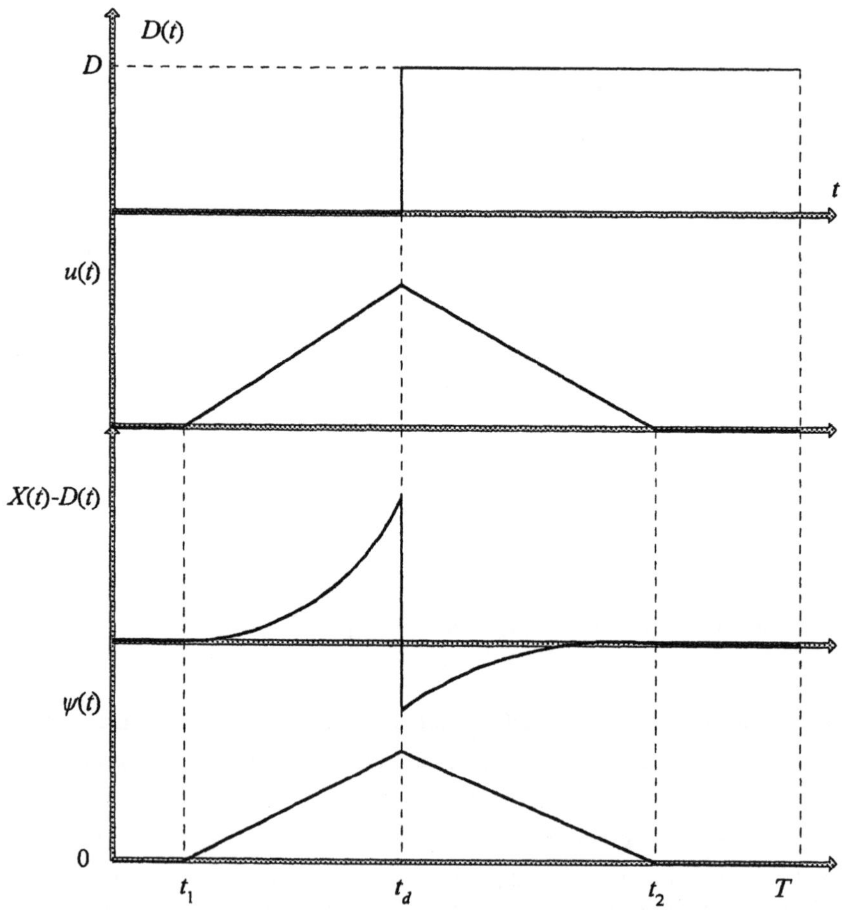

Figure 3.14. Demand $D(t)$, optimal control $u(t)$, state $X(t)$ and co-state $\psi(t)$ variables for the first type solution of the $1|1|u(t)\in[0,\infty]|D,T|Q_u, L_x$ problem.

No Backlog

As in the case with linear performance measure, when backlogs are disallowed, i.e. for the $1|1|u(t)\in[0,\infty], X(t)\geq 0|D,T|Q_u, L_x$ problem, the solution algorithm is simplified, because the demand must be completed by the due date.

Dual Formulation

Now problem (3.58), (3.47) and (3.48) has an additional constraint $X(t)\geq D(t)$ which affects neither Hamiltonian (3.59) nor optimality condition (3.60), but changes the co-state differential equation

$$d\psi(t)=c_x^+dt-d\mu(t),\ \ \psi(T+0)=0, \tag{3.68}$$

where Lagrange multiplier $d\mu(t)\geq 0$ corresponds to the no-backlog constraint and satisfies the complementary slackness condition

$$\int_0^T (X(t)-D(t))d\mu(t)=0. \tag{3.69}$$

Algorithm 3.12

Step 1:

If $t_d<2\sqrt{\dfrac{c_uD}{c_x^+}}$, then go to Step 3; otherwise, proceed to the next step.

Step 2:

Assign $t_1=t_d-2\sqrt{\dfrac{c_uD}{c_x^+}}$ and set the optimal solution as

$u(t)=0,\ \ 0\leq t\leq t_1,$

$u(t)=\dfrac{c_x^+}{2c_u}(t-t_1),\ \ t_1\leq t\leq t_d$ and $u(t)=0,\ \ t_d\leq t\leq T$; STOP.

Step 3:

Set the optimal solution as

$u(t)=\dfrac{1}{2c_u}\left(\dfrac{2c_uD}{t_d}-\dfrac{c_x^+t_d}{2}+c_x^+t\right),\ \ 0\leq t\leq t_d$ and $u(t)=0,\ \ t_d\leq t\leq T$;

STOP.

Discussion

Algorithm 3.12 is of $O(1)$ complexity. It presents two closed form types of the optimal solution possible for the problem with backlog disallowed. In both cases, the demand is fully completed by the due date, while the production is started either from the very beginning of the planning horizon or from a switching point $t_1 = t_d - 2\sqrt{\frac{c_u D}{c_x^+}}$ before which the no-production regime is beneficial.

LEMMA 3.21. (Correctness of Algorithm 3.12) Given problem (3.58), (3.47) and (3.48) with backlog disallowed, the optimal solution is

- if $t_d \geq 2\sqrt{\frac{c_u D}{c_x^+}}$, then

$$u(t) = 0,\ 0 \leq t \leq t_1,\ u(t) = \frac{c_x^+}{2c_u}(t - t_1),\ t_1 \leq t \leq t_d \text{ and } u(t) = 0,\ t_d \leq t \leq T,$$

where $t_1 = t_d - 2\sqrt{\frac{c_u D}{c_x^+}}$;

- otherwise $u(t) = \frac{1}{2c_u}\left(\frac{2c_u D}{t_d} - \frac{c_x^+ t_d}{2} + c_x^+ t\right)$, $0 \leq t \leq t_d$ and $u(t) = 0,\ t_d \leq t \leq T$.

PROOF. To prove the lemma, we construct a solution for the primal and dual problems which is feasible and thus optimal. There can be two types of such solutions. The first one has a no-production regime at the beginning of the planning horizon. The solution which satisfies the primal and dual problem in this case is (see Figure 3.15)

$$X(t) = 0,\ u(t) = 0 \text{ and } \dot{\mu}(t) = 0,\ \psi(t) = 0,\ 0 \leq t \leq t_1;$$

$$X(t) = \int_{t_1}^{t} u(\tau)d\tau,\ u(t) = \frac{\psi(t)}{2c_u},\ t_1 \leq t \leq t_d \text{ and } \psi(t) = c_x^+(t - t_1),\ \dot{\mu}(t) = 0,\ t_1 \leq t < t_d;$$

$$d\psi(t_d) = -d\mu(t_d),\ d\mu(t_d) > 0; \tag{3.70}$$

$$X(t) = D,\ u(t) = 0 \text{ and } \psi(t) = 0,\ \dot{\mu}(t) = 0,\ t_d < t \leq T.$$

To find the only switching point, t_1, we consider this solution at interval $[t_1, t_d]$

$$X(t_d) = \int_{t_1}^{t_d} u(t)dt = \int_{t_1}^{t_d} \frac{\psi(t)}{2c_u}dt = \int_{t_1}^{t_d} \frac{c_x^+(t - t_1)}{2c_u}dt = D.$$

From the last expression we obtain

$$t_1 = t_d - 2\sqrt{\frac{c_u D}{c_x^+}}. \qquad (3.71)$$

Since the switching time point, t_1, determined in (3.71) is always less than t_d, it can be non-feasible only if it is negative, i.e.

$$t_d < 2\sqrt{\frac{c_u D}{c_x^+}}, \quad \psi(0) > 0. \qquad (3.72)$$

In this case, production starts from the very beginning of the planning horizon. This is the second type of the optimal solution which is characterized by the following equations

$$X(t) = \int_{t_1}^{t} u(t)dt,\ u(t) = \frac{\psi(t)}{2c_u},\ 0 \le t \le t_d \text{ and } \psi(t) = \psi(0) + c_x^+ t,\ \dot{\mu}(t) = 0,\ 0 \le t < t_d;$$

$$d\psi(t) = c_x^+ dt - d\mu(t),\ d\mu > 0,\ t = t_d; \qquad (3.73)$$

$$X(t) = D,\ u(t) = 0 \text{ and } \psi(t) = 0,\ \dot{\mu}(t) = 0,\ t_d < t < T.$$

Although there are now no switching points, an initial value of the co-state variable is still unknown. To find this value, we consider the solution at interval $[0, t_d]$ from (3.73)

$$X(t_d) = \int_0^{t_d} u(t)dt = \int_0^{t_d} \frac{\psi(t)}{2c_u} dt = \int_0^{t_d} \frac{\psi(0) + c_x^+ t}{2c_u} dt = D.$$

From the last expression we find

$$\psi(0) = \frac{2c_u D}{t_d} - \frac{c_x^+ t_d}{2}.$$

Note, we assumed $\psi(0) > 0$ in order to find a solution of the second type. By using the found equation for $\psi(0)$, one can readily verify condition (3.72), which identifies the case when a solution of the first type is not feasible. Finally, the optimal control can be immediately set as stated in the lemma by substituting expressions for t_1 and $\psi(0)$ in the corresponding equations.

■

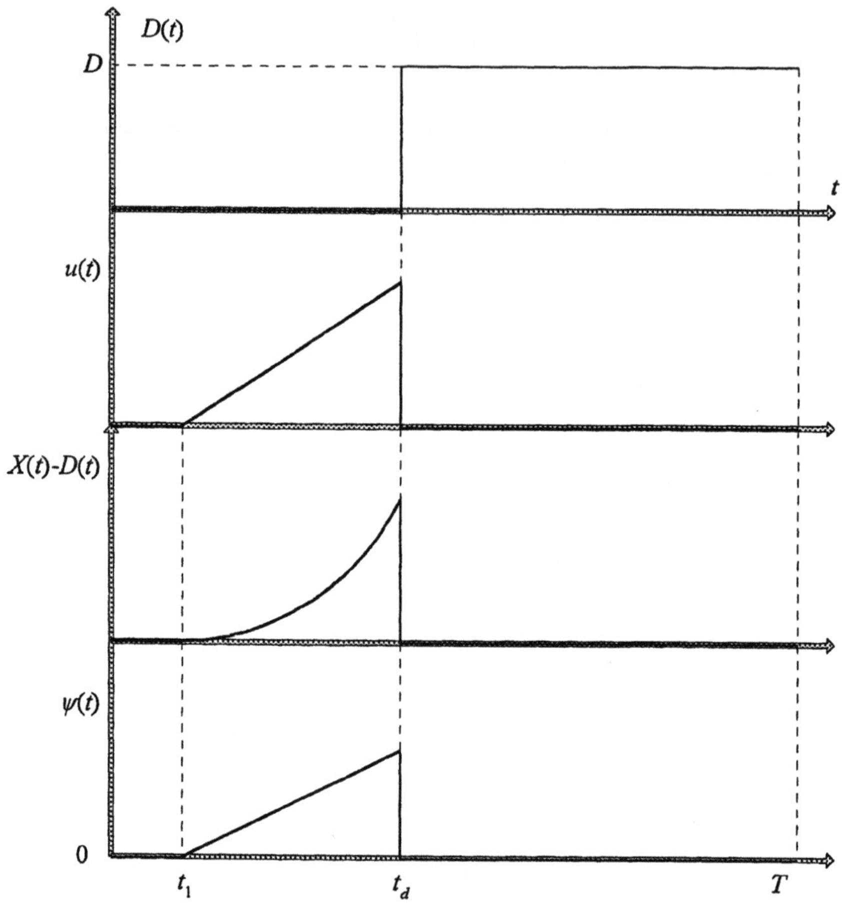

Figure 3.15. Demand $D(t)$, optimal control $u(t)$, state $X(t)$ and co-state $\psi(t)$ variables for the first type solution of the $1|1|u(t)\in[0,\infty], X(t)\geq 0|D,T|Q_u, L_x$ problem.

3.2.4 NON-LINEAR INVENTORY AND PRODUCTION COSTS

Although both costs entering the objective are now non-linear, the problem turns out to be polynomially solvable with and without backlogs.

Backlogs

The problem to be considered is one-machine, one-product-type scheduling with backlogs to minimize the quadratic inventory and production costs, that is, the $1|1|u(t)\in[0,\infty]|D,T|Q_u, Q_x$ problem which is stated as follows

$$J = \int_0^T \left[c_x^+ Z^+(t)^2 + c_x^- Z^-(t)^2 + c_u u(t)^2 \right] dt \to \min, \tag{3.74}$$

subject to (3.47) and (3.48).

Dual Formulation

The Hamiltonian, H, resulting from the maximum principle is

$$H(t) = -\left[c_x^+ Z^+(t)^2 + c_x^- Z^-(t)^2 + c_u u(t)^2 \right] + \psi(t)u(t). \tag{3.75}$$

The co-state variable, $\psi(t)$ in (3.75), satisfies the co-state differential equation (3.56). Maximization of the Hamiltonian as a function of $u(t)$ results, as in the previous section, in

$$u(t) = \begin{cases} \dfrac{\psi(t)}{2c_u}, & \text{if } \psi(t) \geq 0; \\ 0, & \text{if } \psi(t) < 0. \end{cases} \tag{3.76}$$

Algorithm 3.13

Step 1:

Calculate parameters $y_1 = \sqrt{\dfrac{c_x^+}{c_u}}$, $y_2 = \sqrt{\dfrac{c_x^-}{c_u}}$ and

$$A = \frac{D\sqrt{c_x^+ c_x^-}\, sh(y_2(T - t_d))}{y_2 sh(y_2(T - t_d)) sh(y_1 t_d) + y_1 ch(y_2(T - t_d)) ch(y_1 t_d)}.$$

Step 2:

Set the optimal control as

$u(t) = \dfrac{S(t)}{2c_u}$, where

$S(t) = 2A ch(y_1 t)$ for $t \in [0, t_d]$,

$S(t) = 2A ch(y_1 t_d) \dfrac{sh(y_2(T - t))}{sh(y_2(T - t_d))}$ for $t \in [t_d, T]$.

STOP.

Discussion

Algorithm 3.13 gives a closed form of the optimal control in $O(1)$ time for problem (3.74), (3.47) and (3.48). The solution includes two non-linear control functions, increasing and decreasing respectively, which are changed over at the due date t_d.

LEMMA 3.22. Given problem (3.74), (3.47) and (3.48), the co-state variable $\psi(t)$ is non-negative, $\psi(t) \geq 0$ for $t \in [0, T]$.

PROOF. The proof of this lemma is similar to the proof of the lower bound in Lemma 3.8 and therefore is omitted.
■

LEMMA 3.23. (Correctness of Algorithm 3.13) Given problem (3.74), (3.47) and (3.48), the optimal control is

$$u(t) = \frac{Ach(y_1 t)}{c_u} \text{ for } t \in [0, t_d],\ u(t) = \frac{Ach(y_1 t_d)}{c_u} \frac{sh(y_2(T-t))}{sh(y_2(T-t_d))} \text{ for } t \in [t_d, T],$$

where $y_1 = \sqrt{\frac{c_x^+}{c_u}}$, $y_2 = \sqrt{\frac{c_x^-}{c_u}}$,

$$A = \frac{\sqrt{c_x^+ c_x^-} Dsh(y_2(T-t_d))}{y_2 sh(y_2(T-t_d)) sh(y_1 t_d) + y_1 ch(y_2(T-t_d)) ch(y_1 t_d)}.$$

PROOF. Since $D(t)$=0 for $0 \leq t < t_d$, the inventory $X(t)$-$D(t)$ is non-negative at this time interval and the co-state differential equation (3.56) simplifies to

$$\dot{\psi}(t) = 2c_x^+ X(t). \tag{3.77}$$

Next, we prove, by contradiction, that at the remaining part of the planning horizon $t_d < t \leq T$, the optimal solution of the problem is such that $X(t) \leq D$. Indeed, if there exists a point of time at which the cumulative production, $X(t)$, is greater than D, then it is greater than D until the end of the planning horizon even under a no-production regime. According to (3.56), the co-state variable in such a case increases up to zero at t=T. Consequently, it is negative at a terminal interval of time that contradicts Lemma 3.22.

Given $X(t) \leq D$ for $t_d < t \leq T$, the co-state differential equation (3.56) simplifies to

$$\dot{\psi}(t) = 2c_x^- (X(t) - D). \tag{3.78}$$

By differentiation (3.77) and (3.78) and taking into account Lemma 3.22, we obtain

$$\ddot{\psi}(t)=\frac{c_x^+}{c_u}\psi(t), \qquad 0\le t<t_d, \tag{3.79}$$

$$\ddot{\psi}(t)=\frac{c_x^-}{c_u}\psi(t), \quad t_d<t\le T. \tag{3.80}$$

Thus, the system of the state and co-state differential equations is presented as an ordinary, second-order differential equation with the following boundary conditions

$$\dot{\psi}(0)=0, \qquad \psi(T)=0. \tag{3.81}$$

In addition, due to the jump of the demand function $D(t)$ at point t_d, co-state equation (3.56) results in a jump in the derivative of the co-state variable, so that

$$\psi(t_d-0)=\psi(t_d+0),$$
$$\dot{\psi}(t_d-0)=\dot{\psi}(t_d+0)+2c_x^+\big(X(t_d-0)-D(t_d-0)\big)-2c_x^-\big(X(t_d+0)-D(t_d+0)\big). \tag{3.82}$$

Since function $X(t)$ is continuous and demand function $D(t)$ jumps up at point t_d at value D (see (3.49)), the last condition takes the form

$$\dot{\psi}(t_d-0)=\dot{\psi}(t_d+0)+2X(t_d)(c_x^+-c_x^-)+2c_x^-D, \tag{3.83}$$

Due to the boundary conditions (3.81), we solve equations (3.79) and (3.80) separately as follows

$$\psi(t)=2Ach(y_1t) \text{ for } t\in[0,t_d],$$

$$\psi(t)=2Bsh(y_2(T-t)) \text{ for } t\in[t_d,T].$$

By integrating the state equation

$$X(t_d)=\int_0^{t_d}u(t)dt=\frac{1}{2c_u}\int_0^{t_d}\psi(t)dt=\frac{Ash(y_1t_d)}{c_uy_1},$$

and by using conditions (3.82) and (3.83), we obtain a system of two linear equations in two unknowns A and B

$$Ach(y_1t_d)=Bsh(y_2(T-t_d)),$$

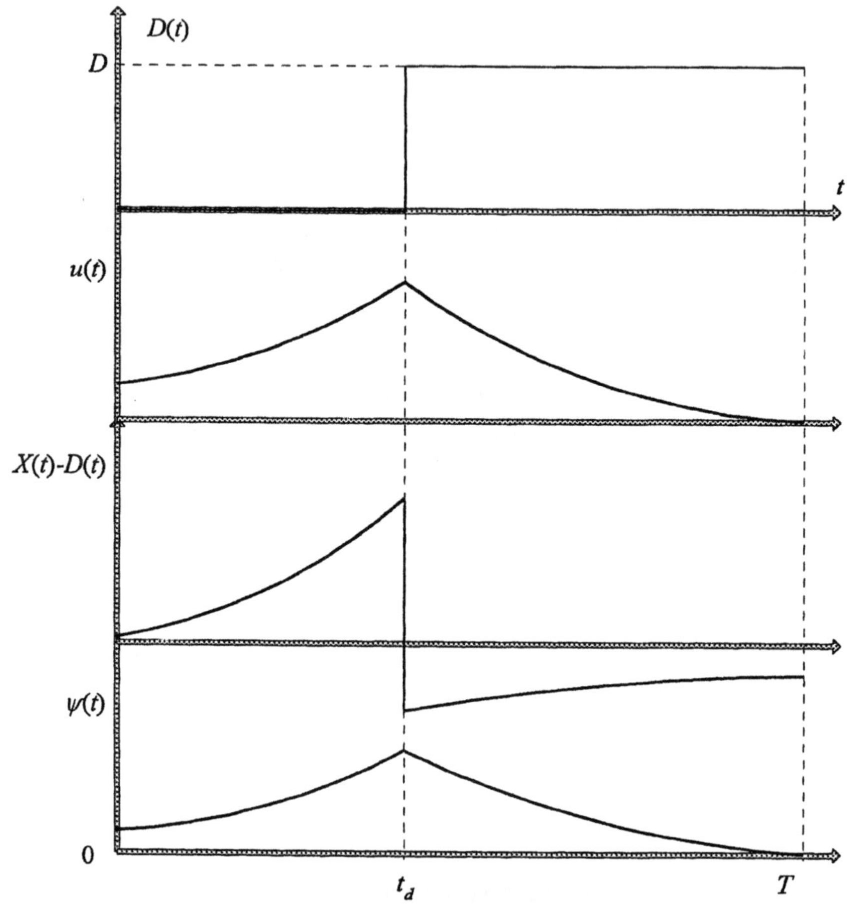

Figure 3.16. Demand $D(t)$, optimal control $u(t)$, state $X(t)$ and co-state $\psi(t)$ variables for the $1|1|u(t)\in[0,\infty]|D,T|\, Q_u, Q_x$ problem.

$$Ay_1 sh(y_1 t_d) = -By_2 ch(y_2 (T-t_d)) + \frac{Ash(y_1 t_d)}{c_u y_1}(c_x^+ - c_x^-) + c_x^- D.$$

The solution of this system determines the optimal behavior of $\psi(t)$ (shown in Figure 3.16), which substituted in (3.76) proves the lemma.

■

Note that if the inventory cost is identical to the backlog cost, $c_x^+ = c_x^-$, the parameters of the optimal control function simplify significantly as

$$y_1 = y_2 = \sqrt{\frac{c_x^+}{c_u}}, \quad A = \frac{c_x^+}{y_1}\frac{Dsh(y_1(T-t_d))}{ch(y_1 T)}.$$

Special Cases of Zero Costs

The problem with no production cost, $1|1|u(t) \in [0,\infty] |D,T| Q_x$, has been discussed in Section 3.2.2. The problem with no inventory cost, $1|1|u(t) \in [0,\infty] |D,T| Q_u, Q_x, c_x^+ = 0$ is solved by Algorithm 3.13 with the following modifications. At Step 1, parameter $A = \dfrac{c_x^- D}{y_2 ch(y_2(T-t_d)) + y_2^2 t_d sh(y_2(T-t_d))}$ and at Step 2, $S(t) = 2Ash(y_2(T-t_d))$ for $t \in [0,t_d]$ and $S(t) = 2Ash(y_2(T-t))$ for $t \in [t_d, T]$. These modifications follow from a constant mode of $\psi(t)$ at $[0,t_d]$ and an exponential mode at $[t_d, T]$.

No Backlog

The $1|1|u(t) \in [0,\infty], X(t) \geq 0 |D,T| Q_u, Q_x$ problem is also solvable for arbitrary cost parameters.

Dual Formulation

Now problem (3.74), (3.47) and (3.48) has an additional constraint $X(t) \geq D(t)$ which affects neither Hamiltonian (3.75) nor optimality condition (3.76), but changes the co-state differential equation

$$d\psi(t) = c_x^+(X(t) - D(t))dt - d\mu(t), \quad \psi(T+0) = 0, \tag{3.84}$$

where Lagrange multiplier $d\mu(t) \geq 0$ corresponds to the no-backlog constraint and satisfies the complementary slackness condition (3.69).

Algorithm 3.14

Step 1:

Calculate $y = \sqrt{\dfrac{c_x^+}{c_u}}$ and set the optimal control as

$u(t) = yD\dfrac{ch(yt)}{sh(yt_d)}$ for $t \in [0, t_d]$,

$u(t) = 0$ for $t \in [t_d, T]$.

STOP.

Example 3.7

Consider the $1|1|u(t) \in [0,\infty], X(t) \geq 0 |D,T| Q_u, Q_x$ problem with the following parameters: $T = 10$ hr, c_u=1\$hr/unit^2, c_x^+=0.2\$/hr.unit^2, c_x^-=1\$/hr.unit^2 and

demand $t_d = 8$ hr and D=1.5unit. At Step 1, Algorithm 3.14 calculates $y = \sqrt{0.2}$ /hr and sets the optimal solution as
$u(t) = 0.0375ch(0.447t)$unit / hr, $0 \le t \le 8$,
$u(t) = 0$, $8 < t \le 10$.
■

Discussion

Algorithm 3.14 is quite simple and provides a closed-form solution for the problem in $O(1)$ time. The main difference in the solution with respect to that of the backlog problem is that the demand is completed by the due date. Before the due date, the control effort increases exponentially. Afterwards, production is not needed.

The following two lemmas prove optimality of the solution that Algorithm 3.14 constructs.

LEMMA 3.24. Given problem (3.74), (3.47) and (3.48), the co-state variable $\psi(t)$ is non-negative, $\psi(t) \ge 0$ for $t \in [0,T]$.

PROOF. The proof of the lemma, similar to the proof of the lower bound in Lemma 3.8, is omitted.
■

LEMMA 3.25. (Correctness of Algorithm 3.14) Given problem (3.74), (3.47) and (3.48) with backlog disallowed, the optimal control is

$$u(t) = yD\frac{ch(yt)}{sh(yt_d)} \text{ for } t \in [0,t_d],\ u(t) = 0 \text{ for } t \in [t_d, T], \text{ where } y = \sqrt{\frac{c_x^+}{c_u}}.$$

PROOF. Let us construct the optimal solution

$$X(t)>0,\ \psi(t) > 0 \text{ for } 0 < t < t_d \text{ and } X(t)=0,\ \psi(t) = 0 \text{ for } t_d < t \le T,$$

as shown in Figure 3.17. Then, the co-state differential equation (3.84) at interval $[0, t_d]$ simplifies to

$$\dot{\psi}(t) = 2c_x^+\left(X(t) - D(t)\right). \tag{3.85}$$

By differentiation (3.85) and substituting (3.47), we obtain

$$\ddot{\psi}(t) = 2c_x^+ u(t). \tag{3.86}$$

Next, by taking into account optimality conditions (3.76) and Lemma 3.24, we find

$$\ddot{\psi}(t) = \frac{c_x^+}{c_u} \psi(t) . \tag{3.87}$$

Thus, the conditions of the lemma allow the system of the state and co-state differential equations to be presented as an ordinary, second-order differential equation (3.87) with the following boundary conditions

$$\dot{\psi}(0) = 0 , \qquad \dot{\psi}(t_d) = 2c_x^+ D . \tag{3.88}$$

The solution of (3.87)-(3.88) proves the lemma. Note, that the found solution meets the conditions $X(t)>0$, $\psi(t) > 0$ for $0 < t < t_d$ assumed in the beginning of the proof.

■

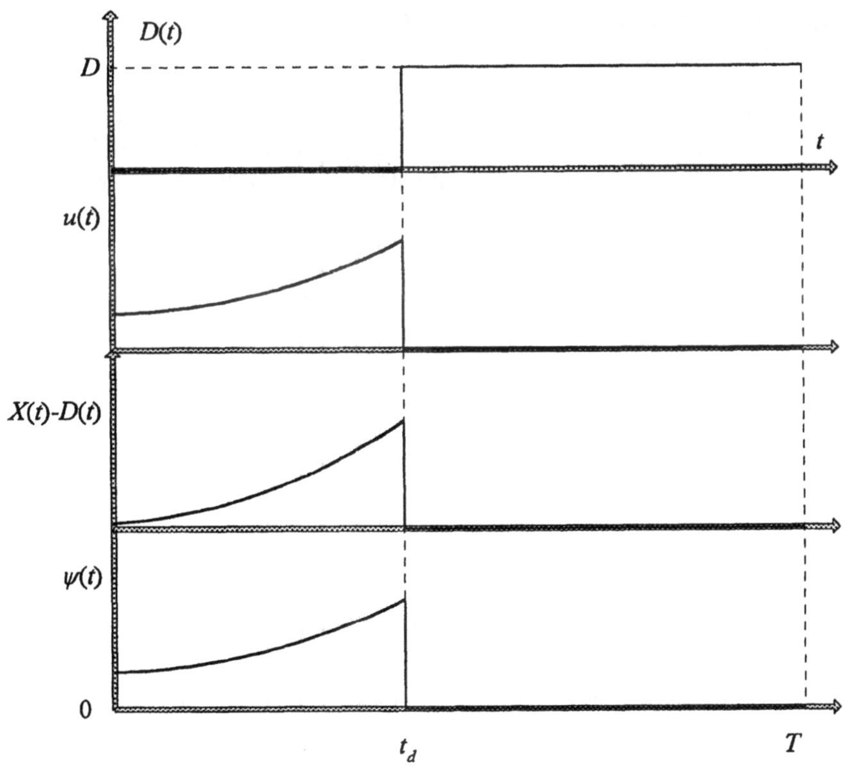

Figure 3.17. Demand $D(t)$, optimal control $u(t)$, state $X(t)$ and co-state $\psi(t)$ variables for the $1|1|u(t) \in [0,\infty], X(t) \geq 0|D,T|Q_u, Q_x$ problem.

3.3 DEMAND OVER INFINITE PLANNING HORIZON

This section addresses the scheduling problems characterized by periodic demand as a function of time specified over an infinite planning horizon. A periodic form of demand causes the optimal solution to one-machine, one-product scheduling problem to be also periodic unless the initial inventory is fixed. When the initial inventory is fixed, there must be a transition time before a solution can enter a periodic steady-state. The period of the optimal solution is considered to be equal to the demand period, T, that implies $X(t)=X(t+T)$. Thus, to provide periodicity of the solution, it is enough to solve the problem over finite planning horizon T with boundary condition $X(0)=X(T)$. Again, the problems we consider differ by inventory-related constraints that allow or disallow backlogs and by the form of the objective function.

3.3.1 LINEAR PERFORMANCE MEASURE

Backlogs

The first problem to be considered is one-machine, one-product-type scheduling with backlogs to minimize a linear cost of inventory and production over period T, that is, the $1|1|u(t)\in[0,\infty], X(0)=X(T)\,|d(t)|\,L_u, L_x$ problem which is stated as follows

$$J=\int_0^T\left[c_x^+X^+(t)+c_x^-X^-(t)+c_u u(t)\right]dt\to\min, \tag{3.89}$$

$$X^+(t)=\max\{X(t),0\},\ X^-(t)=\max\{-X(t),0\}$$

subject to

$$\dot X(t)=u(t)-d(t),\qquad X(0)=X(T), \tag{3.90}$$

$$u(t)\ge 0. \tag{3.91}$$

Dual Formulation

The Hamiltonian, H, resulting from the maximum principle is

$$H(t)=-\left[c_x^+X^+(t)+c_x^-X^-(t)+c_u u(t)\right]+\psi(t)\left[u(t)-d(t)\right]. \tag{3.92}$$

The co-state variable, $\psi(t)$ in (3.92), satisfies the following co-state differential equation with its transversality condition

$$\dot{\psi}(t)=\begin{cases} c_x^+, \text{ if } X(t)>0; \\ -c_x^-, \text{ if } X(t)<0; \\ a,\ a\in[-c_x^-,c_x^+], \text{ if } X(t)=0; \end{cases} \qquad \psi(0)=\psi(T). \tag{3.93}$$

Maximization of the Hamiltonian implies

$$\frac{\partial H(t)}{\partial u(t)}=-c_u+\psi(t)\leq 0.$$

Therefore, the optimality condition is: if $\psi(t)=c_u$ holds at an interval of time, then the control $u(t)$ is non-negative; otherwise, $\psi(t)<c_u$ and no control should be undertaken at this interval, i.e. $u(t)=0$. Note that according to the maximum principle, the periodicity of the state variables causes the periodicity of the co-state variables, $\psi(0)=\psi(T)$, in (3.93).

Algorithm 3.15
Step 1:

Set the optimal solution as

$u(t)=d(t)$ for $0\leq t\leq T$, STOP.

Discussion

From Algorithm 3.15, one can observe that given a periodic demand, there is only one regime of optimal control which is the fractional production $u(t)=d(t)$. This is due to the boundary inventory constraint $X(0)=X(T)$ which necessitates tracking the demand no matter what the production cost coefficient c_u is.

The complexity of Algorithm 3.15 is $O(1)$ time and its correctness is easily verified by solving the dual formulation in the following lemma.

LEMMA 3.26. (Correctness of Algorithm 3.15) Given problem (3.89)-(3.91), the optimal control is $u(t)=d(t)$ for $0\leq t\leq T$.

PROOF. By considering only the control dependent term of the Hamiltonian (3.92), we obtain

$$H_u(t)=u(t)\left[\psi(t)-c_u\right]. \tag{3.94}$$

Expression (3.94) implies that if $\psi(t)<c_u$, then $u(t)=0$, otherwise equation (3.6) must be satisfied, that is, $\psi(t)=c_u$. Then solution $X(t)\equiv 0$, $\psi(t)\equiv c_u$ and $u(t)\equiv d(t)$

always satisfies state (3.90) and co-state (3.93) equations with their boundary conditions as well as their optimality condition (3.6).

■

Special Cases of Zero Costs

Solution Algorithm 3.15 does not include any cost parameters and thus does not change for the special cases of zero production $1|1|u(t)\in[0,\infty], X(0){=}X(T)|d(t)|\ L_x$ and zero inventory $1|1|u(t)\in[0,\infty], X(0){=}X(T)|d(t)|\ L_u, L_x, c_x^+=0$ costs.

No Backlog

For the case when backlogs are disallowed, i.e. for the $1|1|u(t)\in[0,\infty], X(t)\geq 0, X(0){=}X(T)|d(t)|\ L_u, L_x$ problem, the solution algorithm does not require any modification because the optimal inventory level it finds is always non-negative.

3.3.2 NON-LINEAR INVENTORY COST

Next we consider a linear production cost function and a non-linear inventory cost which will be approximated by a quadratic function. It will be shown below that the introduced non-linearity does not affect the optimal control solution at all.

Backlogs

The problem to be considered is one-machine, one-product-type scheduling with backlogs to minimize quadratic inventory and linear production costs over period T, that is, the $1|1|u(t)\in[0,\infty], X(0)=X(T)\,|d(t)|\ L_u, Q_x$ problem which is stated as follows

$$J=\int_0^T\left[c_x^+X^+(t)^2+c_x^-X^-(t)^2+c_u u(t)\right]dt\rightarrow\min, \tag{3.95}$$

subject to (3.90) and (3.91).

Dual Formulation

The Hamiltonian, H, resulting from the maximum principle is

$$H(t)=-\left[c_x^+X^+(t)^2+c_x^-X^-(t)^2+c_u u(t)\right]+\psi(t)\left[u(t)-d(t)\right]. \tag{3.96}$$

The co-state variable, $\psi(t)$ in (3.96), satisfies the following co-state differential equation with its transversality condition

$$\dot{\psi}(t) = \begin{cases} 2c_x^+ X(t), & \text{if } X(t) > 0; \\ 2c_x^- X(t), & \text{if } X(t) \le 0; \end{cases} \quad \psi(0) = \psi(T). \tag{3.97}$$

Maximization of the Hamiltonian implies that if $\psi(t) = c_u$ holds at an interval of time, then the optimal control $u(t)$ is non-negative. Otherwise, $\psi(t) < c_u$ and no control should be undertaken at this interval, i.e. $u(t)=0$.

Algorithm

The algorithm is identical to Algorithm 3.15.

Discussion

Algorithm 3.15 is correct for both linear and quadratic inventory costs. The latter is proved in the following lemma.

LEMMA 3.27. (Correctness of Algorithm 3.15 for the case of the quadratic inventory cost) Given problem (3.95), (3.90) and (3.91), the optimal control is $u(t) = d(t)$, for $0 \le t \le T$.

PROOF. Since the proof is similar to that of Lemma 3.26, it is left as an exercise for the reader.

■

Special Cases of Zero Costs

Solution Algorithm 3.15 does not include any cost parameters and thus does not change for the special cases of zero production $1|1|u(t) \in [0,\infty], X(0)=X(T)|d(t)|\, Q_x$ and zero inventory $1|1|u(t) \in [0,\infty], X(0)=X(T)|d(t)|\, L_u, Q_x, c_x^+ = 0$ costs.

No Backlog

Similar to the case with the linear performance measure, when backlogs are disallowed, i.e. for the $1|1|u(t) \in [0,\infty], X(t) \ge 0, X(0)=X(T)||d(t)|\, L_u, Q_x$ problem, the solution algorithm does not change because the optimal inventory level it finds is non-negative.

3.3.3 NON-LINEAR PRODUCTION COST

In this section we consider a piece-wise linear function of inventory and backlog costs combined with a non-linear production cost function. This change significantly complicates the optimal behavior of the system. In contrast to the trivial solution of the problems with linear production costs (Sections 3.3.1 and 3.3.2), in this section there are only a few solvable cases which are found when special conditions are imposed on the demand. In what follows, we present two solvable cases appearing when demand is either non-lumpy or two-level, piece-wise constant.

Backlogs

The problem to be considered is one-machine, one-product-type scheduling with backlogs to minimize the piece-wise inventory cost and the quadratic production cost over period T, that is, the $1|1|u(t)\in[0,\infty]$,$X(0)=X(T)|d(t)|Q_u, L_x$ problem which is stated as follows

$$J=\int_0^T\left[c_x^+X^+(t)+c_x^-X^-(t)+c_u u(t)^2\right]dt\rightarrow\min, \tag{3.98}$$

subject to (3.90) and (3.91).

Dual Formulation

The Hamiltonian, H, resulting from the maximum principle is

$$H(t)=-\left[c_x^+X^+(t)+c_x^-X^-(t)+c_u u(t)^2\right]+\psi(t)\left[u(t)-d(t)\right]. \tag{3.99}$$

The co-state variable, $\psi(t)$ in (3.99), satisfies the co-state differential equation (3.93).

Maximization of the Hamiltonian implies that if the control function $u(t)$ that satisfies

$$\frac{\partial H}{\partial u(t)}=-2c_u u(t)+\psi(t)=0 \tag{3.100}$$

is non-negative, then it is optimal. Otherwise, no control should be undertaken at point t, i.e. $u(t)=0$.

Algorithm 3.16

Step 1:

If demand function $d(t)$ is continuous, almost everywhere differentiable, and non-lumpy, $-c^-\le 2c_u\dot{d}(t)\le c^+$ for all $t\in[0,T]$, then set the optimal solution as

$u(t)=d(t)$ for $0\le t\le T$, STOP;
otherwise, go to the next step.

Step 2:

If demand function $d(t)$ is two-level piece-wise constant

$d(t)=\begin{cases}d_1, & \text{if } 0\le t\le t_1,\\ d_2, & \text{if } t_1<t\le T,\end{cases}$ and

$\frac{c_u(c_x^+ + c_x^-)}{c_x^+c_x^-}(d_2-d_1)\le t_1\le T-\frac{c_u(c_x^+ + c_x^-)}{c_x^+c_x^-}(d_2-d_1)$, then calculate time points $t_x = \frac{c_u}{c_x^-}(d_2 - d_1)$ and $t_y = \frac{c_u}{c_x^+}(d_2 - d_1)$.

Set the optimal solution as

$u(t) = \frac{1}{2}(d_2 + d_1) - \frac{c_x^-}{2c_u}t$ for $0 \le t \le t_x$;

$u(t) = d_1$ for $t_x \le t \le t_1 - t_y$;

$u(t) = d_1 + \frac{c_x^+}{2c_u}(t - t_1 + t_x)$ for $t_1 - t_y \le t \le t_1 + t_y$;

$u(t) = d_2$ for $t_1 + t_y \le t \le T - t_x$;

$u(t) = \frac{1}{2}(d_2 + d_1) - \frac{c_x^-}{2c_u}(t - T)$ for $T - t_x \le t \le T$;

STOP.

Discussion

Algorithm 3.16 finds a closed form optimal solution for two special cases. Similar to Algorithm 3.15, the first case is characterized by only one regime of optimal control which is the fractional production $u(t)=d(t)$. The other case features a piece-wise linear control in spite of a piece-wise constant demand.

The first solvable case emerges when the demand is non-lumpy, i.e. $-c^- \le 2c_u\dot{d}(t) \le c^+$ over the period T. This is proved in the following lemma.

LEMMA 3.28. (Correctness of Algorithm 3.16 Step 1) Given problem (3.98), (3.90) and (3.91), if demand function $d(t)$ is continuous, almost everywhere differentiable, and $-c^- \le 2c_u\dot{d}(t) \le c^+$ for all $t \in [0, T]$, then the optimal solution is $u(t) = d(t)$ for $0 \le t \le T$.

PROOF. Let the optimal inventory for the problem be zero

$$X(t) \equiv 0 \text{ for } 0 \le t \le T.$$

Since $X(t)$ is constant, $u(t)=d(t)$ must hold for $0 \le t \le T$, and thus, according to (3.100), $\psi(t) = 2c_u d(t)$. This can occur only if $-c^- \le 2c_u\dot{d}(t) \le c^+$ that follows from the third line of co-state differential equation (3.93).

■

The following lemma identifies the optimal solution for the case when demand is a two-level, piece-wise constant function.

LEMMA 3.29. (Correctness of Algorithm 3.16 Step 2) Given problem (3.98), (3.90) and (3.91), if $d(t) = \begin{cases} d_1, \text{ if } 0 \le t \le t_1, \\ d_2, \text{ if } t_1 < t \le T, \end{cases}$ and

$\frac{c_u(c_x^+ + c_x^-)}{c_x^+ c_x^-}(d_2 - d_1) \le t_1 \le T - \frac{c_u(c_x^+ + c_x^-)}{c_x^+ c_x^-}(d_2 - d_1)$, then the optimal solution is

$u(t) = \frac{1}{2}(d_2 + d_1) - \frac{c_x^-}{2c_u} t$ for $0 \le t \le t_x$; $\quad u(t) = d_1$ for $t_x \le t \le t_1 - t_y$;

$u(t) = d_1 + \frac{c_x^+}{2c_u}(t - t_1 + t_x)$ for $t_1 - t_y \le t \le t_1 + t_y$; $u(t) = d_2$ for $t_1 + t_y \le t \le T - t_x$;

$u(t) = \frac{1}{2}(d_2 + d_1) - \frac{c_x^-}{2c_u}(t - T)$ for $T - t_x \le t \le T$,

where $t_x = \frac{c_u}{c_x^-}(d_2 - d_1)$ and $t_y = \frac{c_u}{c_x^+}(d_2 - d_1)$.

PROOF. Since demand is a periodic function, we assume that $d_1 < d_2$ without loss of generality. The case of $d_1 = d_2$ readily satisfies the lemma.

Let there exist time points t_x and t_y such that the optimal inventory to the problem is

$X(t) < 0$ for $0 \le t < t_x$, $\quad X(t) = 0$ for $t_x \le t \le t_1 - t_y$, $\quad X(t) > 0$ for $t_1 - t_y < t < t_1 + t_y$,

$X(t) = 0$ for $t_1 + t_y \le t \le T - t_x$, $\quad X(t) < 0$ for $T - t_x \le t \le T$,

as shown in Figure 3.18. Since $X(t)$ is constant along two intervals $t_x \le t \le t_1 - t_y$ and $t_1 - t_y < t < t_1 + t_y$, then for these intervals $u(t)=d_1$ and $u(t)=d_2$ respectively. Next, with respect to the identified inventory behavior and co-state equation (3.93), the co-state variable for the remaining time intervals evolves linearly

$$\psi(t) = 2c_u d_1 + c_x^+(t - t_1 + t_y) \text{ for } t_1 - t_y < t < t_1 + t_y,$$

$$\psi(t) = c_u(d_2 + d_1) - c_x^- t \text{ for } 0 \le t \le t_x \text{ and}$$

$$\psi(t) = c_u(d_2 + d_1) - c_x^-(t - T) \text{ for } T - t_x \le t \le T.$$

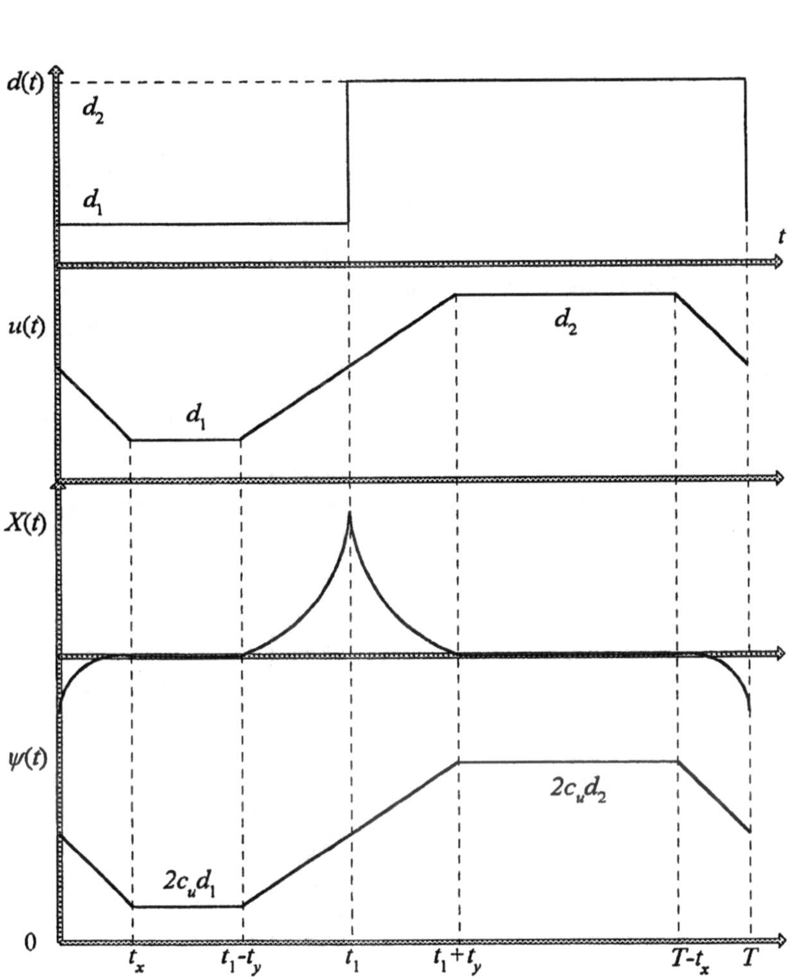

Figure 3.18. Demand $D(t)$, optimal control $u(t)$, state $X(t)$ and co-state $\psi(t)$ variables for the $1|1|u(t)\in[0,\infty],X(0)=X(T)|d(t)|Q_u, L_x$ problem.

By substituting the above co-state variable behavior into $u(t)=\dfrac{\psi(t)}{2c_u}$, the optimal control is obtained as stated in the lemma. Furthermore, according to (3.100), $\psi(t)=2c_u d(t)$ for $t_x \le t \le t_1 - t_y$ and $t_1 + t_y \le t \le T - t_x$. Thus, to find the still unknown switching time points, t_x and t_y, we make use of the state and co-state boundary conditions at point $t_1 + t_y$

$$\psi(t_1+t_y)=2c_u d_1+2c_x^+ t_y=2c_u d_2\,,$$

$$X(t_1+t_y)=\int_{t_1-t_y}^{t_1+t_y}\left[\frac{\psi(t)}{2c_u}-d(t)\right]dt=\int_{t_1-t_y}^{t_1}\left[\frac{\psi(t)}{2c_u}-d_1\right]dt+\int_{t_1}^{t_1+t_y}\left[\frac{\psi(t)}{2c_u}-d_2\right]dt=0\,.$$

Finally, by substituting the corresponding equation for $\psi(t)$ into the last system, we obtain time points t_x and t_y as stated in the lemma. The conditions $t_x\geq 0$ and $t_1-t_y\geq t_x$ are imposed to ensure that the solution constructed in the beginning of the proof is feasible, i.e. all switching points lay within the period and do not overlap.
■

Lemma 3.29 has proved the case of a two-level, piece-wise constant demand. The optimal inventory behavior is characterized by two intervals of time $t_x\leq t\leq t_1-t_y$ and $t_1+t_y\leq t\leq T-t_x$ where production is carried out according to the fractional production regime $u(t)=d_1$ and $u(t)=d_2$ respectively. Evidently, three more solvable cases can be proved by assuming the optimal inventory behavior to contain either one of the fractional production intervals or none of them. The reader is left to develop these cases as an exercise.

Note, the maximum principle does not provide enough information about the optimal solution for exactly solving the $1|1|u(t)\in[0,\infty],X(0)=X(T)|d(t)|Q_u,L_x$ problem with demand as an arbitrary function of time. On the other hand, the numerical approximation algorithms suggested in Section 3.1 cannot be effectively implemented because new boundary conditions $X(0)=X(T)$ and $\psi(0)=\psi(T)$ significantly complicate the solution procedure.

Special Cases of Zero Costs

The problem with no production cost, $1|1|u(t)\in[0,\infty],X(0)=X(T)|d(t)|L_x$ has been solved in Section 3.3.1.

The problem with no inventory cost, $1|1|u(t)\in[0,\infty],X(0)=X(T)|d(t)|Q_u,L_x,c_x^+=0$ is solved with no significant modification by Algorithm 3.16 via substituting $c_x^+=0$ into the optimal solution.

No Backlog

In contrast to the backlog problem, the $1|1|u(t)\in[0,\infty],X(t)\geq 0,X(0)=X(T)|d(t)|Q_u,L_x$ problem is solvable for quite a wide range of demand functions. The only condition which is imposed on the demand is $\frac{1}{T}\int_0^T d(\tau)d\tau - \frac{1}{t}\int_0^t d(\tau)d\tau > \frac{c_x^+}{4c_u}(T-t)$. This implies that the difference between average demand over period T and average demand up to a time point t must exceed $\frac{c_x^+}{4c_u}(T-t)$.

Dual Formulation

Now problem (3.98), (3.90) and (3.91) has an additional constraint $X(t) \geq 0$ which affects neither Hamiltonian (3.99) nor optimality condition (3.100), but changes the co-state differential equation

$$d\psi(t) = c_x^+ dt - d\mu(t), \quad \psi(0+0) = \psi(T+0), \tag{3.101}$$

where Lagrange multiplier $d\mu(t) \geq 0$ corresponds to the no-backlog constraint and satisfies the complementary slackness condition

$$\int_0^T X(t)d\mu(t) = 0. \tag{3.102}$$

Algorithm 3.17

Step 1:

If $\frac{1}{T}\int_0^T d(\tau)d\tau - \frac{1}{t}\int_0^t d(\tau)d\tau > \frac{c_x^+}{4c_u}(T-t)$ for $0 < t < T$, then set the optimal solution as

$u(t) = \frac{1}{T}\int_0^T d(\tau)d\tau - \frac{c_x^+}{4c_u}(T-2t)$, STOP.

Example 3.8

Consider the $1|1|u(t) \in [0,\infty], X(t) \geq 0, X(0)=X(T)|d(t)|Q_u, L_x$ problem with polynomial demand $d(t) = at^n$, $a > 0$, and n is a positive integer, $0 \leq t \leq T$.

At Step 1, Algorithm 3.17 verifies condition

$$\frac{1}{T}\int_0^T d(\tau)d\tau - \frac{1}{t}\int_0^t d(\tau)d\tau > \frac{c_x^+}{4c_u}(T-t)$$

which, by substituting $d(t) = at^n$, results in $\frac{a}{n+1}\sum_{k=0}^{n-1} T^k t^{n-k-1} > \frac{c_x^+}{4c_u}$. This inequality holds for each t if $a > \frac{(n+1)c_x^+}{4c_u T^{n-1}}$. Thus, if $a > \frac{(n+1)c_x^+}{4c_u T^{n-1}}$, the algorithm sets the optimal solution as

$$u(t) = \frac{aT^n}{n+1} - \frac{c_x^+}{4c_u}(T-2t),\ 0 \le t \le T.$$

■

Discussion

Algorithm 3.17 presents a closed form solution of the problem with backlog disallowed. Specifically, the optimal control is a linearly increasing function in time with inventory starting and ending at zero. Note, even if the demand function is analytically integrable, checking the only condition at Step 1 strongly depends on the form of the demand function. For example, if this function is K-interval, piece-wise constant, this check, and thus Algorithm 3.17, require $O(K)$ time that follows from Subroutine to Algorithm 3.2.

LEMMA 3.30. (Correctness of Algorithm 3.17) Given problem (3.98), (3.90), (3.91) with backlogs disallowed, if $\frac{1}{T}\int_0^T d(\tau)d\tau - \frac{1}{t}\int_0^t d(\tau)d\tau > \frac{c_x^+}{4c_u}(T-t)$, $0 < t < T$ then the optimal control is $u(t) = \frac{1}{T}\int_0^T d(\tau)d\tau - \frac{c_x^+}{4c_u}(T-2t)$.

PROOF. Let the optimal inventory be $X(0) = X(T) = 0$ and $X(t) > 0$ for $0 < t < T$. Then $\psi(t) = \psi(0+0) + c_x^+ t$ and $u(t) = \frac{\psi(t)}{2c_u}$ (see Figure 3.19). To find the inventory level over the period, we integrate the state equation (3.90)

$$X(t) = \frac{1}{2c_u}(\psi(0+0)t + \frac{c_x^+}{2}t^2) - \int_0^t d(\tau)d\tau. \tag{3.103}$$

Next, the initial value for the co-state variable is obtained from (3.103) by setting $X(T)=0$

$$\psi(0+0) = \frac{2c_u}{T}\int_0^T d(t)dt - \frac{c_x^+}{2}T. \tag{3.104}$$

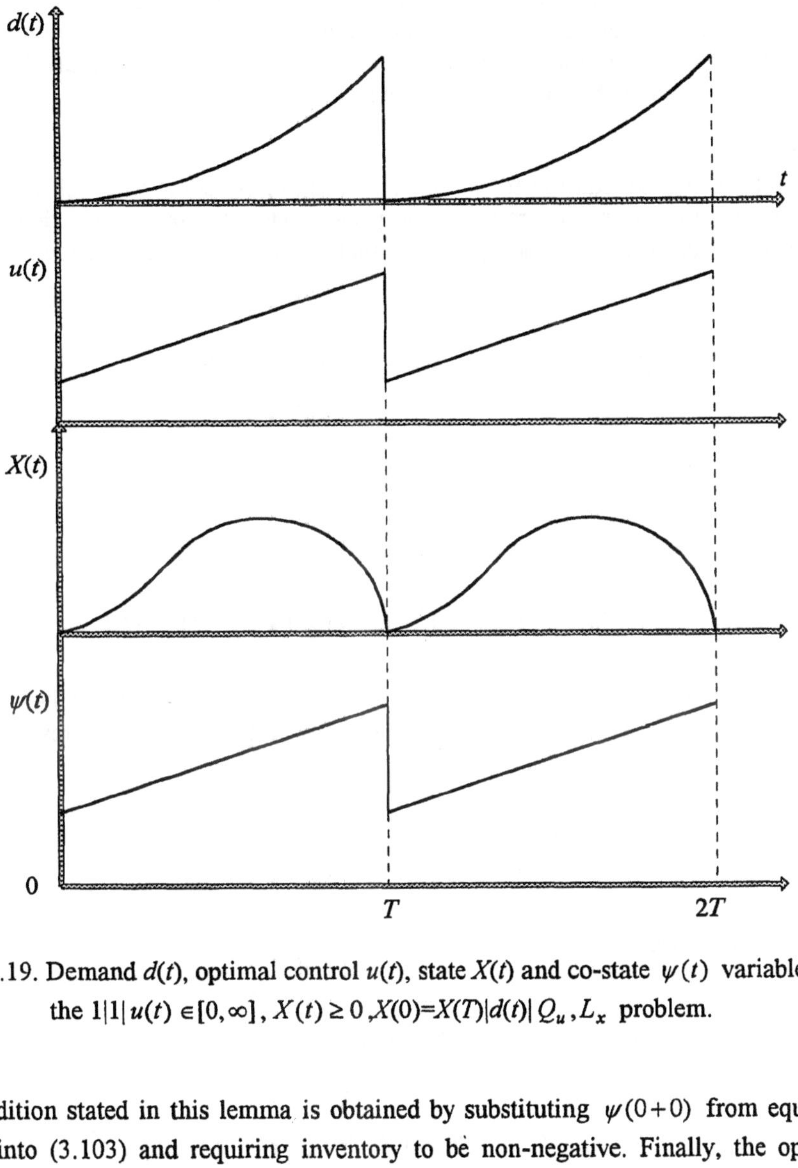

Figure 3.19. Demand $d(t)$, optimal control $u(t)$, state $X(t)$ and co-state $\psi(t)$ variables for the $1|1|u(t)\in[0,\infty], X(t)\geq 0, X(0)=X(T)|d(t)|Q_u, L_x$ problem.

The condition stated in this lemma is obtained by substituting $\psi(0+0)$ from equation (3.104) into (3.103) and requiring inventory to be non-negative. Finally, the optimal control $u(t)=\dfrac{\psi(0+0)+c_x^+ t}{2c_u}$ is obtained by substituting $\psi(0+0)$ from (3.104).

■

3.3.4 NON-LINEAR INVENTORY AND PRODUCTION COSTS

Both costs entering the objective are now non-linear. This further complicates the optimal behavior of the system. However, when demand is a polynomial function of time, the problem is found to be analytically solvable.

Backlogs

The problem to be considered is one-machine, one-product-type scheduling with backlogs to minimize the quadratic inventory and production costs over period T, that is, the $1|1|u(t)\in[0,\infty],X(0)=X(T)|d(t)|Q_u,Q_x$ problem which is stated as follows

$$J=\int_0^T\left[c_x^+X^+(t)^2+c_x^-X^-(t)^2+c_u u(t)^2\right]dt\rightarrow\min, \tag{3.105}$$

subject to (3.90) and (3.91).

Dual Formulation

The Hamiltonian, H, resulting from the maximum principle is

$$H(t)=-\left[c_x^+X^+(t)^2+c_x^-X^-(t)^2+c_u u(t)^2\right]+\psi(t)\left[u(t)-d(t)\right]. \tag{3.106}$$

The co-state variable, $\psi(t)$ in (3.106), satisfies the co-state differential equation (3.97). Maximization of the Hamiltonian as a function of $u(t)$ results, as in the previous section, in

$$u(t)=\begin{cases}\dfrac{\psi(t)}{2c_u}, \text{ if } \psi(t)\geq 0;\\ 0, \text{ if } \psi(t)<0.\end{cases} \tag{3.107}$$

Algorithm 3.18

Step 1:

If the demand function is a polynomial of order P, $d(t)=\sum_{p=0}^{p=P}a_p t^p$,

and $c_x^+=c_x^-$, then calculate parameters $y=\sqrt{\dfrac{c_x^+}{c_u}}$, $b_P=2c_u a_P$,

$$b_{P-1}=2c_u a_{P-1};\quad b_p=2c_u a_p+\frac{(p+1)(p+2)}{y^2}b_{p+2}\quad \text{for}\quad p=0,\ldots,P\text{-}2;$$

$$A=-\frac{\sum_{p=0}^{p=P-1}T^p(yb_p-(p+1)b_{p+1})}{2y\left(1+e^{yT}\right)}\quad \text{and } B=\frac{\sum_{p=0}^{p=P-1}T^p(yb_p-(p+1)b_{p+1})}{2y\left(1+e^{-yT}\right)}.$$

Set the optimal solution as

$$u(t)=\frac{1}{2c_u}\left(Ae^{yt}+Be^{-yt}+\sum_{p=0}^{p=P}b_p t^p\right);\ \text{STOP.}$$

Discussion

Algorithm 3.18 provides a closed form of the optimal control for the demand function expressed as a polynomial, when the inventory cost is equal to the backlog cost. The complexity of this step is *O*(*P*) due to calculation of coefficients b_p.

LEMMA 3.31. Given problem (3.105), (3.90) and (3.91), the co-state variable $\psi(t)$ is bounded as $\psi(t)\geq 0$ for $t\in[0,T]$.

PROOF. The proof of the lemma is identical to the proof of the lower bound in Lemma 3.8.

■

LEMMA 3.32. (Correctness of Algorithm 3.18) Given problem (3.105), (3.90) and (3.91), if the demand function is a polynomial of order *P*, $d(t)=\sum_{p=0}^{p=P}a_p t^p$,

and $c_x^+=c_x^-$, then the optimal control is

$$u(t)=\frac{S(t)}{2c_u},$$

where $S(t)=Ae^{yt}+Be^{-yt}+\sum_{p=0}^{p=P}b_p t^p$ for all $t\in[0,T]$, $y=\sqrt{\frac{c_x^+}{c_u}}$, $b_P=2c_u a_P$,

$$b_{P-1}=2c_u a_{P-1};\ b_p=2c_u a_p+\frac{(p+1)(p+2)}{y^2}b_{p+2}\ \text{for } p=0,\ldots,P\text{-}2;$$

$$A = -\frac{\sum_{p=0}^{p=P-1} T^p (yb_p - (p+1)b_{p+1})}{2y\left(1+e^{yT}\right)} \quad \text{and} \quad B = \frac{\sum_{p=0}^{p=P-1} T^p (yb_p - (p+1)b_{p+1})}{2y\left(1+e^{-yT}\right)}.$$

PROOF. By using condition $c_x^+ = c_x^-$, the co-state differential equation (3.97) simplifies

$$\dot{\psi}(t) = 2c_x^+ X(t). \tag{3.108}$$

By differentiation (3.108) and substituting (3.90), we obtain

$$\ddot{\psi}(t) = 2c_x^+ (u(t) - d(t)). \tag{3.109}$$

Next, by taking into account optimality condition (3.107) and Lemma 3.31, we find

$$\ddot{\psi}(t) = \frac{c_x^+}{c_u} \psi(t) - 2c_x^+ d(t).$$

Thus, the conditions of the lemma allow the system of state and co-state differential equations (3.90) and (3.108) to be presented as a single, second-order ordinary differential equation

$$\ddot{\psi}(t) = \frac{c_x^+}{c_u} \psi(t) - 2c_x^+ \sum_{p=0}^{p=P} a_p t^p, \qquad \psi(0) = \psi(T), \qquad \dot{\psi}(0) = \dot{\psi}(T).$$

The analytical solution of this equation, denoted as $S(t)$, is discussed in Section 2.3.1. The optimal control is found by substituting the solution for $\psi(t)$ into (3.107).
■

Note, the maximum principle does not provide enough information about the optimal solution for exactly solving the $1|1|u(t) \in [0,\infty], X(0)=X(T)|d(t)|Q_u, Q_x$ problem with demand as an arbitrary function of time. On the other hand, the numerical approximation algorithms suggested in Section 3.1 cannot be effectively implemented because new boundary conditions $X(0)=X(T)$ and $\psi(0) = \psi(T)$ significantly complicate the solution procedure.

Special Cases of Zero Costs

The problem with no production cost, i.e. the $1|1|u(t) \in [0,\infty], X(0)=X(T)|d(t)|Q_x$ problem has been solved in Section 3.3.2. The problem with no inventory cost, i.e. the $1|1|u(t)\in[0,\infty]$, $X(0)=X(T)|d(t)|Q_u, Q_x, c_x^+=0$ problem can be solved for arbitrary demand by setting a constant production rate equal to the average demand, $u(t)=\frac{1}{T}\int_0^T d(t)dt$, $t\in[0,T]$. This solution follows from the fact that $\psi(t)$ must be constant ($\psi(t)$ is periodic, $\psi(0)=\psi(T)$, and cannot increase because $c_x^+=0$). The value of $X(0)$ is chosen such that no backlog can occur along the period, $X(0)=\max_{t\in[0,T]}\left[\int_0^t d(\tau)d\tau-\frac{t}{T}\int_0^T d(\tau)d\tau\right]$.

No Backlog

To solve the $1|1|u(t)\in[0,\infty]$,$X(0)=X(T)$, $X(t)\geq 0\,|d(t)|\,Q_u, Q_x$ problem with backlogs disallowed, we again reduce it to a combinatorial search for a feasible path on a 2-weighted graph.

Dual Formulation

Now problem (3.105), (3.90) and (3.91) has an additional constraint, $X(t)\geq 0$, which affects neither Hamiltonian (3.106) nor optimality condition (3.107), but changes the co-state differential equation

$$d\psi(t)=c_x^+X(t)dt-d\mu(t),\quad \psi(T+0)=0, \tag{3.110}$$

where Lagrange multiplier $d\mu(t)\geq 0$ corresponds to the no-backlog constraint and satisfies the complementary slackness condition (3.102).

Algorithm 3.19

Step 1:

Select time intervals $\tau_k=[t_{left}^k, t_{right}^k]$ at each of which the demand function decreases, $t_{right}^k<t_{left}^{k+1}$, and number them beginning from k=2. In addition, define τ_1 and τ_K as the intervals which are degenerated to a point: $\tau_1=\{0\}$ and $\tau_K=\{T\}$, where K is the total number of obtained intervals.

Step 2 and Step 3 are identical to Steps 2 and 3 of Algorithm 3.4.

Subroutine 1 to Algorithm 3.19

(i) Set time points $t_x=t_{left}^n$, $t_y=t_{right}^n$, $t_{out}=t_{left}^n$.

(ii) Integrate the following system of differential equations

$$\dot{X}(t)=\frac{\psi(t)}{2c_u}-d(t),\quad X(t_{out})=0,$$

$$\dot{\psi}(t)=2c_x^+X(t),\qquad \psi(t_{out})=2c_u d(t_{out}),$$

over the interval $[t_{out}, t_{right}^m]$.

If $X(t)<0$ for any $t<t_{left}^m$, then no arc links nodes τ_n and τ_m; STOP.

Otherwise, set $t_{out}=t_{right}^n$ and integrate the above system.

If $X(t)>0$ for all $t\in\tau_m$, then no arc links nodes τ_n and τ_m; STOP.

Otherwise go to (iii)

(iii) Integrate the system from (ii) with $t_{out}=0.5(t_x+t_y)$;

If $X(t)<0$ for any $t<t_{left}^m$ then $t_y=t_{out}$ and go to (iv),

otherwise set time point $t_{in}\in\tau_m$ such that $X(t_{in})=0$.

If $\psi(t_{in})>2c_u d(t_{in})$, then $t_x=t_{out}$ and go to (iv);

if $\psi(t_{in})<2c_u d(t_{in})$, then $t_y=t_{out}$ and go to (iv).

(iv) If $t_y-t_x\le\varepsilon$, then set $arc_{nm}=(\tau_n,\tau_m)\in A$ and its two weights $a(arc_{nm})=t_{out}$ and $b(arc_{nm})=t_{in}$; STOP.

Otherwise go to (iii).

Discussion

Since Algorithm 3.19 is very similar to Algorithm 3.4 that locates intervals of time at which fractional production $u(t)=d(t)$ occurs, it approximates the globally optimal solution with the same accuracy and complexity. At Step 1, the condition of selecting time intervals τ_k is changed, while the other two steps are identical to those of Algorithm 3.4. Step 2 invokes four subroutines which are now slightly changed. Namely, the no-backlog condition is additionally checked and the co-state differential equation, integrated iteratively in the subroutines, is modified, as presented in procedure (ii) of Subroutine 1 to Algorithm 3.19. Subroutines 2-4 to Algorithm 3.19 are omitted since they are modified accordingly.

Note, that according to (3.102), $d\mu(t)=0$ at time intervals of positive inventories. Although jumps of the measure function, $d\mu(t)>0$, at points $t=t_{out}$ and $t=t_{in}$ are formally possible, its value is zero, which is proved similarly to the proof of Lemma 3.12.

3.3.5 MAINTAINING STABLE INVENTORY

Backlogs

The problem to be considered is one-machine, one-product-type production control to maintain stable inventory under the demand when its change in time is unknown, that is, the $1|1|u(t)\in[0,\infty]|d(t)|\overline{X}$ problem which is stated as follows

$$\sup_{0\le t<\infty} |X(t)| < \infty, \tag{3.111}$$

$$\dot{X}(t) = u(t) - d(t), \quad X(0) = 0, \tag{3.112}$$

$$u(t) \ge 0. \tag{3.113}$$

In contrast to the optimization problems, maintaining stable inventory is a problem of feedback control. Feedback control is a control rule given as a function of inventory, rather than a function of time as was the case in the previous sections. Generally, feedback control rules may involve several components with adjustable positive parameters G and the desired positive inventory level $\overline{X}$. In principle, it is possible to suggest a large number of feedback rules to maintain stable inventory. In what follows, we present three feedback control-based solutions which are useful in practice and easy to implement.

Feedback Rule 3.1

Proportional feedback control:

$$\text{at every } t \text{ set } \quad u(X(t)) = \begin{cases} G(\overline{X} - X(t)), & \text{if } X(t) \le \overline{X}, \\ 0, & \text{if } X(t) > \overline{X}. \end{cases}$$

Feedback Rule 3.2

Proportional plus integral feedback control:

at every t set

$$u(X(t)) = \begin{cases} G_1(\overline{X} - X(t)) + G_2 \int_0^t (\overline{X} - X(\tau))d\tau, & \text{if } X(t) \le \overline{X}, \\ 0, & \text{if } X(t) > \overline{X}. \end{cases}$$

Feedback Rule 3.3
Impulse feedback control:

$$\text{set } u(X(t)) = \begin{cases} 0, & \text{if } X(t) > 0, \\ \infty, & \int_{t-0}^{t+0} u(\tau)d\tau = G_3\overline{X}, \text{ if } X(t) = 0, \end{cases} \quad \text{at every } t.$$

Discussion

Feedback Rules 3.1-3.3 are of $O(1)$ complexity at every time t. Although these rules are simple to use, the stability of inventories is not guaranteed when dealing with arbitrary demands. In other words, given a specific feedback rule, either inventory or backlog level can grow unlimitedly. Therefore, our objective is to determine a class of demand functions for which inventory stability is guaranteed. The following three lemmas identify demand-related conditions sufficient for the inventory to be bounded when the three respective feedback rules are applied.

LEMMA 3.33. (Stability of Feedback Rule 3.1) Given problem (3.111)-(3.113), if the proportional feedback control is applied with $\overline{X} \geq 0$ and demand is bounded, $d(t) \leq B$, then the inventory is bounded as $\min\left\{0, \overline{X} - \frac{B}{G}\right\} \leq X(t) \leq \overline{X}$.

PROOF. First note that the inventory level can never exceed $\overline{X}$. Indeed, by definition of the proportional feedback control, once $X(t)$ reaches $\overline{X}$, production stops and the demand again depletes inventory below $\overline{X}$, i.e. the upper inventory bound is

$$X(t) \leq \overline{X}. \tag{3.114}$$

To estimate the lower bound of $X(t)$, we substitute the proportional feedback control into state equation (3.112)

$$\dot{X}(t) = G(\overline{X} - X(t)) - d(t), \quad X(0) = 0. \tag{3.115}$$

According to Section 2.3.1, the general form of the solution for the linear differential equation (3.115) is

$$X(t) = e^{-Gt}\int_0^t e^{G\tau}(G\overline{X} - d(\tau))d\tau. \tag{3.116}$$

An estimate for the lower bound of $X(t)$ is obtained by setting $d(t) \equiv B$ in (3.116)

$$X(t) \ge e^{-Gt} \int_0^t e^{G\tau} (G\overline{X} - B) d\tau .$$

The last inequality can be presented in the following form

$$X(t) \ge e^{-Gt} (G\overline{X} - B) \int_0^t e^{G\tau} d\tau \ge \min\{0, \overline{X} - \frac{B}{G}\} .$$

■

LEMMA 3.34. (Stability of Feedback Rule 3.2) Given problem (3.111)-(3.113), if the proportional plus integral feedback control is applied with $\overline{X} \ge 0$ and demand is bounded, $d(t) \le B$, then the inventory is bounded as $\min\left\{0, \overline{X} - \frac{B}{G_1}\right\} \le X(t) \le \overline{X}$.

PROOF. First, as with the previous lemma, the inventory level can never exceed $\overline{X}$ since $X(0) = 0$ and $\overline{X} > 0$. Indeed, by definition of the proportional feedback control, once $X(t)$ reaches $\overline{X}$, production stops and the demand again depletes inventory below $\overline{X}$, i.e. the upper inventory bound is

$$X(t) \le \overline{X} . \tag{3.117}$$

Next, since Rule 3.2 determines the control, which is always greater than that determined by Rule 3.1, the lower bound of $X(t)$ found in Lemma 3.33 is valid.

■

LEMMA 3.35. (Stability of Feedback Rule 3.3) Given problem (3.111)-(3.113), if the impulse feedback control is applied, then the inventory is bounded as $0 \le X(t) \le G_3\overline{X}$.

PROOF. According to the impulse feedback control and (3.112), at time t=0, as well as at all points where $X(t)$=0, the inventory level jumps up to the value of $G_3\overline{X}$. At the intervals of time where $X(t)$>0, the inventory decreases with respect to the state equation

$$\dot{X}(t) = -d(t) . \tag{3.118}$$

This behavior thus ensures that $0 \le X(t) \le G_3\overline{X}$.

■

Example 3.9

Consider the $1|1|u(t)\in[0,\infty]|d(t)|\overline{X}$ problem with constant demand $d(t)=d$. Feedback Rules 3.1-3.3 are applied with unit parameters G, G_1, G_2, G_3 and $\overline{X}>\frac{d}{G}$. Since the demand is bounded with *B*=*d*, all three feedback rules provide stable inventories bounded as $0\le X(t)\le\overline{X}$, as proven in Lemmas 3.33-3.35.

In addition, if Rules 3.1 and 3.2 are applied, the inventory level asymptotically converges to $\overline{X}-\frac{d}{G}$ and $\overline{X}$ respectively. Specifically, under Rule 3.1, the inventory level evolves according to (3.116)

$$X(t)=e^{-Gt}\int_0^t e^{G\tau}(G\overline{X}-d)d\tau=(G\overline{X}-d)\frac{1-e^{-Gt}}{G}. \tag{3.119}$$

The asymptotic convergence is found from (3.119) as

$$\lim_{t\to\infty}X(t)=\overline{X}-\frac{d}{G}. \tag{3.120}$$

To determine the asymptotic convergence of $X(t)$ under Rule 3.2, we substitute the proportional plus integral feedback control into state equation (3.112)

$$\dot{X}(t)=G_1(\overline{X}-X(t))+G_2\int_0^t(\overline{X}-X(\tau))d\tau-d. \tag{3.121}$$

By introducing a new variable

$$Y(t)=\int_0^t(\overline{X}-X(\tau))d\tau,$$

we present (3.121) as a two-dimensional, first-order linear differential equation

$$\frac{d}{dt}\begin{vmatrix}Y(t)\\ \dot{Y}(t)\end{vmatrix}=A\begin{vmatrix}Y(t)\\ \dot{Y}(t)\end{vmatrix}+f,\ Y(0)=0,\ \dot{Y}(0)=\overline{X}, \tag{3.122}$$

where matrices $A=\begin{vmatrix}0 & 1\\ -1 & -1\end{vmatrix}$ and $f=\begin{vmatrix}0\\ d\end{vmatrix}$.

According to Section 2.3.1, the general form of the solution for the linear differential equation (3.122) is

$$\hat{Y}(t)=\begin{vmatrix}Y(t)\\ \dot{Y}(t)\end{vmatrix}=e^{At}\left(\begin{vmatrix}0\\ \overline{X}\end{vmatrix}+\int_0^t e^{-A\tau}fd\tau\right). \tag{3.123}$$

Matrix exponent e^{At} tends to zero as t tends to infinity because eigenvalues $\lambda_{1,2}$ of matrix A have a negative real component

$$\lambda_{1,2}=-\frac{1}{2}\left(1\pm\sqrt{-3}\right).$$

The asymptotic convergence of $\hat{Y}(t)$ is, thus, found by integrating (3.123) as

$$\lim_{t\to\infty}\hat{Y}(t)=\lim_{t\to\infty}e^{At}\left(\begin{vmatrix}0\\ \overline{X}\end{vmatrix}+A^{-1}(I-e^{-At})\begin{vmatrix}0\\ d\end{vmatrix}\right)=$$

$$=\lim_{t\to\infty}e^{At}\left(\begin{vmatrix}0\\ \overline{X}\end{vmatrix}+A^{-1}\begin{vmatrix}0\\ d\end{vmatrix}\right)-A^{-1}\begin{vmatrix}0\\ d\end{vmatrix}=-A^{-1}\begin{vmatrix}0\\ d\end{vmatrix}=\begin{vmatrix}d\\ 0\end{vmatrix}.$$

By considering the second component of $\hat{Y}(t)$, $\dot{Y}(t)=\overline{X}-X(t)$, we conclude that

$$\lim_{t\to\infty}X(t)=\overline{X}.$$

Figure 3.20 presents inventory convergence under Rules 3.1 and 3.2. From this figure one can also observe that the inventory level is periodic and therefore does not converge if Rule 3.3 is applied.

■

No Backlog

For the case when backlogs are disallowed, i.e. for the $1|1|u(t)\in[0,\infty], X(t)\geq 0\,|d(t)|\,\overline{X}$ problem, Feedback Rule 3.3 is readily applicable because it provides no backlog (see Lemma 3.35). On the other hand, Feedback Rules 3.1 and 3.2 do not necessarily prevent backlogs, especially if the demand is lumpy. However, from equation (3.116), it readily follows that if Rule 3.1 is applied, no backlog can occur when demand is bounded as

$d(t) \le G\overline{X}$ at every point of time. Finally, the same bound $d(t) \le G_1\overline{X}$ ensures no backlog, if Rule 3.2 is applied, because the control it determines is always greater than that determined by Rule 3.1.

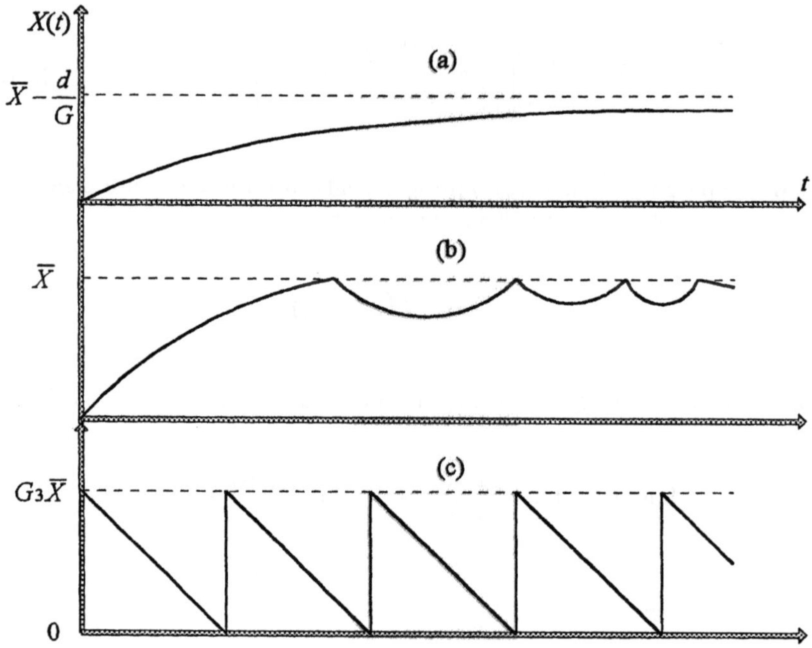

Figure 3.20. Evolution of inventory $X(t)$ under constant demand and Rules 3.1 (a), 3.2 (b) and 3.3 (c) for the $1|1|u(t) \in [0,\infty]\,|d(t)|\,\overline{X}$ problem.

3.4 VARIABLE PLANNING HORIZON

This section addresses the scheduling problems characterized by a certain amount of product to be completed by the end of a planning horizon. Unlike the previous sections, the length of the planning horizon is unknown *a priori*. Thus, the objective in such cases is to minimize the total cost of inventory and production as well as the penalty for prolonging the planning horizon. The penalty for prolonging the planning horizon is further referred to as the cost of the planning horizon. Specifically, when the inventory and production costs are negligible, the objective transforms into the makespan widely used in combinatorial scheduling.

3.4.1 LINEAR PERFORMANCE MEASURE

We start with one-machine, one-product-type scheduling to minimize the total linear cost of inventory, production and the planning horizon, that is, the $1|1|u(t)\in[0,\infty], X(T)=X_T \,||\, L_u, L_x, T$ problem which is stated as follows

$$J=\int_0^T \left[c_x^+ X(t)+c_u u(t)\right]dt+c_T T \rightarrow \min,$$

subject to

$$\dot{X}(t)=u(t), \qquad X(0)=0, \qquad X(T)=X_T, \tag{3.124}$$

$$u(t)\geq 0. \tag{3.125}$$

Note, that the above statement is characterized by non-negative inventory levels along the overall planning horizon because the planning horizon ends up as soon as the required amount of the product $X(T)=X_T$ is completed. Thus, the notion of backlog is inapplicable to problems with variable planning horizon.

Next, the objective can be simplified by substituting the state equation (3.124)

$$J=\int_0^T \left[c_x^+ X(t)+c_u u(t)\right]dt+c_T T=c_u X_T+\int_0^T c_x^+ X(t)dt+c_T T.$$

Since the total production cost $c_u X_T$ is a constant which does not influence the optimization, the objective of the problem takes the following form

$$J = \int_0^T c_x^+ X(t)dt + c_T T \to \min . \tag{3.126}$$

The fact that the total production cost is independent of the control function has resulted from linearity of the production cost and fixed terminal inventory.

Dual Formulation

The Hamiltonian, H, resulting from the maximum principle is

$$H(t) = -c_x^+ X(t) + \psi(t)u(t) = c_T . \tag{3.127}$$

The co-state variable, $\psi(t)$ in (3.127), satisfies the following co-state differential equation with no transversality conditions

$$\dot{\psi}(t) = c_x^+ . \tag{3.128}$$

Maximization of the Hamiltonian implies

$$\frac{\partial H(t)}{\partial u(t)} = \psi(t) \le 0 . \tag{3.129}$$

Therefore, the optimality condition is: if $\psi(t) = 0$ holds at an interval of time, then the control effort $u(t)$ is arbitrary; otherwise, $\psi(t) < c_u$ and no control should be undertaken at this interval, i.e. $u(t)=0$. Furthermore, according to equation (3.128), $\psi(t)$ linearly increases. Therefore, with respect to the optimality condition, it can reach zero only at $t=T$. Thus, no control should be undertaken $u(t)=0$ when $t<T$ because $\psi(t) < 0$.

Algorithm 3.20

Step 1:

Set $T=0$.

If $X_T > 0$, then

set the optimal solution as

$$u(0) = \infty , \quad \int_0^{T+0} u(t)dt = X_T , \quad \text{STOP;}$$

Otherwise, $u(0)=0$; STOP.

Discussion

From Algorithm 3.20, one can observe that the new terminal constraint, $X(T) = X_T$, dictates the required amount of the product to be carried out immediately, as fast as possible, regardless of the inventory and production costs.

The complexity of Algorithm 3.20 is $O(1)$ and its correctness is easily verified by solving the dual formulation in the following lemma.

LEMMA 3.36. (Correctness of Algorithm 3.20) For problem (3.124)-(3.126), the optimal planning horizon T=0 and the optimal control is

- if $X_T > 0$ then $u(0) = \infty$, $\int_0^{T+0} u(t)dt = X_T$;
- otherwise $u(0)$=0.

PROOF. The proof is by contradiction. Let T>0, then, as shown in the dual formulation, $\psi(t) < 0$ and $u(t)$=0 for $0 \le t < T$. As a result, there is no production $X(t)$=0 at this time interval and equation (3.127) never holds.

Thus T=0, and required amount X_T of the product must be completed instantly at this single time point t=T=0.
■

Special Cases of Zero Costs

The $1|1|u(t) \in [0,\infty], X(T) = X_T \| L_x, T$ and $1|1|u(t) \in [0,\infty], X(T) = X_T \| L_u, L_x, T, c_x^+ = 0$ problems are solvable by Algorithm 3.20, since neither production nor inventory cost coefficients are involved in the algorithm.

The problem with the zero planning horizon cost, i.e. the $1|1|u(t) \in [0,\infty], X(T) = X_T \| L_u, L_x, T, c_T = 0$ problem, has an infinite number of optimal solutions. Indeed, the length of the planning horizon can now be set at any time point and the optimal control, determined by Algorithm 3.20, is applied at this time point without changing the objective value.

3.4.2 NON-LINEAR INVENTORY COST

Consider a linear production cost function and non-linear inventory cost, which is approximated by a quadratic function. It is shown below that the introduced non-linearity does not affect the optimal solution at all.

The problem to be considered is one-machine, one-product-type scheduling to minimize quadratic inventory as well as linear production and planning horizon costs, that is, the $1|1|u(t)\in[0,\infty], X(T)=X_T \| L_u, Q_x, T$ problem which is stated as follows

$$J=\int_0^T\left[c_x^+X(t)^2+c_uu(t)\right]dt+c_TT\rightarrow\min,$$

subject to (3.124) and (3.125).

As in the previous section, the total production cost is constant and thus does not influence the optimization. Therefore, the objective simplifies to

$$J=\int_0^T c_x^+X(t)^2\,dt+c_TT\rightarrow\min. \tag{3.130}$$

Dual Formulation

The Hamiltonian, H, resulting from the maximum principle is

$$H(t)=-c_x^+X(t)^2+\psi(t)u(t)=c_T. \tag{3.131}$$

The co-state variable, $\psi(t)$ in (3.131), satisfies the following co-state differential equation with no transversality condition

$$\dot{\psi}(t)=2c_x^+X(t). \tag{3.132}$$

Maximization of the Hamiltonian implies that if the control function $u(t)$ that satisfies (3.129) is non-negative, then it is optimal. Otherwise, no control should be undertaken at point t, i.e. $u(t)=0$.

Algorithm

The algorithm is identical to Algorithm 3.20.

Discussion

Algorithm 3.20 is correct for both linear and quadratic inventory costs. The latter is proven in the following lemma.

LEMMA 3.37. (Correctness of Algorithm 3.20 for the case of the quadratic inventory cost) Given problem (3.130), (3.124) and (3.125), then T=0 and the optimal control is

- if $X_T > 0$ then $u(0) = \infty$, $\int_0^{T+0} u(t)dt = X_T$;
- otherwise $u(0)$=0.

PROOF. The proof is by contradiction. Let $T > 0$, then, according to the state differential equation (3.124), $X(t)$ is a non-decreasing function of time. With respect to the co-state differential equation (3.132), $\psi(t)$ is also a non-decreasing function. Moreover, according to (3.132), once production starts, $\psi(t)$ increases. Therefore, with respect to condition (3.129), the co-state variable can reach zero only at t=T, when no control is exercised, $u(t)$=0. Therefore, $X(t)$=0 when $0 \le t < T$ because $\psi(t) < 0$. Consequently, equation (3.131) never holds ($c_T \neq 0$).

Thus T=0, and required amount X_T of the product must be completed instantly at this single time point t=T=0 as was the case with the linear inventory cost proven in Lemma 3.36.

■

Special Cases of Zero Costs

As with the linear inventory cost problem, zero production, inventory or planning horizon costs do not influence the optimal solution and, therefore no modification to Algorithm 3.20 is required.

3.4.3 NON-LINEAR PRODUCTION COST

In this section, we consider the linear inventory and planning horizon costs when the production cost is non-linear. The production cost function is such that additional control efforts may become increasingly more expensive. This change influences the optimal solution, which, in contrast to what has been described in the previous sections, takes on finite values over a non-zero planning horizon.

The problem to be considered is one-machine, one-product-type scheduling to minimize quadratic production cost as well as linear inventory and planning horizon costs, that is, the $1|1|u(t) \in [0,\infty], X(T) = X_T \| Q_u, L_x, T$ problem which is stated as follows

$$J = \int_0^T \left[c_x^+ X(t) + c_u u(t)^2\right] dt + c_T T \to \min, \tag{3.133}$$

subject to (3.124) and (3.125).

Dual Formulation

The Hamiltonian, H, resulting from the maximum principle is

$$H(t) = -\left[c_x^+ X(t) + c_u u(t)^2\right] + \psi(t)u(t) = c_T . \tag{3.134}$$

The co-state variable, $\psi(t)$ in (3.134), satisfies the co-state differential equation (3.128).

Maximization of the Hamiltonian implies that if the control function $u(t)$ that satisfies

$$\frac{\partial H}{\partial u(t)} = -2c_u u(t) + \psi(t) = 0 \tag{3.135}$$

is non-negative, then it is optimal. Otherwise, no control should be undertaken at point t, i.e. $u(t)$=0.

Algorithm 3.21
Step 1:

Calculate $T = \frac{2}{c_x^+}\left(\sqrt{c_u c_T + c_x^+ c_u X_T} - \sqrt{c_u c_T}\right)$ and set the optimal solution as

$u(t) = \frac{1}{2c_u}\left(2\sqrt{c_u c_T} + c_x^+ t\right)$ for $0 \le t \le T$, STOP.

Discussion

Algorithm 3.21 determines a closed form solution. Its optimality is proven in the following lemma. This solution is characterized by the production, which is exercised from the very beginning of the planning horizon and grows linearly until the required amount of the product is completed.

LEMMA 3.38. (Correctness of Algorithm 3.21) Given problem (3.133), (3.124) and (3.125), the optimal control and the length of the planning horizon are

$$u(t) = \frac{1}{2c_u}\left(2\sqrt{c_u c_T} + c_x^+ t\right) \text{ and } T = \frac{2}{c_x^+}\left(\sqrt{c_u c_T + c_x^+ c_u X_T} - \sqrt{c_u c_T}\right).$$

PROOF. First, we prove that production must start from the very beginning of the planning horizon. The proof is by contradiction. Let there exist a time point t_1 >0, such that $u(t) = 0$ for $0 \le t \le t_1$. Then, according to the state differential equation (3.124), $X(t)$=0 for $0 \le t \le t_1$ and equation (3.134) is never satisfied at $0 \le t \le t_1$ because $c_T \ne 0$. Thus, the production starts from the beginning of the planning horizon and, with respect to (3.124), (3.128) and (3.135), is described as follows (see Figure 3.21)

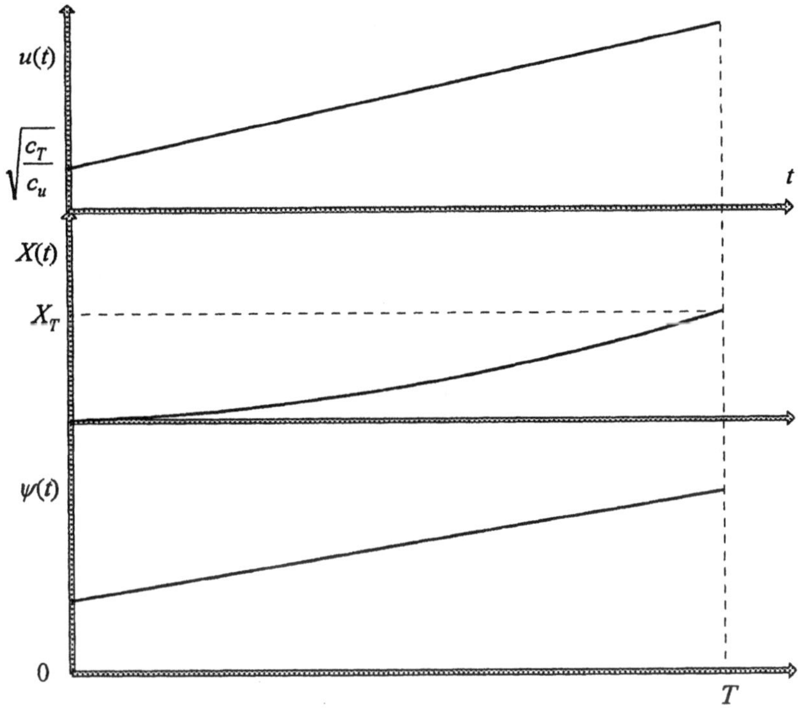

Figure 3.21. Optimal control $u(t)$, state $X(t)$ and co-state $\psi(t)$ variables for the $1|1|u(t)\in[0,\infty], X(T)=X_T \| Q_u, L_x, T$ problem.

$$X(t)=\int_0^t u(\tau)d\tau\,,\quad \psi(t)=\psi(0)+c_x^+ t\,,\quad u(t)=\frac{\psi(t)}{2c_u}\,. \tag{3.136}$$

By taking into account the terminal inventory condition and (3.124), we find

$$X(T)=\frac{T}{2c_u}\left(\psi(0)+\frac{c_x^+ T}{2}\right)=X_T\,, \tag{3.137}$$

$$\psi(0)=2\sqrt{c_u c_T}\,, \tag{3.138}$$

and thus

$$T=\frac{2}{c_x^+}\left(\sqrt{c_u c_T + c_x^+ c_u X_T}-\sqrt{c_u c_T}\right). \tag{3.139}$$

Finally, optimal control $u(t) = \frac{\psi(t)}{2c_u}$ is readily obtained by replacing $\psi(t)$ with its expressions (3.136) and (3.138).

■

Special Cases of Zero Costs

First note that the $1|1|u(t) \in [0, \infty], X(T) = X_T \| L_x, T$ problem has been solved in Section 3.4.1. However, the case of zero inventory cost, i.e. the $1|1|u(t) \in [0, \infty], X(T) = X_T \| Q_u, L_x, T, c_x^+ = 0$ problem, deserves separate consideration. Although, the change in the primal and dual formulations is straightforwardly obtained by substituting $c_x^+ = 0$ in the corresponding equations, solving these formulations results in a new form of the optimal control function.

Algorithm 3.22

Step 1:

Calculate $T = X_T \sqrt{\frac{c_u}{c_T}}$ and set the optimal solution as

$u(t) = \frac{X_T}{T}$ for $0 \le t \le T$, STOP.

Discussion

Algorithm 3.22 presents a closed form solution whose optimality is proved in the following lemma. This optimal solution is characterized by constant-in-time production, until the required amount of the product is completed.

LEMMA 3.39. (Correctness of Algorithm 3.22) Given problem (3.124), (3.125) and (3.133) with $c_x^+ = 0$, the optimal solution is $u(t) = \frac{X_T}{T}$ and $T = X_T \sqrt{\frac{c_u}{c_T}}$.

PROOF. First, we prove that production must start from the very beginning of the planning horizon. The proof is by contradiction. Let there exist a time point $t_1 > 0$, such that $u(t) = 0$ for $0 \le t \le t_1$. Then, according to the state differential equation (3.124), $X(t)=0$ for $0 \le t \le t_1$ and equation (3.134) is never satisfied at $0 \le t \le t_1$ because $c_T \ne 0$. Thus, the production starts from the beginning of the planning horizon.

Secondly, by taking into account $c_x^+ = 0$, the co-state differential equation (3.128) results in

$$\dot{\psi}(t) = 0 .$$

Therefore, with respect to (3.124) and (3.135), the optimal solution is described as follows

$$X(t)=\int_0^t u(\tau)d\tau\,,\quad \psi(t)=\psi(0)\,,\quad u(t)=\frac{\psi(0)}{2c_u}\,. \tag{3.140}$$

By taking into account the terminal inventory condition and (3.124), we find

$$X(T)=\frac{T}{2c_u}\psi(0)=X_T\ ; \tag{3.141}$$

$$c_u\frac{X_T^2}{T^2}=c_T\ . \tag{3.142}$$

The optimal control value and the length of the planning horizon stated in this lemma are immediately obtained from equations (3.141) and (3.142).
■

3.4.4 NON-LINEAR INVENTORY AND PRODUCTION COSTS

In this section we show that, although inventory and production costs are now considered non-linear, the problem is still analytically solvable. This is to say, analytical expressions for the optimal control and planning horizon are available.

The problem to be considered is one-machine, one-product-type scheduling to minimize the quadratic costs of inventory and production as well as the linear cost of the planning horizon, that is, the $1|1|u(t)\in[0,\infty],X(T)=X_T\,\|\,Q_u,Q_x,T$ problem which is stated as follows

$$J=\int_0^T\left[c_x^+X(t)^2+c_uu(t)^2\right]dt+c_TT\to\min\,, \tag{3.143}$$

subject to (3.124) and (3.125).

Dual Formulation

The Hamiltonian, H, resulting from the maximum principle is

$$H(t)=-\left[c_x^+X(t)^2+c_uu(t)^2\right]+\psi(t)u(t)=c_T\,. \tag{3.144}$$

The co-state variable $\psi(t)$ in (3.144) satisfies the co-state differential equation (3.132). Maximization of the Hamiltonian as a function of $u(t)$ results, as in the previous section, in

$$u(t) = \begin{cases} \dfrac{\psi(t)}{2c_u}, \text{ if } \psi(t) \geq 0; \\ 0, \text{ if } \psi(t) < 0. \end{cases} \tag{3.145}$$

Algorithm 3.23

Step 1:

Calculate the following parameters

$$y = \sqrt{\frac{c_x^+}{c_u}}, \ B = 2X_T^2 c_u^2 \frac{c_x^+}{c_T}, \ T = \frac{1}{y}\ln\left(B + \sqrt{B^2 + 1}\right), \ A = \frac{X_T\sqrt{c_x^+ c_u}}{sh(yT)}.$$

Step 2:

Set the optimal solution as

$$u(t) = \frac{A}{c_u} ch(yt) \text{ for } 0 \leq t \leq T, \text{ STOP.}$$

Example 3.10

Consider the $1|1|u(t) \in [0, \infty], X(T) = X_T \| Q_u, Q_x, T$ problem with the following parameters: X_T=2unit, c_u=1\$hr/unit^2, c_x^+=0.2\$/hr.unit^2, c_x^-=1\$/hr.unit^2 and c_T=1\$/hr.

At Step 1, Algorithm 3.23 calculates $y = \sqrt{0.2}$ /hr, B=1.6, T=2.8hr and A=0.66\$/unit. Step 2 sets the optimal solution as
$u(t) = 0.66ch(0.447t), \ 0 \leq t \leq 2.8$.

■

Discussion

Algorithm 3.23 presents a closed form solution whose optimality is proved in the following lemma. This solution is characterized by production control, which is a non-linear in time function.

LEMMA 3.40. (Correctness of Algorithm 3.23) Given problem (3.143), (3.124) and (3.125), the optimal solution is $u(t) = \dfrac{A}{c_u} ch(yt)$ for $0 \leq t \leq T$,

where $T = \dfrac{1}{y}\ln\left(B + \sqrt{B^2+1}\right)$, $y = \sqrt{\dfrac{c_x^+}{c_u}}$, $A = \dfrac{X_T\sqrt{c_x^+ c_u}}{sh(yT)}$ and $B = 2X_T^2 c_u^2 \dfrac{c_x^+}{c_T}$.

PROOF. First, since the proof that the production starts from the beginning of the planning horizon is identical to that for Lemma 3.39, it is omitted here. Thus, $\psi(t)>0$, $u(t)>0$ and $X(t)>0$ for $0<t\le T$.

Secondly, by taking into account (3.145), (3.124) and differentiating the co-state equation (3.132), we find

$$\ddot{\psi}(t)=\frac{c_x^+}{c_u}\psi(t)\,,\ \dot{\psi}(0)=0\,,\ \dot{\psi}(T)=2c_x^+X_T\,. \tag{3.146}$$

According to Section 2.3.1, the solution to the second order differential equation (3.146) is (see Figure 3.22)

$$\psi(t)=2Ach(yt)\,, \tag{3.147}$$

where

$$y=\sqrt{\frac{c_x^+}{c_u}}\ \text{ and }\ A=\frac{X_T\sqrt{c_x^+c_u}}{sh(yT)}\,. \tag{3.148}$$

Equations (3.145) and (3.147) determine the optimal control over the planning horizon. To find the length of the planning horizon, we substitute (3.147), $u(t)=\frac{\psi(t)}{2c_u}$ and $X(t)=\int_0^t u(\tau)d\tau$ into (3.144) and get

$$\frac{A^2}{c_u}=c_T\,.$$

The optimal length of the planning horizon stated in this lemma is readily obtained from the last equation.

■

Special Cases of Zero Costs

The special case of zero production cost, i.e. the $1|1|u(t)\in[0,\infty],X(T)=X_T\|Q_x,T$ problem has been solved in Section 3.4.2. The special case of zero inventory cost, i.e. the $1|1|u(t)\in[0,\infty],X(T)=X_T\|Q_u,Q_x,T,c_x^+=0$ problem is solved by setting $u(t)=\frac{X_T}{T}$, $t\in[0,T]$. This solution follows from the fact that $\psi(t)$ must be constant since $c_x^+=0$. The optimal value of T is obtained from (3.144) as $T=X_T\sqrt{\frac{c_u}{c_T}}$.

Finally, the special case of zero planning horizon cost, i.e. the $1|1|u(t) \in [0,\infty], X(T) = X_T \| Q_u, Q_x, T, c_T = 0$ problem has no optimal solution. Indeed, by letting $c_T \to 0$ in Algorithm 3.23, one can observe that T tends to infinity and optimal control $u(t)$ tends to zero at each t.

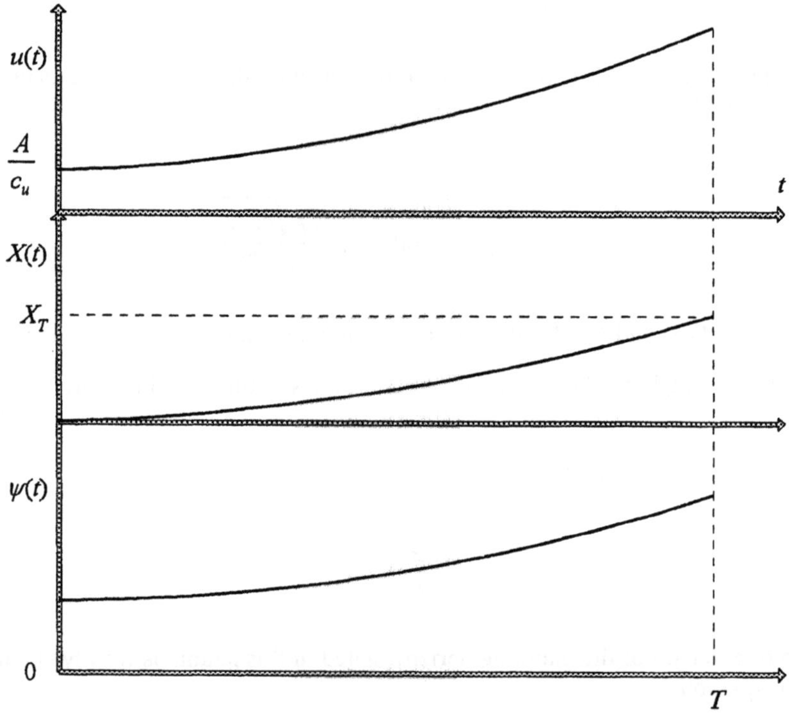

Figure 3.22. Optimal control $u(t)$, state $X(t)$ and co-state $\psi(t)$ variables for the $1|1|u(t) \in [0,\infty], X(T) = X_T \| Q_u, Q_x, T$ problem.

4

LIMITED CAPACITY PROBLEMS

4.1 DEMAND OVER FINITE PLANNING HORIZON

This section addresses capacitated scheduling problems characterized by demand as a function of time specified over a finite planning horizon. The problems differ by inventory related constraints, allowing or disallowing, backlogs and by the form of the objective function.

4.1.1 LINEAR PERFORMANCE MEASURE

Backlogs

The first problem to be considered is one-machine, one-product-type scheduling with backlogs and bounded production rate to minimize the total linear cost of inventory and production, that is, the $1|1|u(t)\in[0,U]\,|d(t),T|\,L_u, L_x$ problem which is stated as follows

$$J=\int_0^T\left[c_x^+X^+(t)+c_x^-X^-(t)+c_u u(t)\right]dt\rightarrow\min, \tag{4.1}$$

$$X^+(t)=\max\{X(t),0\},\ X^-(t)=\max\{-X(t),0\}$$

subject to

$$\dot{X}(t)=u(t)-d(t),\qquad X(0)=0, \tag{4.2}$$

$$0\le u(t)\le U. \tag{4.3}$$

Dual Formulation

Problem (4.1)-(4.3) is stated in the form of optimal control, where $u(t)$ is a control variable and $X(t)$ is a state variable whose behavior is described by the state differential equation (4.2). The Hamiltonian, H, resulting from the maximum principle is as follows

$$H(t) = -\left[c_x^+ X^+(t) + c_x^- X^-(t) + c_u u(t)\right] + \psi(t)\left[u(t) - d(t)\right]. \tag{4.4}$$

The co-state variable, $\psi(t)$ in (4.4), satisfies the following co-state differential equation with the corresponding transversality condition

$$\dot{\psi}(t) = \begin{cases} c_x^+, \text{ if } X(t) > 0; \\ -c_x^-, \text{ if } X(t) < 0; \\ a,\ a \in [-c_x^-, c_x^+], \text{ if } X(t) = 0; \end{cases} \qquad \psi(T) = 0. \tag{4.5}$$

Maximization of the Hamiltonian with respect to admissible controls $u(t)$ implies that

$$u(t) = \begin{cases} U, \text{ if } \psi(t) > c_u, \\ 0, \text{ if } \psi(t) < c_u, \\ b,\ b \in [0, U], \text{ if } \psi(t) = c_u. \end{cases} \tag{4.6}$$

The first line in the optimality condition (4.6) defines production at the maximum rate (full production regime). The second line which is similar to that found for the corresponding uncapacitated scheduling problem, defines the no-production regime. Finally, the last condition in (4.6) presents the fractional production which allows production at an intermediate rate to track the demand.

Algorithm 4.1

Step 1:

Run Algorithm 3.1. If $u(t) \le U$ for $0 \le t \le T$, then the found solution is optimal; STOP.
Otherwise, go to the next step.

Step 2:

If there exists a point of time $\bar{t} \in [0, T]$ such that $d(t) < U$ for $0 \le t \le \bar{t}$ and $d(t) > U$ for $\bar{t} < t \le T$, then define function

$$f(t_1) = \int_{t_1}^{t_x} d(t)dt - U \frac{c_x^-(T - t_1) - c_u}{c_x^+ + c_x^-},$$

where $t_x = \dfrac{c_x^+ t_1 + c_x^- T - c_u}{c_x^+ + c_x^-}$.

If function $f(t_1) > 0$ at interval [0, $\bar{t}$], then set $\hat{t} = 0$; otherwise,

if function $f(t_1) < 0$ at interval [0, $\bar{t}$], then set $\hat{t} = T - \dfrac{c_u}{c_x^-}$; otherwise

$\hat{t} = t_1$, where $t_1 \in [0, \bar{t}]$ is a root of equation $f(t_1) = 0$.

Set the optimal solution as

$u(t) = d(t)$ for $0 \le t < \hat{t}$,

$u(t) = U$ for $\hat{t} \le t < T - \dfrac{c_u}{c_x^-}$,

$u(t) = 0$ for $T - \dfrac{c_u}{c_x^-} \le t \le T$; STOP.

Otherwise, go to the next step.

Step 3:

If there exists a point of time $\bar{t} \in [0, T]$ such that $d(t)>U$ for $0 \le t < \bar{t}$ and $d(t)<U$ for $\bar{t} \le t \le T$, then define function

$$f(t_1) = \int_0^{t_1} d(t)dt - Ut_1 .$$

If equation $f(t_1) = 0$ has a root $t_1 \ge T - \dfrac{c_u}{c_x^-}$, then set $\hat{t} = T - \dfrac{c_u}{c_x^-}$;

otherwise, set $\hat{t} = t_1$.

Set the optimal solution as

$u(t) = U$ for $0 \le t < \hat{t}$,

$u(t) = d(t)$ for $\hat{t} \le t < T - \dfrac{c_u}{c_x^-}$,

$u(t) = 0$ for $T - \dfrac{c_u}{c_x^-} \le t < T$; STOP.

Example 4.1

Consider the $1|1|u(t) \in [0, U]\,|d(t),T|\,L_u, L_x$ problem with parameters T=10hr, U=3unit/hr, c_u=1\$/unit, c_x^+=0.2\$/hr.unit, c_x^-=1\$/hr.unit and linearly growing demand $d(t)$=(2+0.15t)unit/hr.

Step 1 of Algorithm 4.1 runs Algorithm 3.1 which returns $u(t)$=2+0.15t for $0 \le t \le 9$. Since condition $u(t) \le U$ does not hold for $6.67 \le t \le 9$, Algorithm 4.1 proceeds to Step 2 where time point $t_x = 0.167t_1 + 7.5$ is calculated and function $f(t_1) = \int_{t_1}^{t_x} d(t)dt - U \frac{c_x^-(T-t_1) - c_u}{c_x^+ + c_x^-}$ =-0.073 t_1^2 +1.021 t_1 -3.281 is defined. Because $\bar{t} = 6.67$hr and root $t_1 = 5$ of equation $f(t_1) = 0$ is within the interval [0,6.67], Step 2 sets the optimal solution as

$u(t) = (2 + 0.15t)$ unit/hr for $0 \le t < 5$,
$u(t) = 3$ unit/hr for $5 \le t < 9$,
$u(t) = 0$ for $9 \le t \le 10$.
■

Algorithm 4.2
Step 1:

Set accuracy ξ required for computing the objective of the optimal solution. Select time intervals $\tau_k = [t_{left}^k, t_{right}^k]$ at which $d(t) \le U$, $t_{right}^k < t_{left}^{k+1}$, and number them beginning from k=2. In addition, define τ_1 and τ_K as the intervals which are degenerated to a point, $\tau_1 = \{0\}$ and $\tau_K = \{T\}$, where K is the total number of obtained intervals. Calculate accuracy of integration $\varepsilon = \frac{\xi}{UK(c_u + T\max\{c_x^+, c_x^-\})}$.

Step 2 and Step 3 are identical to Steps 2 and 3 of Algorithm 3.4.

Subroutine 1 to Algorithm 4.2

(i) Set time points $t_x = t_{left}^n$, $t_y = t_{right}^n$, $t_{out} = t_{left}^n$.

(ii) Integrate the following system of differential equations

$$\dot{X}(t) = \begin{cases} U - d(t), & \text{if } \psi(t) \ge c_u, \\ -d(t), & \text{if } \psi(t) < c_u, \end{cases} \qquad X(t_{out}) = 0,$$

$$\dot{\psi}(t) = \begin{cases} c_x^+, & \text{if } X(t) \ge 0, \\ -c_x^-, & \text{if } X(t) < 0, \end{cases} \qquad \psi(t_{out}) = c_u.$$

over the interval $[t_{out}, t_{right}^m]$.

If $X(t) < 0$ for all $t \in \tau_m$, then no arc links nodes τ_n and τ_m; STOP.

Otherwise, set $t_{out} = t^n_{right}$ and integrate the above system.

If $X(t) > 0$ for all $t \in \tau_m$, then no arc links nodes τ_n and τ_m; STOP. Otherwise go to (iii)

(iii) Integrate the system from (ii) with $t_{out} = 0.5(t_x + t_y)$ and set time point $t_{in} \in \tau_m$ such that $X(t_{in}) = 0$.

If $\psi(t_{in}) > c_u$, then $t_x = t_{out}$ and go to (iv);

if $\psi(t_{in}) < c_u$, then $t_y = t_{out}$ and go to (iv).

(iv) If $t_y - t_x \le \varepsilon$, then set $arc_{nm} = (\tau_n, \tau_m) \in A$ and its three weights $a(arc_{nm}) = t_{out}$, $b(arc_{nm}) = t_{in}$ and

$$c(arc_{nm}) = \int_{t_{out}}^{t_{in}} c_x^+ X^+(t) + c_x^- X^-(t) + c_u u(t)dt$$; go to (v).

Otherwise go to (iii).

(v) If there exist arcs arc_{jn} such that $|b(arc_{jn}) - a(arc_{nm})| < \varepsilon$, then set $a(arc_{nm}) = \max_j b(arc_{jn})$. For each arc arc_{jn} entering node *n*, update weights

$$SD^n_{jm} = \begin{cases} \int_{b(arc_{jn})}^{a(arc_{nm})} c_u d(t)dt, & \text{if } a(arc_{nm}) \ge b(arc_{jn}); \\ -1, & \text{otherwise.} \end{cases}$$ STOP.

Otherwise, STOP.

Algorithm 4.3

Step 1:

If demand is $\overline{K}$-interval, single-mode, piece-wise constant and there exist time points $\bar{t}_1$ and $\bar{t}_2$, such that

$d(t) < U$ for $0 \le t < \bar{t}_1$, $d(t) > U$ for $\bar{t}_1 < t < \bar{t}_2$

and $d(t) < U$ for $\bar{t}_2 < t \le T$,

then denote two time intervals $\tau_2 = [0, \bar{t}_1]$ and $\tau_3 = [\bar{t}_2, T]$. In addition, define τ_1 and τ_4 as the intervals which are degenerated to a point, $\tau_1 = \{0\}$ and $\tau_4 = \{T\}$. Assign K=4.

Step 2:

Find $\bar{k}_1$ and $\bar{k}_2$ such that

$\bar{k}_1 = \max\{k | t_k < \bar{t}_1\}$, $\bar{k}_2 = \min\{k | t_{k-1} > \bar{t}_2\}$.

Step 3 and Step 4 are identical to Steps 2 and 3 of Algorithm 3.5 respectively.

Subroutine 1 to Algorithm 4.3

(i) Assign $k_1 = 1$, $k_2 = \bar{k}_1 + 1$, $k_3 = \bar{k}_2$.

(ii) Calculate time points

$$t_{in} = \frac{(U - d_{k_2})c_x^+ y_1 - (U + d_{k_1})(c_x^+ + c_x^-)y_2}{(U - d_{k_2})(U - d_{k_3})c_x^+ - (U + d_{k_1})(Uc_x^+ + d_{k_2}c_x^-)},$$

$$t_{out} = \frac{(U - d_{k_3})t_{in} - y_1}{(U + d_{k_1})}, \quad t_x = \frac{c_x^- t_{in} + c_x^+ t_{out}}{c_x^+ + c_x^-},$$

where $y_1 = d_{k_1}t_{k_1} - d_{k_3}t_{k_3-1} + \sum_{j=k_1+1}^{j=k_3-1} d_j(t_j - t_{j-1})$,

$$y_2 = d_{k_2}t_{k_2} - d_{k_3}t_{k_3-1} + \sum_{j=k_2+1}^{j=k_3-1} d_j(t_j - t_{j-1}).$$

(iii) If $t_{out} \in [t_{k_1-1}, t_{k_1}]$, $t_x \in [t_{k_2-1}, t_{k_2}]$, $t_{in} \in [t_{k_3-1}, t_{k_3}]$, then set $arc_{nm} = (\tau_n, \tau_m) \in A$ and its two weights $a(arc_{nm}) = t_{out}$, $b(arc_{nm}) = t_{in}$. Set the optimal control $u(t) = U$ for $t_{out} \le t \le t_{in}$, STOP. Otherwise go to (iv).

(iv) If $k_1 < \bar{k}_1$, then assign $k_1 = k_1 + 1$ and go to (ii).

Otherwise, set $k_1 = 1$ and if $k_2 < \bar{k}_2 - 1$, then assign $k_2 = k_2 + 1$ and go to (ii).

Otherwise, set $k_2 = 1$ and if $k_3 < \bar{K}$, then assign $k_3 = k_3 + 1$ and go to (ii).

Otherwise, no arc links nodes *n* and *m*; STOP.

Discussion

This section presents three algorithms. Algorithms 4.1 and 4.3 yield exact optimal solutions of the $1|1|u(t) \in [0,U]|d(t),T|\, L_u, L_x$ problem for a few special classes of the demand function. Algorithm 4.2 approximates the optimal solution with a given accuracy for arbitrary demand.

Specifically, Algorithm 4.1 Step 1 invokes Algorithm 3.1 developed for the corresponding uncapacitated problem. If demand is weak, $d(t) \le U$, the found solution is optimal. In addition, when the unit production cost is greater than the unit backlog cost $c_u > c_x^- T$, no production at all is optimal regardless of the demand. The other two steps of Algorithm 4.1 cope with the demand which exceeds capacity U at one part of the planning horizon and falls below the capacity at the other. This causes the optimal control function to switch twice at most. The first switching is at point t_1, where the full production, $u(t)=U$, and the fractional production, $u(t)=d(t)$, change over. The second switching is at point $T-\frac{c_u}{c_x^-}$. Beyond this point it is not worthwhile to exercise any additional control, as the cumulative effect of the additional backlog will not be very significant. The same effect of the no-production regime towards the end of the planning horizon was observed also in unlimited capacity problems studied in Sections 3.1.1 and 3.1.2.

The complexity of Algorithm 4.1 is $O(1)$ only if the root of equation $f(t_1)=0$ can be found analytically. This, of course, is not always the case. If demand is K-interval, piece-wise constant (in addition to the demand conditions of Steps 2 and 3), then the root of equation $f(t_1)=0$ and hence the exact optimal solution is available in $O(K)$ time. If demand is piece-wise constant and conditions of Steps 2 and 3 are relaxed to allow single-mode demand, the problem is still exactly solvable by Algorithm 4.3 in $O(K^3)$ time. However, if demand is an arbitrary complex function of time, only a numerical approximation of the optimal solution is possible as provided by Algorithm 4.2.

LEMMA 4.1. (Correctness of Algorithm 4.1) Given problem (4.1)-(4.3), the optimal control is

- if there exists a point of time $\bar{t} \in [0,T]$ such that $d(t)<U$ for $0 \le t \le \bar{t}$ and $d(t)>U$ for $\bar{t} < t \le T$, then

$$u(t)=d(t) \text{ for } 0 \le t < \hat{t},\; u(t)=U \text{ for } \hat{t} \le t < T-\frac{c_u}{c_x^-},\; u(t)=0 \text{ for } T-\frac{c_u}{c_x^-} \le t < T;$$

where $\hat{t}=0$ if function $f(t_1)=\int_{t_1}^{t_x} d(t)dt - U\frac{c_x^-(T-t_1)-c_u}{c_x^+ + c_x^-} > 0$ at interval $[0,\bar{t}\,]$,

$\hat{t} = T - \frac{c_u}{c_x^-}$ if function $f(t_1) < 0$ at interval $[0, \bar{t}\,]$, and $\hat{t} = t_1$, $f(t_1) = 0$ if $t_1 \in [0, \bar{t}\,]$,

$$t_x = \frac{c_x^+ t_1 + c_x^- T - c_u}{c_x^+ + c_x^-};$$

- if there exists a point of time $\bar{t} \in [0, T]$ such that $d(t)>U$ for $0 \le t < \bar{t}$ and $d(t)<U$ for $\bar{t} \le t \le T$, then

  $u(t) = U$ for $0 \le t < \hat{t}$, $u(t) = d(t)$ for $\hat{t} \le t < T - \frac{c_u}{c_x^-}$, $u(t) = 0$ for $T - \frac{c_u}{c_x^-} \le t < T$,

  where $\hat{t} = \max\{t_1, T - \frac{c_u}{c_x^-}\}$ and t_1 is a root of equation $\int_0^{t_1} d(t)dt - Ut_1 = 0$.

PROOF. To prove the first part of the lemma, we need to find the control value at the fractional production characterized by $\psi(t) = c_u$ at an interval of time. By differentiating this condition, we obtain

$$\dot{\psi}(t) = 0. \tag{4.7}$$

From (4.5) it immediately follows that equation (4.7) can be met only if $X(t)=0$. By differentiating the last equality and substituting within it the state equation (4.2), we finally obtain

$$u(t) = d(t). \tag{4.8}$$

Given $\frac{c_u}{c_x^-} < T$ (see Lemma 3.1) and a point of time $\bar{t} \in [0, T]$ such that $d(t)<U$ for $0 \le t \le \bar{t}$ and $d(t)>U$ for $\bar{t} < t \le T$, we next prove that the optimal control is characterized by three regimes and at most two switching time points where the regimes change over in the only sequence: fractional production, full production and then no-production, as illustrated in Figure 4.1. (Either the fractional or full production may vanish). This proof is by contradiction. Let us assume that full production must be carried out prior to the fractional production

$$u(t) = U \text{ for } t \in [0, t_1],\ \ u(t) = d(t) \text{ for } t \in [t_1, t_2].$$

Then, according to (4.5) we find

$$\psi(t) \ge c_u \text{ for } t \in [0, t_1],\ \ \psi(t) = c_u \text{ for } t \in [t_1, t_2].$$

and consequently

$$\psi(t_1) = c_u,\ X(t_1) = 0\,. \tag{4.9}$$

Since the demand is less than U up to $\bar{t}$, then $X(t) > 0$ for $t \in (0, t_1)$ and $t_1 > \bar{t}$. Thus, according to (4.5) $\psi(t)$ increases at $t \in [0, t_1]$ and $\psi(t_1) > c_u$ that contradicts (4.9). Note, that similar to the proofs of Lemmas 3.1 and 3.2, the transversality condition necessitates that the last regime will always be no-production.

Consider now a solution which consists of the proven sequence of regimes: fractional production at $[0, t_1]$, full production at $[t_1, t_2]$ and no-production at $[t_2, T]$. With respect to (4.5), it follows that

$$\psi(t) = c_u \text{ for } t \in [0, t_1],\ \ \psi(t) \ge c_u \text{ for } t \in [t_1, t_2] \text{ and } \psi(t) \le c_u \text{ for } t \in [t_2, T]\,.$$

This behavior of the co-state variable implies that there exists a point t_x, $t_1 \le t_x \le t_2$, so that

$$X(t)=0 \text{ for } t \in [0, t_1],\ X(t)>0 \text{ for } t \in (t_1, t_x) \text{ and } X(t)<0 \text{ for } t \in (t_x, T]\,.$$

Thus, three unknown time points t_1, t_x and t_2 must satisfy the following three equations

$$X(t_x) = 0,\ \ \psi(t_2) = c_u \text{ and } \psi(T) = 0\,. \tag{4.10}$$

By integrating the state and co-state differential equations over interval $[t_1, T]$, expressions (4.10) transform into the following system of three algebraic equations in the three unknowns

$$U(t_x - t_1) = \int_{t_1}^{t_x} d(t)dt\,,\quad c_x^+(t_x - t_1) = c_x^-(t_2 - t_x) \text{ and } c_u - c_x^-(T - t_2) = 0\,.$$

Solution of the last equation proves the second part of the lemma. The third part of the lemma is proved similarly.

■

Since the maximum principle does not provide enough information about the optimal solution for the $1|1|u(t) \in [0, U]\,|d(t), T|\,L_u, L_x$ problem with demand as an arbitrary function of time, Algorithm 4.2 has been presented to approximate the optimal solution with an *a priori* given accuracy.

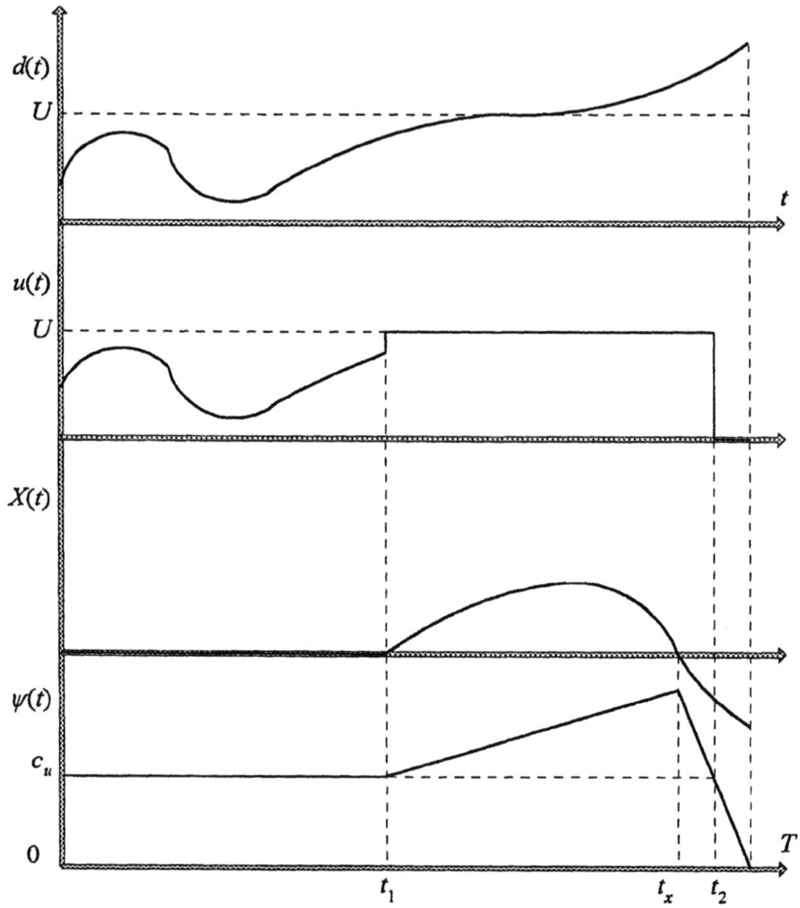

Figure 4.1. Demand $d(t)$, optimal control $u(t)$, state $X(t)$ and co-state $\psi(t)$ variables for a special case of the $1|1|\, u(t) \in [0,U]\, |d(t),T|\, L_u, L_x$ problem treated in Step 2 of Algorithm 4.1.

The following lemma states that even if the demand function is arbitrary, that is the conditions of Lemma 4.1 are not satisfied, then the general form of the optimal solution has a quite certain structure induced by time intervals where demand $d(t)$ is less than capacity U (see Figure 4.2). Only at these intervals can the fractional regime $u(t)=d(t)$ occur and provide zero inventories.

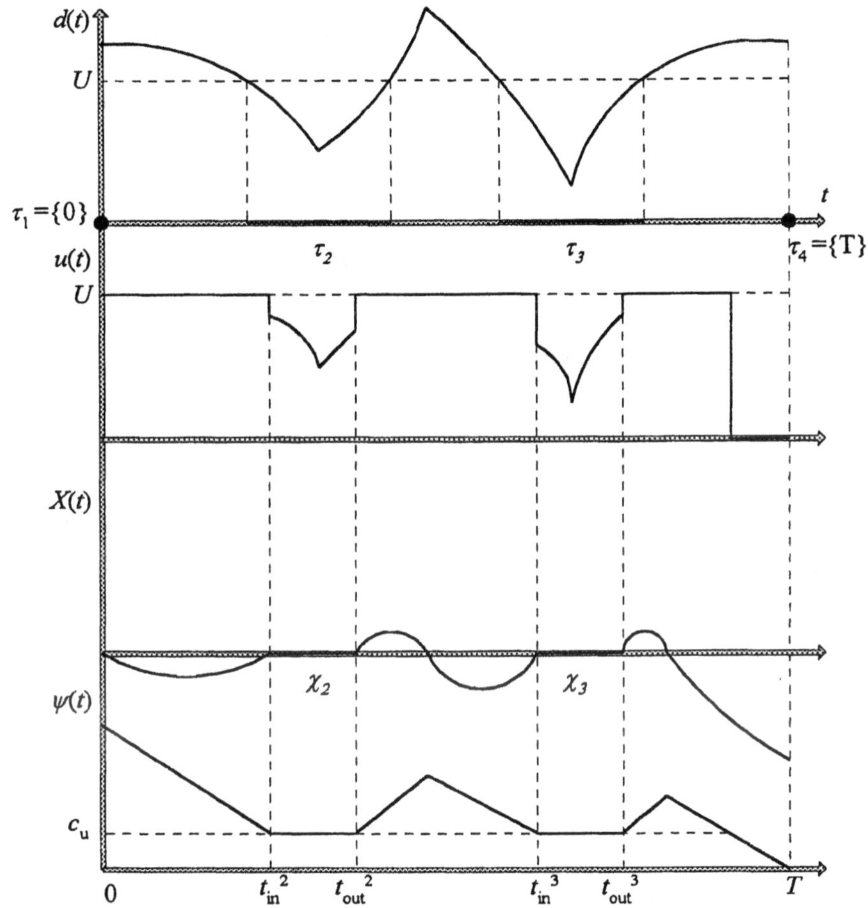

Figure 4.2. Demand $d(t)$, optimal control $u(t)$, state $X(t)$ and co-state $\psi(t)$ variables for the $1|1|u(t)\in[0,U]\,|d(t),T|\,L_u, L_x$ problem with arbitrary demand.

LEMMA 4.2. (Correctness of Algorithm 4.2 Step 1) Given problem (4.1)-(4.3), if $d(t)\leq U$ at K intervals of time τ_k, $k=1,\ldots,K$, then the optimal solution contains at most K time intervals of zero inventory, $\chi_k=\{t\in\tau_k \,|\, X(t)=0\}$.

PROOF. The proof follows from Lemma 4.1 and the fact that at an interval of time where $d(t)\leq U$ does not hold, $X(t)$ cannot be kept at zero.
■

Algorithm 4.2 makes use of Lemma 4.2 to locate intervals τ_k, $k = 1,...,K$ and to link them so that the solution of the dual problem will be feasible. This is accomplished by construction of a graph where the nodes are these intervals and the arcs are links between pairs of nodes. Thus, as with the $1|1|u(t) \in [0,\infty]|d(t),T|Q_u, L_x$ problem, continuous-time scheduling is again reduced to a combinatorial problem of finding a shortest feasible path in the graph.

As was the case with Algorithm 3.4, the dichotomous search is applied to set the arcs of the graph by calculating roots of specially constructed slack functions. Efficiency of the search is due to the monotonicity of the slack functions.

Definition 4.1. Given a pair of nodes τ_n and τ_m, $n,m \neq 1,K$, F_6 is a function which maps $t_{out} \to \psi(t_{in}) - c_u$, where $t_{out} \in \tau_n$, $t_{in} \in \tau_m$ and $X(t_{in}) = 0$.

Definition 4.2. Given a pair of nodes τ_1 and τ_m, $m \neq 1,K$, F_7 is a function which maps $\psi(0) \to \psi(t_{in}) - c_u$, where $t_{in} \in \tau_m$ and $X(t_{in}) = 0$.

Definition 4.3. Given a pair of nodes τ_1 and τ_K, F_8 is a function which maps $\psi(0) \to \psi(T)$.

Definition 4.4. Given a pair of nodes τ_n and τ_K, $n \neq 1,K$, F_9 is a function which maps $t_{out} \to \psi(T)$, where $t_{out} \in \tau_n$.

Subroutines 1-4 to Algorithm 4.2, which are invoked at Step 2, locate roots of the slack functions F_6 - F_9 respectively. The following lemma states the monotonicity of the defined slack functions.

LEMMA 4.3. Slack functions F_6 - F_9 are non-decreasing in their arguments.

PROOF. Since the proof is similar to that of Lemma 3.7, it is omitted here. ■

The following lemma sets bounds for the co-state variable within which the dichotomous search is performed in Subroutines 2 and 3.

LEMMA 4.4. Given problem (4.1)-(4.3), the co-state variable $\psi(t)$ is bounded as $\psi(t) \in [0,\hat{\psi}]$ for $t \in [0,T]$, where $\hat{\psi} = c_x^- T$.

PROOF. Since the proof is similar to that of Lemma 3.8, it is omitted here. ■

Lemmas 4.3 and 4.4 prove the correctness of Algorithm 4.2 Step 2. The last step uses the sequence-dependent, shortest path procedure described in Section 3.1.3.

THEOREM 4.1. The problem (4.1)-(4.3) is solvable with accuracy ξ in $O\left(\max\{K^4, \frac{K^3 R}{\xi}\log\frac{KR}{\xi}\}\right)$ time, where K is the number of time intervals at which $d(t) \le U$ and $R = TU\left(c_u + T\max\{c_x^+, c_x^-\}\right)$.

PROOF. First note that Algorithm 4.2 links time intervals where $d(t) \le U$ so that the primal and dual formulations of the problem are satisfied with a given accuracy ε. Since Algorithm 4.2 is based on the same procedures as Algorithm 3.4, its complexity is determined by (3.25)

$$O\left(\max\{K^4, K^2\frac{T}{\varepsilon}\log\frac{T}{\varepsilon}\}\right).$$

However, accuracy of the objective ξ that the algorithm provides is different. To estimate the accuracy, we vary the objective function (4.1)

$$\xi \le \max\{c_x^+, c_x^-\}\int_0^T |\Delta X(t)|dt + c_u\int_0^T |\Delta u(t)|dt. \tag{4.11}$$

Each link between two fractional regimes results in a maximum error of $\int_{t_{out}}^{t_{in}} |\Delta u(t)|dt = \varepsilon U$ in the second term of (4.11). Consequently, the maximum error in the first term for each link is $\int_0^T |\Delta X(t)|dt = \int_0^T\int_0^t |\Delta u(\tau)|d\tau dt \le \varepsilon UT$.

By taking into account that there can be at most K links over the planning horizon, the worst accuracy estimate (4.11) is

$$\xi = c_u UK(1+gT)\varepsilon, \tag{4.12}$$

where $g = \dfrac{\max\{c_x^+, c_x^-\}}{c_u}$.

To obtain the complexity and accuracy of Algorithm 4.2, we express ε from (4.12), substitute it into (3.25) and introduce parameter R to simplify the presentation.

■

According to Theorem 4.1, Algorithm 4.2 is of pseudopolynomial complexity which is due to multipliers T and ε entering (3.25). However, if demand is single-mode, piece-wise constant, the algorithm becomes strongly polynomial. This is due to the fact that the approximate integration which is carried out in Subroutines 1-4 to Algorithm 4.2 is reduced to solving systems of algebraic equations in the corresponding subroutines of Algorithm 4.3. Such modification has been presented only for Subroutine 1. Modification of the other three subroutines is left as an exercise for the reader.

In what follows we state the correctness and complexity of Algorithm 4.3.

LEMMA 4.5. If demand is K-interval, single-mode, piece-wise constant, then the dynamics of $\psi(t)$ is also single-mode, i.e. there exists a time point $t_\psi \in [0,T]$ such that $\psi(t)$ does not decrease before t_ψ and does not increase after it.

PROOF. Since the proof is similar to that of Lemma 3.9, it is omitted here. ■

LEMMA 4.6. (Rule for linking nodes in graph G) Let two adjacent fractional regimes $u(t)=d(t)$ occur at intervals $\chi_2 \subseteq \tau_2$ and $\chi_3 \subseteq \tau_3$, and demand be $\overline{K}$-interval, single-mode, piece-wise constant, $d(t)<U$ for $0 \le t < \bar{t}_1$, $d(t)>U$ for $\bar{t}_1 < t < \bar{t}_2$ and $d(t)<U$ for $\bar{t}_2 < t \le T$. If there exist indexes $1 \le k_1 \le \bar{k}_1$, $\bar{k}_1 < k_2 < \bar{k}_2$, and $\bar{k}_2 \le k_3 \le \overline{K}$, $\bar{k}_1 = \max\{k | t_k < \bar{t}_1\}$, $\bar{k}_2 = \min\{k | t_{k-1} > \bar{t}_2\}$ and points $t_{out} \in [t_{k_1-1}, t_{k_1}]$, $t_x \in [t_{k_2-1}, t_{k_2}]$, $t_{in} \in [t_{k_3-1}, t_{k_3}]$, such that

$$t_{in} = \frac{(U-d_{k_2})c_x^+ y_1 - (U+d_{k_1})(c_x^+ + c_x^-)y_2}{(U-d_{k_2})(U-d_{k_3})c_x^+ - (U+d_{k_1})(Uc_x^+ + d_{k_2}c_x^-)},$$

$$t_{out} = \frac{(U-d_{k_3})t_{in} - y_1}{(U+d_{k_1})}, \qquad t_x = \frac{c_x^- t_{in} + c_x^+ t_{out}}{c_x^+ + c_x^-},$$

where

$$y_1 = d_{k_1}t_{k_1} - d_{k_3}t_{k_3-1} + \sum_{j=k_1+1}^{j=k_3-1} d_j(t_j - t_{j-1}), \; y_2 = d_{k_2}t_{k_2} - d_{k_3}t_{k_3-1} + \sum_{j=k_2+1}^{j=k_3-1} d_j(t_j - t_{j-1}),$$

then the optimal control between these fractional regimes is
$u(t) = U$ for $t_{out} \le t \le t_{in}$.

PROOF. According to Lemma 4.2, there exists a time point t_x between intervals τ_2 and τ_3, such that $X(t_x)$ equals zero. Then, according to (4.7) and (4.9)

$$X(t_x) = X(t_{in}) = X(t_{out}), \; \psi(t_{in}) - \psi(t_{out}) = 0, \tag{4.13}$$

where t_{out} is the terminal point of the first fractional regime, $t_{out} \in \tau_2$, and t_{in} is the starting point of the other fractional regime, $t_{in} \in \tau_3$.

By integrating the system of the state and co-state differential equations over interval $[t_{out}, t_{in}]$ with three boundary conditions (4.13), we find three corresponding algebraic equations in three unknowns t_x, t_{out} and t_{in}

$$d_{k_1}(t_{k_1} - t_{out}) + d_{k_2}(t_x - t_{k_2-1}) + \sum_{j=k_1+1}^{j=k_2-1} d_j(t_j - t_{j-1}) = U(t_x - t_{out}),$$

$$d_{k_2}(t_{k_2} - t_x) + d_{k_3}(t_{in} - t_{k_3-1}) + \sum_{j=k_2+1}^{j=k_3-1} d_j(t_j - t_{j-1}) = U(t_{in} - t_x),$$

$$c_x^+(t_x - t_{out}) = c_x^-(t_{in} - t_x).$$

The solution of this system proves the lemma.
■

Lemma 4.6 proves correctness of Step 2 of Algorithm 4.3 with modified subroutines as compared to Algorithm 3.5. Step 3 of Algorithm 4.3 involves the coloring procedure for constructing a feasible path. Correctness of this procedure is proved in Lemma 3.11.

THEOREM 4.2. If demand is single-mode, K-interval piece-wise constant function of time, then problem (4.1)-(4.3) is solvable in $O(K^3)$ time.

PROOF. At Step 1, Algorithm 4.2 identifies exactly two intervals where the fractional production may occur. In Step 2, the algorithm constructs a graph with four nodes and invokes one of the four subroutines which sets an arc between a pair of nodes in $O(K^3)$ time. Step 2 is repeated $O(1)$ times because $|A| \le 4^2$. Thus, this step totally needs $O(K^3)$ time.

Since the graph has only four nodes, the coloring procedure executed in the last step of the algorithm requires $O(1)$ time.
■

Special Cases of Zero Costs

The problems to be considered here are characterized by either no production cost or no inventory cost, that is the $1|1|u(t)\in[0,U]\,|d(t),T|\,L_x$ and $1|1|u(t)\in[0,U]\,|d(t),T|\,L_u, L_x, c_x^+=0$ problems. The solution algorithms are straightforwardly obtained by substituting either $c_u=0$ or $c_x^+=0$ into the corresponding conditions and formulas of Algorithms 4.1-4.3. The only significant computational simplification is in Algorithm 4.2 for the case of zero production cost. The shortest path procedure in this case becomes sequence-independent. This results in an improved complexity estimate, $O\left(\frac{K^3R}{\xi}\log\frac{KR}{\xi}\right)$.

No Backlog

For the $1|1|u(t)\in[0,U], X(t)\geq 0\,|d(t),T|\,L_u, L_x$ problem to have a feasible solution, it is necessary that the average demand must not exceed capacity U

$$\frac{1}{t}\int_0^t d(\tau)d\tau \leq U \quad \text{for } 0<t\leq T. \tag{4.14}$$

In contrast to the backlog problem considered in the previous section, the no-backlog problem is polynomially solvable for any demand satisfying (4.14). Moreover, if the demand does not exceed the capacity at every point of time, $d(t)\leq U$, then a closed form optimal solution is simply to produce according to the fractional production $u(t)=d(t)$.

Dual Formulation

When backlogs are disallowed, problem (4.1)-(4.3) has the additional constraint $X(t)\geq 0$. This affects neither Hamiltonian (4.4) nor optimality conditions (4.6), but changes the co-state differential equation

$$d\psi(t)=c_x^+dt-d\mu(t),\quad \psi(T+0)=0, \tag{4.15}$$

where the Lagrange multiplier $d\mu(t)\geq 0$ corresponds to the no-backlog constraint and satisfies the complementary slackness condition

$$\int_0^T X(t)d\mu(t)=0. \tag{4.16}$$

Algorithm 4.4
Step 1:

If $d(t) \le U$, then set the optimal solution as $u(t) = d(t)$ for $0 \le t \le T$. Otherwise go to the next step.

Step 2:

Select time intervals $\tau_k = [t_{left}^k, t_{right}^k]$ at which $d(t) \le U$, $t_{right}^k < t_{left}^{k+1}$, and number them beginning from k=2. In addition, define τ_1 and τ_K as the intervals which are degenerated to a point, $\tau_1 = \{0\}$ and $\tau_K = \{T\}$, where K is the total number of obtained intervals.

Step 3 and Step 4 are identical to Steps 2 and 3 of Algorithm 3.5 respectively.

Subroutine 1 to Algorithm 4.4

(i) Define

$$f(t_{out}) = \int_{t_{out}}^{t_{left}^m} d(t)dt - U(t_{left}^m - t_{out}),$$

as a function of t_{out} .

(ii) If equation $f(t_{out}) = 0$ has a root $t_{out} \in \tau_n$, then set $arc_{nm} = (\tau_n, \tau_m) \in A$ and its two weights $a(arc_{nm}) = t_{out}$, $b(arc_{nm}) = t_{left}^m$. Set the optimal control as

$u(t) = U$ for $t_{out} \le t \le t_{left}^m$, STOP.

Otherwise, no arc links nodes n and m; STOP.

Discussion

Algorithm 4.4 Step 1 determines a closed form solution for the problem characterized by a demand function not exceeding the maximum production rate U. Steps 2-4 of Algorithm 4.4 locate intervals of the fractional production regime $u(t)=d(t)$ and link them by intervals of the full production regime. This is accomplished, as with Algorithm 3.5, by constructing a graph with nodes as the fractional production intervals and arcs as the full production intervals. A root-finding procedure when setting an arc between two fractional regimes is carried out by Subroutine 1 to Algorithm 4.4. Subroutines 2-4 link the beginning of the planning horizon with a fractional regime; a fractional regime with the end of the planning horizon; and the beginning with the end of the planning horizon respectively. Since these subroutines are organized similarly to Subroutine 1, their detailed description is left as an exercise for the reader. The total complexity of the algorithm is $O(K^3)$ if the root of function $f(t_{out})$ can be found analytically in Subroutines 1-4.

Lemma 4.7 below proves the condition when the fractional production is optimal over the entire planning horizon.

LEMMA 4.7. (Correctness of Algorithm 4.4 Step 1) Given problem (4.1)-(4.3) with backlogs disallowed, if $d(t) \le U$ for $0 \le t \le T$, then the optimal control is $u(t) = d(t)$ for $0 \le t \le T$.

PROOF. It is easy to observe that if $d(t) \le U$ for $0 \le t \le T$, the solution

$$u(t) = d(t),\ X(t) = 0,\ \psi(t) = c_u \ \text{ for } 0 \le t \le T,$$

$$d\mu(t) = 0 \ \text{ for } 0 \le t < T,\ d\mu(T) = c_u \ \text{ and } \ \psi(T+0) = 0$$

always meets the state (4.2)-(4.3) and co-state (4.4), (4.6), (4.15)-(4.16) conditions and, thus, it is optimal.
■

The following lemma proves the correctness of Algorithm 4.4 Steps 2 and 3 for the case of linking two fractional regimes by Subroutine 1.

LEMMA 4.8. (Rule for linking nodes in graph *G*) Given two adjacent fractional regimes $u(t)=d(t)$ at intervals $\chi_n \subseteq \tau_n$ and $\chi_m \subseteq \tau_m$, if t_{out} is a root of the following equation

$$\int_{t_{out}}^{t_{left}^m} d(t)dt - U(t_{left}^m - t_{out}) = 0, \tag{4.17}$$

then the optimal control between these fractional production regimes is

$$u(t) = U \ \text{ for } t_{out} \le t \le t_{left}^m.$$

PROOF. According to the proof of Lemma 4.7, the necessary condition for the fractional regime to occur is $d(t) \le U$. Therefore, Step 2 of the algorithm identifies time intervals where this condition holds. Furthermore, given fractional regimes at $\chi_n \subseteq \tau_n$ and $\chi_m \subseteq \tau_m$, according to (4.6), (4.15) and (4.16), $X(t) = 0$ and $\psi(t) = c_u$ at χ_n and χ_m.

Consider the solution between the fractional regimes at χ_n and χ_m in the following form

$$X(t) = U(t - t_{out}) - \int_{t_{out}}^{t} d(\tau)d\tau \text{, for } t_{out} \le t \le t_{in}, \tag{4.18}$$

$$\psi(t) = c_u + c_x^+(t - t_{out}) \text{ for } t_{out} \le t \le t_{in},\ d\mu(t_{in}) = c_x^+(t_{in} - t_{out}), \tag{4.19}$$

where t_{out} is the terminal point of the first fractional regime, $t_{out} \in \chi_n$ and t_{in} is the starting point of the other fractional regime, $t_{in} \in \chi_m$.

Next, we prove by contradiction that point t_{in} coincides with the left boundary of interval τ_m, i.e. $t_{in} = t_{left}^m$ (see Figure 4.3). Indeed, if it is not the case, then at interval $[t_{left}^m, t_{in})$ $X(t) > 0$. Consequently, $\psi(t)$ increases up to the value of c_u, i.e. $\psi(t) \le c_u$. This contradicts the co-state variable behavior (4.19). Once $t_{in} = t_{left}^m$, the only unknown parameter when linking the two intervals is t_{out}. It is calculated from (4.18) as a root of equation $X(t_{left}^m) = 0$, as stated in the lemma.

■

Step 3 of Algorithm 4.4 involves the coloring procedure for constructing a feasible path in a 2-weighted graph. The correctness of the procedure has been proved in Lemma 3.11.

4.1.2 NON-LINEAR INVENTORY COST

Next, we consider a linear production cost function and non-linear inventory and backlog costs, which will be approximated by a quadratic function. It will be shown below that introducing non-linearity complicates the solution and reduces the class of the demand functions for which the exact optimal control can be found in polynomial time. However, this non-linearity does not significantly affect the algorithm which approximates the optimal solution for arbitrary demands.

Backlogs

The problem to be considered is one-machine, one-product-type scheduling with backlogs and bounded production rate to minimize the quadratic cost of inventory and linear production cost, that is, the $1|1|u(t) \in [0,U]\,|d(t),T|\,L_u, Q_x$ problem which is stated as follows

$$J = \int_0^T \left[c_x^+ X^+(t)^2 + c_x^- X^-(t)^2 + c_u u(t)\right]dt \to \min, \tag{4.20}$$

$$X^+(t) = \max\{X(t),0\},\ X^-(t) = \max\{-X(t),0\}$$

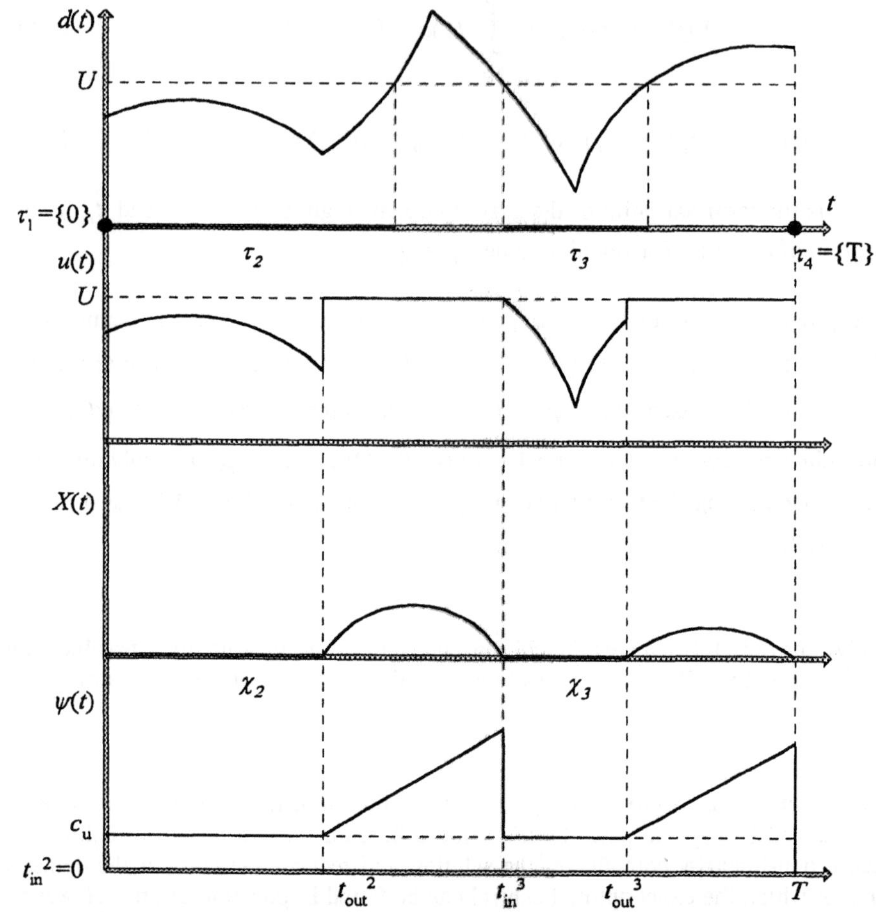

Figure 4.3. Demand $d(t)$, optimal control $u(t)$, state $X(t)$ and co-state $\psi(t)$ variables for the $1|1|\,u(t)\in[0,U]\,,X(t)\geq 0\,|d(t),T|\,L_u\,,L_x$ problem.

subject to (4.2) and (4.3).

Dual Formulation

The Hamiltonian, H, resulting from the maximum principle is

$$H(t) = -\left[c_x^+ X^+(t)^2 + c_x^- X^-(t)^2 + c_u u(t)\right] + \psi(t)\left[u(t) - d(t)\right]. \tag{4.21}$$

The co-state variable, $\psi(t)$ in (4.21), satisfies the following co-state differential equation with the corresponding transversality condition

$$\dot{\psi}(t) = \begin{cases} 2c_x^+ X(t), & \text{if } X(t) \geq 0; \\ 2c_x^- X(t), & \text{if } X(t) < 0; \end{cases} \qquad \psi(T) = 0\,. \tag{4.22}$$

The quadratic inventory and backlog costs do not influence the control-dependent terms of the Hamiltonian. Therefore, maximization of the Hamiltonian at every point of time, with respect to the maximum principle, results in the same optimality conditions (4.6) obtained in the previous section.

Algorithm 4.5

Step 1: Run Algorithm 3.2. If $u(t) \leq U$ for $0 \leq t \leq T$, then the found solution is optimal; STOP.
Otherwise, go to the next step.

Step 2:

If demand is pressing, $\frac{1}{t}\int_0^t d(\tau)d\tau > U$ for $0 < t < T$, then

calculate t_1 as a root of the following function

$$\frac{c_u}{2c_x^-} = -Ut_1(T-t_1) + (T-t_1)\int_0^{t_1} d(t)dt + \int_{t_1}^{T}\int_{t_1}^{t} d(\tau)d\tau dt\,.$$

Set the optimal solution as
$u(t) = U$ for $0 \leq t < t_1$,
$u(t) = 0$ for $t_1 \leq t \leq T$; STOP.

Example 4.2

Consider the $1|1|u(t) \in [0,U]\,|d(t),T|\,L_u, Q_x$ problem with parameters T=10hr, U=3unit/hr, c_u=1\$/unit, c_x^+=0.2\$/hr.unit^2, c_x^-=1\$/hr.unit^2 and linearly decreasing demand $d(t)$=(3.5-0.1t)unit/hr.

Step 1 of Algorithm 4.5 runs Algorithm 3.2 which returns $u(t)$=3.5-0.1t for $0 \leq t \leq 9.37$. Since condition $u(t) \leq U$ does not hold for $0 \leq t < 5$, Algorithm 4.5 proceeds to Step 2 where it verifies that the demand is pressing, $\frac{1}{t}\int_0^t d(\tau)d\tau = 3.5 - 0.05t > U$. Then, the root of equation

$$\frac{c_u}{2c_x^-} = -Ut_1(T-t_1)+(T-t_1)\int_0^{t_1} d(t)dt + \int_{t_1}^{T}\int_{t_1}^{t} d(\tau)d\tau dt$$

is calculated as $t_1 = 9.46$ and the optimal solution is set as
$u(t) = 3$ unit/hr for $0 \le t < 9.46$,
$u(t) = 0$ for $9.46 \le t \le 10$.
■

Algorithm 4.6
Step 1:

Set accuracy ξ required for computing the objective of the optimal solution. Select time intervals $\tau_k = [t_{left}^k, t_{right}^k]$ at which $d(t) \le U$, $t_{right}^k < t_{left}^{k+1}$, and number them beginning from k=2. In addition, define τ_1 and τ_K as the intervals which are degenerated to a point, $\tau_1 = \{0\}$ and $\tau_K = \{T\}$, where K is the total number of obtained intervals. Calculate accuracy of integration

$$\varepsilon = \frac{\xi}{UK(c_u + 2T\max\{c_x^+ UT, c_x^- \int_0^T d(t)dt\})}.$$

Step 2 and Step 3 are identical to Steps 2 and 3 of Algorithm 3.4.

Subroutine 1 to Algorithm 4.6

(i) Set time points $t_x = t_{left}^n$, $t_y = t_{right}^n$, $t_{out} = t_{left}^n$.

(ii) Integrate the following system of two differential equations

$$\dot{X}(t) = \begin{cases} U - d(t), & \text{if } \psi(t) \ge c_u, \\ -d(t), & \text{if } \psi(t) < c_u, \end{cases} \qquad X(t_{out}) = 0,$$

$$\dot{\psi}(t) = \begin{cases} 2c_x^+ X(t), & \text{if } X(t) \ge 0, \\ 2c_x^- X(t), & \text{if } X(t) < 0, \end{cases} \qquad \psi(t_{out}) = c_u,$$

over the interval $[t_{out}, t_{right}^m]$.

If $X(t) < 0$ for all $t \in \tau_m$, then no arc links nodes τ_n and τ_m; STOP.

Otherwise, set $t_{out} = t^n_{right}$ and integrate the above system.

If $X(t) > 0$ for all $t \in \tau_m$, then no arc links nodes τ_n and τ_m; STOP.

Otherwise go to (iii)

(iii) Integrate the system from (ii) with $t_{out} = 0.5(t_x + t_y)$ and set time point $t_{in} \in \tau_m$ such that $X(t_{in}) = 0$.

If $\psi(t_{in}) > c_u$, then $t_x = t_{out}$ and go to (iv);

if $\psi(t_{in}) < c_u$, then $t_y = t_{out}$ and go to (iv).

(iv) If $t_y - t_x \le \varepsilon$, then set $arc_{nm} = (\tau_n, \tau_m) \in A$ and its three weights

$a(arc_{nm}) = t_{out}$, $b(arc_{nm}) = t_{in}$ and

$c(arc_{nm}) = \int\limits_{t_{out}}^{t_{in}} c_x^+ X^+(t)^2 + c_x^- X^-(t)^2 + c_u u(t)dt$; go to (v).

Otherwise go to (iii).

(v) If there exist arcs arc_{jn} such that $|b(arc_{jn}) - a(arc_{nm})| < \varepsilon$, then set $a(arc_{nm}) = \max_j b(arc_{jn})$. For each arc arc_{jn} entering node n, update weights

$$SD^n_{jm} = \begin{cases} \int\limits_{b(arc_{jn})}^{a(arc_{nm})} c_u d(t)dt, & \text{if } a(arc_{nm}) \ge b(arc_{jn}); \\ -1, & \text{otherwise.} \end{cases} \quad \text{STOP.}$$

Otherwise, STOP.

Discussion

This section presents two algorithms. Algorithm 4.5 yields the exact optimal solution of the $1|1|u(t) \in [0,U]\,|d(t),T|\,L_u, Q_x$ problem for a few special cases of the demand function. Algorithm 4.6 approximates the optimal solution with a given accuracy for arbitrary demands.

Specifically, Algorithm 4.5 Step 1 invokes Algorithm 3.2 developed for the corresponding uncapacitated problem. If demand is weak, $d(t) \le U$, the found solution is optimal. In addition, when the backlog cost related to the demand over the planning horizon is less than the production cost, no production is optimal regardless of the demand. The other step of Algorithm 4.5 copes with the case when the demand is pressing (the demand averaged up to any point of time is greater than the production capacity). For this type of demand, there is only one point of time, t_1, at which the control function changes over from the full production, $u(t)=U$, to the no-production regime.

The complexity of Algorithm 4.5 is $O(1)$ if the root of equation $f(t_1)=c_u$ can be found analytically. For example, if demand is K-interval piece-wise constant (in addition to the demand requirements of Steps 1 and 2), then the exact solution is available in $O(K)$ time. However, if demand is an arbitrary complex function of time, only a numerical approximation of the optimal solution is possible as provided by Algorithm 4.6.

The correctness of Algorithm 4.5 is proved in the following lemma.

LEMMA 4.9. (Correctness of Algorithm 4.5) Given problem (4.20), (4.2) and (4.3), the optimal control is determined as follows

- if equation $c_u = 2c_x^- \int_{t_1}^{T}\int_{t_1}^{t} d(\tau)d\tau dt$ has a root $t_1 \in [0,T]$ and $d(t) \le U$, then
  $u(t)=d(t)$ for $0 \le t < t_1$ and $u(t)=0$ for $t_1 \le t \le T$;

- if equation $c_u = 2c_x^- \int_{t_1}^{T}\int_{t_1}^{t} d(\tau)d\tau dt$ does not have a root $t_1 \in [0,T]$, then
  $u(t)=0$ for $0 \le t \le T$;

- if equation $\dfrac{c_u}{2c_x^-} = -Ut_1(T-t_1) + (T-t_1)\int_0^{t_1} d(t)dt + \int_{t_1}^{T}\int_{t_1}^{t} d(\tau)d\tau dt$ has a root $t_1 \in [0,T]$ and $\dfrac{1}{t}\int_0^{t} d(\tau)d\tau > U$ for $0<t<T$, then
  $u(t)=U$ for $0 \le t < t_1$ and $u(t)=0$ for $t_1 \le t \le T$.

PROOF. First we find the control effort at the fractional production characterized by $\psi(t)=c_u$ at an interval of time. By differentiating this condition, we obtain

$$\dot{\psi}(t)=0. \tag{4.23}$$

From (4.22), it immediately follows that equation (4.23) can be met only if $X(t)=0$. By differentiating this equality and substituting the state equation (4.2), we obtain

$$u(t) = d(t). \tag{4.24}$$

Note, that the fractional production (4.24) can occur at an interval of time only if the demand is weak at the interval, $d(t) \le U$.

Next, consider the solution which includes two time intervals characterized by the fractional production $u(t) = d(t)$, $X(t)$=0 for $0 \le t \le t_1$ and no-production $u(t)$=0, $X(t)$<0 for $t_1 < t \le T$ regimes respectively. According to conditions (4.6), $\psi(t_1) = c_u$ and the single switching point, t_1, can be found by integrating the corresponding equation from (4.22)

$$\psi(t) = -2c_x^- \int_t^T X(\tau)d\tau, \quad t_1 \le t \le T. \tag{4.25}$$

By taking into account that $X(t) = -\int_{t_1}^{t} d(\tau)d\tau$ for $t_1 < t \le T$, we finally find an algebraic equation for the unknown switching time point, t_1

$$\psi(t_1) = c_u = 2c_x^- \int_{t_1}^{T}\int_{t_1}^{t} d(\tau)d\tau dt. \tag{4.26}$$

Note, that the right-hand side of equation (4.26) is a non-increasing function in t_1. Thus, if there is no root of (4.26) with respect to t_1, then $\psi(0) < c_u$, $\dot{\psi}(t) < 0$, $X(t) \le 0$ and the feasible solution for the stated primal and dual problems is $u(t)$=0 over the entire planning horizon.

To prove the last part of the lemma, consider the solution which includes two time intervals characterized by full, $u(t) = U$ for $0 \le t \le t_1$, and no-production, $u(t)$=0 for $t_1 < t \le T$ respectively, and negative inventory levels $X(t) < 0$ for $0 < t \le T$ (see Figure 4.4). Since condition $\frac{1}{t}\int_0^t d(\tau)d\tau > U$ ensures negative inventories, the presented solution is feasible if (4.26) has a root falling within the planning horizon. Consequently, according to conditions (4.6), $\psi(t_1) = c_u$ and point t_1 can be determined from (4.25). This is accomplished by substituting $X(t) = X(t_1) - \int_{t_1}^{t} d(\tau)d\tau$ for $t_1 < t \le T$ in (4.25). The resulting algebraic equation for the unknown t_1 is

$$\psi(t_1) = c_u = -2c_x^- \int_{t_1}^{T} \left(X(t_1) - \int_{t_1}^{t} d(\tau)d\tau \right) dt . \tag{4.27}$$

Substitution of $X(t_1) = \int_0^{t_1} \big(U - d(t)\big)dt$ into (4.27) proves the last part of the lemma.

■

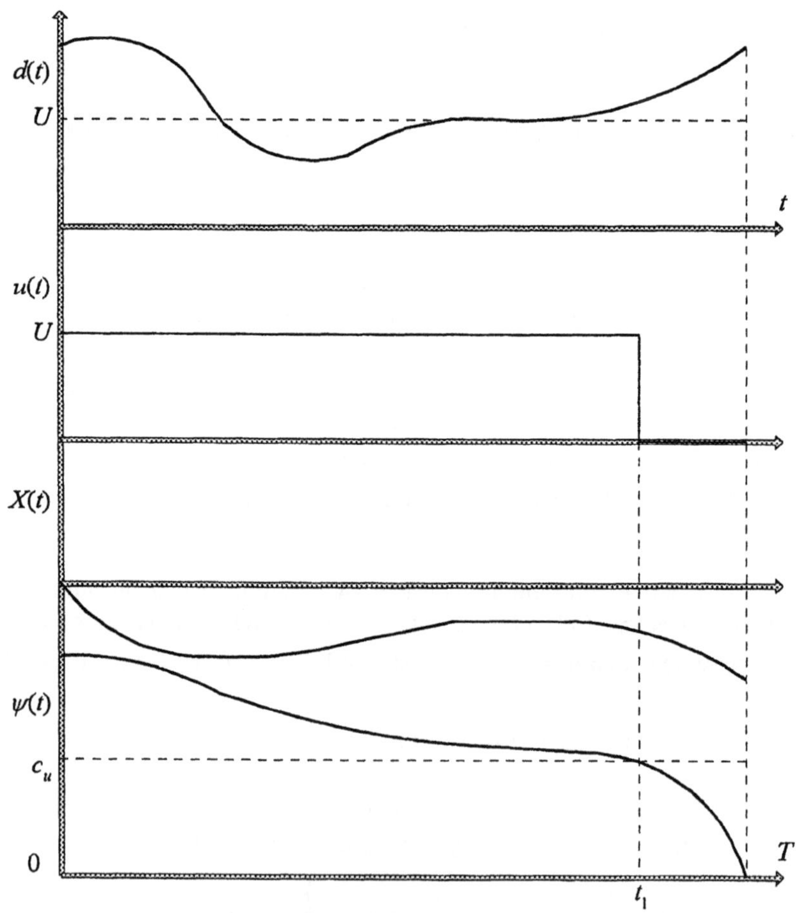

Figure 4.4. Demand $d(t)$, optimal control $u(t)$, state $X(t)$ and co-state $\psi(t)$ variables for the $1|1|u(t) \in [0,U]\,|d(t),T|\,L_u, Q_x$ problem with pressing demand.

Since the maximum principle does not provide enough information about the optimal solution for the $1|1|u(t)\in[0,U]\,|d(t),T|\,L_u,Q_x$ problem with demand as an arbitrary function of time, Algorithm 4.6 has been presented to approximate the optimal solution with an *a priori* given accuracy.

Algorithm 4.6 is very similar to Algorithm 4.2, which locates the intervals where the demand is weak, $d(t)\le U$, and links them so that the solution of the dual problem will be feasible (see Figure 4.5). This is accomplished by construction of a graph in which the nodes are these intervals and the arcs are links between pairs of nodes. Thus, similar to the $1|1|u(t)\in[0,U\,|d(t),T|\,L_u,L_x$ problem, the continuous-time scheduling is again reduced to a combinatorial problem of finding a shortest feasible path in this graph.

Therefore, with no further elaboration, we state Lemmas 4.10-4.12 analogous to Lemmas 4.2-4.4 to justify the correctness of Algorithm 4.6. The proofs of Lemmas 4.10-4.12 are omitted because they are similar to the proofs of Lemmas 3.6, 3.7 and 3.13 respectively.

LEMMA 4.10. (Correctness of Algorithm 4.6 Step 1) Given problem (4.20), (4.2)-(4.3), if $d(t)\le U$ at K intervals of time τ_k, $k=1,\dots,K$, then the optimal solution contains at most K time intervals of zero inventory, $\chi_k=\{t\in\tau_k\,|X(t)=0\}$.

LEMMA 4.11. Slack functions F_6, F_7, F_8 and F_9 are non-decreasing functions of their arguments.

LEMMA 4.12. Given problem (4.20), (4.2)-(4.3), the co-state variable $\psi(t)$ is bounded as $\psi(t)\in[0,\hat{\psi}]$ for $t\in[0,T]$, where $\hat{\psi}=2c_x^-\int_0^T\int_0^t d(\tau)d\tau dt$.

Thus, the correctness of Algorithm 4.6 Step 2 immediately follows from Lemmas 4.11 and 4.12. The last step of the algorithm uses the sequence-dependent, shortest path procedure described in Section 3.1.3. Theorem 4.3 concludes the discussion of the problem by determining the complexity of solving the $1|1|u(t)\in[0,U]\,|d(t),T|\,L_u,Q_x$ problem.

THEOREM 4.3. The problem (4.20), (4.2)-(4.3) is solvable with accuracy ξ in $O\left(\max\{K^4,\frac{K^3R}{\xi}\log\frac{KR}{\xi}\}\right)$ time, where K is the number of time intervals at which $d(t)\le U$ and $R=TU\left(c_u+2T\max\{c_x^+UT,c_x^-D(T)\}\right)$.

PROOF. Given integration accuracy ε, the complexity of Algorithm 4.6 is determined by (3.25)

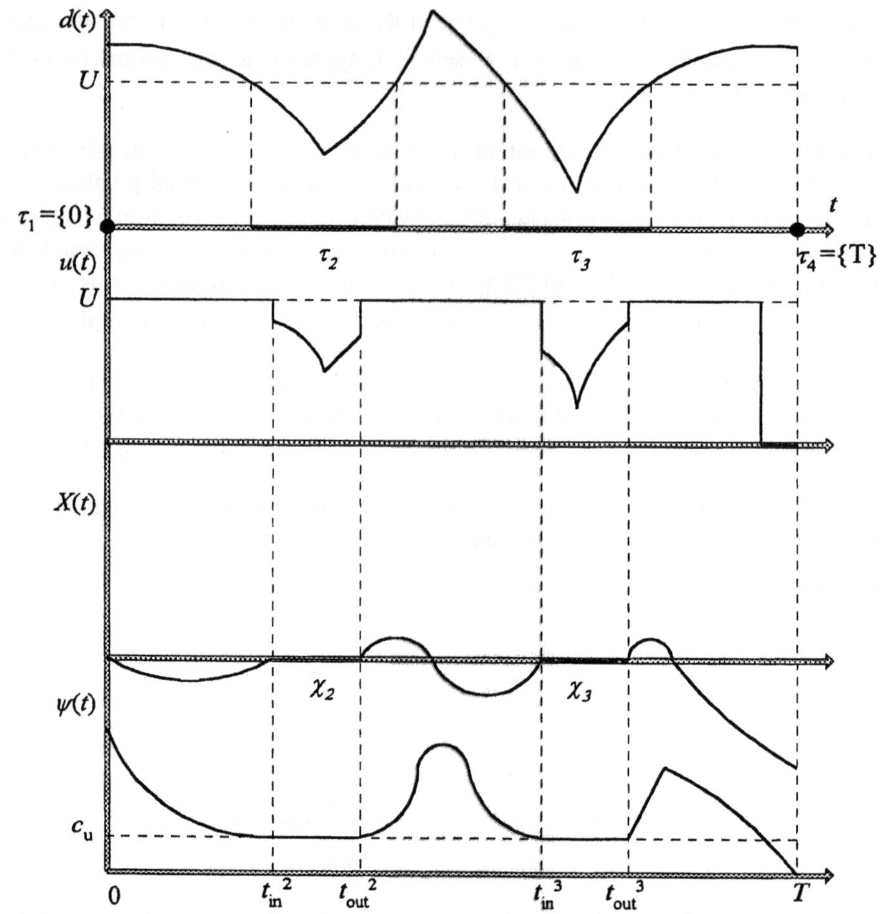

Figure 4.5. Demand $d(t)$, optimal control $u(t)$, state $X(t)$ and co-state $\psi(t)$ variables for the $1|1|u(t) \in [0,U]\,|d(t),T|\,L_u, Q_x$ problem with arbitrary demand.

$$O\left(\max\{K^4, K^2 \frac{T}{\varepsilon} \log \frac{T}{\varepsilon}\right)$$

However, the accuracy of the objective ξ that the algorithm provides is different. Namely, by varying the objective function (4.20) we determine

$$\xi \le 2\max\{c_x^+ \max_t X^+(t), c_x^- \max_t X^-(t)\}\int_0^T |\Delta X(t)|dt + c_u \int_0^T |\Delta u(t)|dt\,. \qquad (4.28)$$

The maximum inventory and backlog are easy to estimate as

$$\max_t X^+(t) = UT \text{ and } \max_t X^-(t) = \int_0^T d(t)dt = D(T)\,.$$

Next, each link between two fractional regimes results in a maximum error of $\int_{t_{out}}^{t_{in}} |\Delta u(t)|dt = \varepsilon U$ due to the second term of (4.28). Consequently, the maximum error induced by the first term is

$$\int_0^T |\Delta X(t)|dt = \int_0^T \int_0^t |\Delta u(\tau)|d\tau dt \le \varepsilon UT\,.$$

By taking into account that there can be at most K links over the planning horizon, the worst accuracy estimate (4.28) is as follows

$$\xi = c_u UK(1+gT)\varepsilon\,, \qquad (4.29)$$

where $g = \frac{2}{c_u}\max\{c_x^+ UT, c_x^- D(T)\}$. Finally, to obtain the complexity and accuracy of Algorithm 4.6, we express ε from (4.29), substitute it into (3.25) and introduce parameter R to simplify the presentation.
■

Special Cases of Zero Costs

Algorithm 4.5 simplifies when solving the problem with zero production cost, i.e. the $1|1|u(t) \in [0,U]\,|d(t),T|\,Q_x$ problem. Indeed, Step 1 results in $t_1 = T$ for arbitrary demand, so that no switching points occur. As a result, either fractional production (Step 2) or full production (Step 3) becomes optimal over the entire planning horizon. Algorithm 4.6 also simplifies in the case of zero production cost, since the shortest path procedure becomes no longer sequence-dependent. This results in an improved complexity estimate $O\left(\frac{K^3R}{\xi}\log\frac{KR}{\xi}\right)$. The $1|1|u(t) \in [0,U]\,|d(t),T|\,L_u, Q_x, c_x^+ = 0$ problem is solved by Algorithms 4.5 and 4.6 with no significant modifications.

No Backlog

When backlogs are disallowed, i.e. for the $1|1|u(t)\in[0,U],X(t)\geq 0|d(t),T|L_u,Q_x$ problem to have a feasible solution, it is necessary that average demand must not exceed capacity U as determined in (4.14).

It turns out that no-backlog problem is polynomially solvable if the demand satisfies (4.14). Moreover, if demand is weak, $d(t)\leq U$, then the closed form optimal solution is just to produce according to the fractional production $u(t)$=$d(t)$.

Dual Formulation

Now problem (4.20), (4.2)-(4.3) has the additional constraint $X(t)\geq 0$ which affects neither Hamiltonian (4.21) nor optimality conditions (4.6), but changes the co-state differential equation

$$d\psi(t)=c_x^+X(t)dt-d\mu(t),\quad \psi(T+0)=0,$$

where Lagrange multiplier $d\mu(t)\geq 0$ corresponds to the no-backlog constraint and satisfies the complementary slackness condition (4.16).

Algorithm

The algorithm is identical to Algorithm 4.4.

Discussion

Applicability of Algorithm 4.4 is explained below by proving Lemmas 4.7-4.9 for the case of the quadratic inventory cost Q_x. As a result, the complexity $O(K^3)$ of the algorithm does not change.

LEMMA 4.13. (Correctness of Algorithm 4.4 Step 1 for the Q_x case) Given problem (4.20), (4.2)-(4.3), if $d(t)\leq U$ for $0\leq t\leq T$, then the optimal control is $u(t)=d(t)$ for $0\leq t\leq T$.

PROOF. The proof is identical to that of Lemma 4.7. ■

LEMMA 4.14. (Rule for linking nodes in graph G for the Q_x case) Let two adjacent fractional regimes $u(t)$=$d(t)$ be at intervals $\chi_n\subseteq\tau_n$ and $\chi_m\subseteq\tau_m$, and t_{out} be a root of the following equation

$$\int_{t_{out}}^{t_{left}^m} d(t)dt-U(t_{left}^m-t_{out})=0.$$

Then the optimal control between these fractional production regimes is $u(t) = U$ for $t_{out} \le t \le t_{left}^{m}$.

PROOF. According to the proof of Lemma 4.7, the necessary condition for the fractional regime to occur is $d(t) \le U$. Therefore, Step 2 of the algorithm identifies time intervals where this condition holds. Furthermore, given fractional regimes at intervals χ_n and χ_m, according to (4.6) and (4.16), $X(t) = 0$ and $\psi(t) = c_u$ at these intervals.

Consider the solution between these fractional regimes in the following form

$$X(t) = U(t - t_{out}) - \int_{t_{out}}^{t} d(\tau)d\tau \text{, for } t_{out} \le t \le t_{in}, \tag{4.30}$$

$$\psi(t) = c_u + 2c_x^+ \int_{t_{out}}^{t} X(\tau)d\tau \quad \text{for } t_{out} \le t \le t_{in}, \quad d\mu(t_{in}) = 2c_x^+ \int_{t_{out}}^{t_{in}} X(\tau)d\tau, \tag{4.31}$$

where t_{out} is the terminal point of the first fractional regime, $t_{out} \in \tau_n$ and t_{in} is the starting point of the other fractional regime, $t_{in} \in \tau_m$.

Next, we prove by contradiction that point t_{in} coincides with the left boundary of interval τ_m, i.e. $t_{in} = t_{left}^{m}$. Indeed, if it is not the case, then at interval $[t_{left}^{m}, t_{in})$ $X(t) > 0$. Consequently, $\psi(t)$ increases up to the value of c_u. Therefore, it is less than c_u that contradicts the co-state variable behavior (4.31). Once $t_{in} = t_{left}^{m}$, the only unknown parameter when linking the two intervals is t_{out}. It is calculated from (4.30) as a root of the equation $X(t_{left}^{m}) = 0$.

■

THEOREM 4.4. If demand satisfies condition (4.14), then the problem (4.20), (4.2)-(4.3) with backlog disallowed is solvable in $O(K^3)$ time.

PROOF. The proof is similar to that of Theorem 4.2. ■

4.1.3 NON-LINEAR PRODUCTION COST

In this section, we consider a piece-wise linear function of inventory and backlog costs and a non-linear production cost function. This change complicates the optimal behavior of the system so that the solvable cases are found only for either pressing demand or weak, non-lumpy demand.

Backlogs

The problem to be considered is one-machine, one-product-type scheduling with backlogs and bounded production rate to minimize the piece-wise inventory cost and the quadratic production cost, that is, the $1|1|u(t)\in[0,U|d(t),T]Q_u,L_x$ problem which is stated as follows

$$J=\int_0^T\left[c_x^+X^+(t)+c_x^-X^-(t)+c_u u(t)^2\right]dt\to\min, \tag{4.32}$$

subject to (4.2) and (4.3).

Dual Formulation

The Hamiltonian, H, resulting from the maximum principle is

$$H(t)=-\left[c_x^+X^+(t)+c_x^-X^-(t)+c_u u(t)^2\right]+\psi(t)\left[u(t)-d(t)\right]. \tag{4.33}$$

The co-state variable, $\psi(t)$ in (4.33), satisfies the co-state differential equation (4.5).

Maximization of the Hamiltonian with respect to admissible controls $u(t)$ implies that

$$u(t)=\begin{cases}U, & \text{if } \psi(t)>2c_uU,\\ 0, & \text{if } \psi(t)<0,\\ \dfrac{\psi(t)}{2c_u}, & \text{if } 0\le\psi(t)\le 2c_uU.\end{cases} \tag{4.34}$$

The first two conditions in (4.34) are similar to those found for the case of linear production cost and determine full production and no-production regimes respectively. The last condition presents the partial production regime which determines production at an intermediate rate less than the capacity U.

Algorithm 4.7

Step 1:

Run Algorithms 3.3-3.5. If $u(t)\le U$ for $0\le t\le T$, then the found solution is optimal; STOP.

Otherwise, go to the next step.

Step 2:

If the demand is pressing, $\frac{1}{t}\int_0^t d(\tau)d\tau \geq U$ for $0 < t < T$,

then calculate $t_1 = T - \frac{2c_u U}{c_x^-}$ and set the optimal solution as

$u(t) = U$ for $0 \leq t \leq t_1$

$u(t) = \frac{c_x^-}{2c_u}(T - t)$ for $t_1 \leq t \leq T$, STOP;

otherwise go to the next step.

Step 3:

If the demand function $d(t)$ is single-mode, i.e. it is non-decreasing before a time point $\hat{t}$ and non-increasing after $\hat{t}$, then find $t_2 \in [0, T]$

as the root of the equation $\frac{1}{t_2}\int_0^{t_2} d(t)dt = U$.

If $Tc_x^- - t_2(c_x^- + c_x^+) > 2c_u U$, then set the optimal solution as

$u(t) = U$ for $0 \leq t \leq t_1$,

$u(t) = \frac{c_x^-}{2c_u}(T - t)$ for $t_1 \leq t \leq T$; STOP.

Otherwise, calculate time points t_2 and t_3 from the following system of two algebraic equations

$$(T - t_2)c_x^- = c_x^+(t_2 - t_3) + 2c_u U, \qquad \frac{1}{t_2}\int_0^{t_2} d(t)dt = U - \frac{c_x^+}{4c_u}\frac{t_3^2}{t_2}.$$

Set the optimal solution as

$u(t) = U - \frac{c_x^+}{2c_u}(t_3 - t)$ for $0 \leq t < t_3$,

$u(t) = U$ for $t_3 \leq t < t_1$,

$u(t) = \frac{c_x^-}{2c_u}(T - t)$ for $t_1 \leq t \leq T$; STOP.

Example 4.3

Consider the $1|1|u(t) \in [0, U]|d(t), T|Q_u, L_x$ problem with parameters T=10hr, U=3unit/hr, c_u=1\$hr/unit^2, c_x^+=0.2\$/hr.unit, c_x^-=1\$/hr.unit and quadratic demand $d(t)$=(2.5+0.02(t-6)^2)unit/hr.

Step 1 of Algorithm 4.7 runs Algorithm 3.3 which returns optimal control exceeding U from the very beginning of the planning horizon. Therefore, the algorithm proceeds to Step 2 where it verifies that the demand is pressing, $\frac{1}{t}\int_0^t d(\tau)d\tau = 2.5 + 0.02(\frac{t^2}{3} - 6t + 72) > U$ for $0 \le t \le 10$. Then it calculates

$t_1 = T - \frac{2c_u U}{c_x^-} = 4$ hr and sets the optimal solution as

$u(t) = 3$ unit/hr for $0 \le t < 4$,

$u(t) = 0$ for $4 \le t \le 10$.

■

Algorithm 4.8

Step 1:

Set accuracy ξ required for computing the optimal solution. Select time intervals $\tau_k = [t_{left}^k, t_{right}^k]$ at which $d(t) \le U$ and $-c_x^- \le 2c_u \dot{d}(t) \le c_x^+$, $t_{right}^k < t_{left}^{k+1}$, and number them beginning from k=2. In addition, define τ_1 and τ_K as the intervals which are degenerated to a point, $\tau_1 = \{0\}$ and $\tau_K = \{T\}$, where K is the total number of obtained intervals. Calculate accuracy of integration

$$\varepsilon = \left(KT \max\{c_x^+, c_x^-\} \left(U + \frac{T \max\{c_x^+, c_x^-\}}{2c_u} \right) \right)^{-1} \xi .$$

Step 2 and Step 3 are identical to Steps 2 and 3 of Algorithm 3.4.

Subroutine 1 to Algorithm 4.8

(i) Set time points $t_x = t_{left}^n$, $t_y = t_{right}^n$, $t_{out} = t_{left}^n$.

(ii) Integrate twice, for $\dot{\psi}(t_{out}) = c_x^+$ and $\dot{\psi}(t_{out}) = -c_x^-$, the following system of differential equations

$$\dot{X}(t) = \begin{cases} \frac{\psi(t)}{2c_u} - d(t), & \text{if } 0 \le \psi(t) \le 2c_u U, \\ U - d(t), & \text{if } 2c_u U \le \psi(t), \end{cases} \qquad X(t_{out}) = 0,$$

$$\dot{\psi}(t) = \begin{cases} c_x^+, & \text{if } X(t) > 0, \\ -c_x^-, & \text{if } X(t) < 0, \end{cases} \qquad \psi(t_{out}) = 2c_u d(t_{out}),$$

over interval $[t_{out}, t^{m}_{right}]$.

If $X(t) < 0$ for all $t \in \tau_m$, then no arc links nodes τ_n and τ_m; STOP.

Otherwise, set $t_{out} = t^{n}_{right}$ and integrate the above system twice, for $\dot{\psi}(t_{out}) = c_x^+$ and $\dot{\psi}(t_{out}) = -c_x^-$.

If $X(t) > 0$ for all $t \in \tau_m$, then no arc links nodes τ_n and τ_m; STOP.

Otherwise go to (iii).

(iii) Integrate the system from (ii) twice, for $\dot{\psi}(0) = c_x^+$ and $\dot{\psi}(0) = -c_x^-$ with $t_{out} = 0.5(t_x + t_y)$ and set time point $t_{in} \in \tau_m$ such that $X(t_{in}) = 0$.

If $\psi(t_{in}) > 2c_u d(t_{in})$, then $t_x = t_{out}$ and go to (iv);

if $\psi(t_{in}) < 2c_u d(t_{in})$, then $t_y = t_{out}$ and go to (iv).

(iv) If $t_y - t_x \le \varepsilon$, then set $arc_{nm} = (\tau_n, \tau_m) \in A$ and its three weights

$a(arc_{nm}) = t_{out}$, $b(arc_{nm}) = t_{in}$ and

$$c(arc_{nm}) = \int_{t_{out}}^{t_{in}} c_x^+ X^+(t) + c_x^- X^-(t) + c_u u(t)^2 \, dt \text{ ; go to (v).}$$

Otherwise go to (iii).

(v) If there exist arcs arc_{jn} such that $\left|b(arc_{jn}) - a(arc_{nm})\right| < \varepsilon$, then set $a(arc_{nm}) = \max_j b(arc_{jn})$. For each arc arc_{jn} entering node n, update weights

$$SD^{n}_{jm} = \begin{cases} \displaystyle\int_{b(arc_{jn})}^{a(arc_{nm})} c_u d(t)^2 \, dt, & \text{if } a(arc_{nm}) \ge b(arc_{jn}); \\ -1, & \text{otherwise.} \end{cases} \quad \text{STOP.}$$

Otherwise, STOP.

Discussion

The two algorithms presented in this section solve the $1|1|u(t) \in [0,U]\,|d(t),T|\,Q_u, L_x$ problem for different forms of the demand function. Algorithms 4.7 Step 1 invokes Algorithms 3.3-3.5 developed for different cases of the corresponding uncapacitated problem. If demand is weak, $d(t) \le U$, the found solution is optimal. Steps 2 and 3 give exact optimal solutions for pressing and single-mode demands respectively. If, however, the demand function is arbitrary, Algorithm 4.8 approximates the optimal solution with a given accuracy in pseudopolynomial time.

The following lemmas prove the first two solvable cases.

LEMMA 4.15. (Correctness of Algorithm 4.7 Step 2) Given problem (4.32), (4.2) and (4.3), if demand is pressing, $\frac{1}{t}\int_0^t d(\tau)d\tau \geq U$ for $0 < t < T$, then the optimal solution is

$$u(t) = U \text{ for } 0 \leq t \leq t_1, \quad u(t) = \frac{c_x^-}{2c_u}(T-t) \text{ for } t_1 \leq t \leq T,$$

where $t_1 = T - \frac{2c_u U}{c_x^-}$.

PROOF. Since the demand is pressing, the inventory level is negative within the planning horizon, $0 < t < T$. Consequently, with respect to the co-state equation (4.5)

$$\psi(t) = c_x^-(T-t) \text{ and } X(t) \leq 0 \text{ for } 0 \leq t \leq T. \tag{4.35}$$

Now, from optimality condition (4.34), we obtain the optimal control, which includes the full production regime changed over by the partial production regime at a point t_1, satisfying switching condition $\psi(t_1) = 2c_u U$. By substituting this condition into (4.35), we obtain the unknown point t_1, as stated in the lemma. The optimal control is found from (4.34)

$$u(t) = U \text{ for } 0 \leq t \leq t_1, \quad u(t) = \frac{c_x^-}{2c_u}(T-t) \text{ for } t_1 \leq t \leq T. \tag{4.36}$$

■

LEMMA 4.16. (Correctness of Algorithm 4.7 Step 3) Given problem (4.32), (4.2) and (4.3), $t_1 = T - \frac{2c_u U}{c_x^-}$, if the demand function $d(t)$ is single-mode, then the optimal solution is

- if $Tc_x^- - t_2(c_x^- + c_x^+) > 2c_u U$, then

$$u(t) = U \text{ for } 0 \leq t \leq t_1, \quad u(t) = \frac{c_x^-}{2c_u}(T-t) \text{ for } t_1 \leq t \leq T,$$

where t_2 is the root of equation $\frac{1}{t_2}\int_0^{t_2} d(t)dt = U$;

• otherwise,

$u(t) = U - \frac{c_x^+}{2c_u}(t_3 - t)$ for $0 \le t < t_3$, $u(t) = U$ for $t_3 \le t < t_1$, $u(t) = \frac{c_x^-}{2c_u}(T - t)$ for $t_1 \le t \le T$,

where t_2 and t_3 satisfy the following system of two algebraic equations

$$(T - t_2)c_x^- = c_x^+(t_2 - t_3) + 2c_u U, \quad \frac{1}{t_2}\int_0^{t_2} d(t)dt = U - \frac{c_x^+}{4c_u}\frac{t_3^2}{t_2}.$$

PROOF. Consider a solution which consists of partial, $u(t) = \frac{\psi(t)}{2c_u}$, full $u(t)=U$ and again partial production regimes changed over at points t_3 and t_1 respectively (see Figure 4.6). Since the demand is single-mode, there exists a time point t_2 such that

$$X(t) > 0, \quad \psi(t) = 2c_u U + c_x^+(t - t_3) \quad \text{for } 0 < t < t_2,$$
$$X(t) < 0, \quad \psi(t) = 2c_u U - c_x^-(t - t_1) \quad \text{for } t_2 < t < T. \tag{4.37}$$

For the solution (4.37) to be feasible, it is neccssary that $X(t_2) = 0$, that is

$$U(t_2 - t_3) + \frac{1}{2c_u}\int_0^{t_3}\left(2c_u U + c_x^+(t - t_3)\right)dt - \int_0^{t_2} d(t)dt = 0. \tag{4.38}$$

Next, by considering the co-state variable in (4.37) at point t_2, we find

$$c_x^+(t_2 - t_3) = -c_x^-(t - t_1). \tag{4.39}$$

If $\psi(0) < 2c_u U$, equations (4.38) and (4.39) result in the system of two algebraic equations in unknowns t_2 and t_3, as stated in the second condition of the lemma. Since the first condition is a special case in which the first partial production regime vanishes, i.e.

$$\psi(0) \ge 2c_u U, \tag{4.40}$$

it is proven similarly.

■

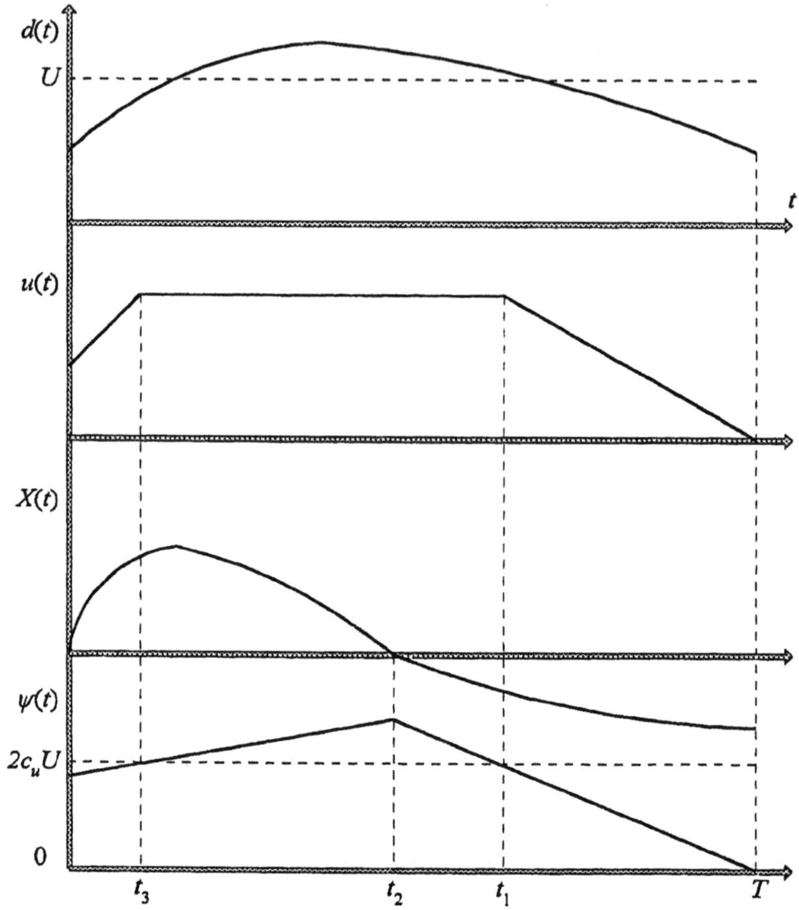

Figure 4.6. Demand $d(t)$, optimal control $u(t)$, state $X(t)$ and co-state $\psi(t)$ variables for the case of the $1|1|u(t)\in[0,U]\,|d(t),T|\,Q_u, L_x$ problem with single-mode demand.

Since the maximum principle does not provide enough information about the optimal solution for the $1|1|u(t)\in[0,U]\,|d(t),T|\,Q_u, L_x$ problem with demand as an arbitrary function of time, Algorithm 4.8 has been presented to approximate the optimal solution with an *a priori* given accuracy.

The following lemma states that even if the demand function is arbitrary, that is, if the conditions of Lemmas 4.15 and 4.16 are not satisfied, then the general form of the optimal solution has a very definite structure characterized by controls providing zero inventories only at those time intervals where the demand is weak and non-lumpy.

LEMMA 4.17. (Correctness of Algorithm 4.8 Step 1) Given problem (4.32), (4.2) and (4.3), if $-c^- \le 2c_u \dot{d}(t) \le c^+$ and $d(t) \le U$ at K intervals of time τ_k, $k = 1,...,K$, then the optimal solution contains at most K time intervals of zero inventory, $\chi_k = \{t \in \tau_k | X(t) = 0\}$.

PROOF. The proof immediately follows from two facts: $d(t) \le U$ is the necessary condition for tracking the demand and $X(t)$ cannot be zero at any sub-interval of the interval of time where $-c^- \le 2c_u \dot{d}(t) \le c^+$ does not hold. The detailed proof of the lemma is left as an exercise for the reader.
■

Similar to Algorithm 3.4, Algorithm 4.8 makes use of the conditions $d(t) \le U$ and $-c^- \le 2c_u \dot{d}(t) \le c^+$ of Lemma 4.17 to locate the intervals, where these conditions are met, i.e. the fractional production $u(t) = d(t)$ is optimal, and to link them so that the solution of the dual problem will be feasible. This is accomplished by construction of a graph in which the nodes represent these intervals and the arcs are links between pairs of nodes.

The links between nodes τ_n and τ_m are calculated numerically with the aid of the slack functions F_1 - F_4, as it is the case in Algorithm 3.4. The only change is in the state differential equation used in the subroutines to Algorithm 4.8. This change is due to the fact that the optimal control is now determined by either fractional or full production. Since the change is the same for all subroutines of Algorithm 4.8, only the first subroutine to this algorithm is presented.

The following two lemmas state monotonicity of the slack functions and bounds for the co-state variable. Both are used in the dichotomous search carried out by the algorithm's subroutines.

LEMMA 4.18. Given problem (4.32), (4.2) and (4.3), slack functions F_1, F_2, F_3 and F_4 are non-decreasing functions of their arguments.

PROOF. The proof is similar to that of Lemma 3.7. ■

LEMMA 4.19. Given problem (4.32), (4.2) and (4.3), the co-state variable $\psi(t)$ is bounded as $\psi(t) \in [0, \hat{\psi}]$ for $t \in [0, T]$, where $\hat{\psi} = c_x^- T$.

PROOF. The proof is similar to that of Lemma 3.8. ■

The correctness of Algorithm 4.8 Step 2 immediately follows from Lemmas 4.18 and 4.19, while the last step of the algorithm uses the sequence-dependent, shortest path procedure discussed in Section 3.1.3.

THEOREM 4.5. Problem (4.32), (4.2) and (4.3) is solvable with accuracy ξ in $O\left(\max\{K^4, \frac{K^3R}{\xi}\log\frac{KR}{\xi}\}\right)$ time, where $R = T^2\max\{c_x^+, c_x^-\}\left(U + \frac{T\max\{c_x^+, c_x^-\}}{2c_u}\right)$ and K is the number of time intervals at which $-c^- \le 2c_u\dot{d}(t) \le c^+$ and $d(t) \le U$.

PROOF. Given integration accuracy ε, the complexity of the algorithm is determined by (3.25)

$$O\left(\max\{K^4, K^2\frac{T}{\varepsilon}\log\frac{T}{\varepsilon}\right).$$

However, the accuracy of the objective, ξ, that the algorithm provides is different. Namely, by varying the objective function (4.32) we determine

$$\xi \le T\max\{c_x^+, c_x^-\}\max_t|\Delta X(t)| + 2Tc_u\max_t u(t)\max_t|\Delta u(t)|. \tag{4.41}$$

The value of $\max_t u(t)$ is evidently U. The value of $\max_t|\Delta u(t)|$ is estimated from (4.5) as $\frac{1}{2c_u}\max_t|\Delta\psi(t)|$, where the error of the co-state variable in the dichotomous search is $\max_t|\Delta\psi(t)| = \varepsilon K\max\{c_x^+, c_x^-\}$ for each link at Step 2 of the algorithm. The estimate of $\max_t|\Delta\psi(t)|$ is due to the fact that every point t^*, where $X(t^*)=0$ (the maximum number of such points is K), induces increments in the co-state variable error by at most $\varepsilon\max\{c_x^+, c_x^-\}$, as illustrated in Figure 3.7. The value of $\max_t|\Delta X(t)|$ is just $T\max_t|\Delta u(t)| = \frac{\varepsilon K\max\{c_x^+, c_x^-\}}{2c_u}T$. By substituting the found estimates into (4.41), the worst accuracy estimate, ξ, is obtained as a function of ε

$$\xi = KT\max\{c_x^+, c_x^-\}\left(U + \frac{T\max\{c_x^+, c_x^-\}}{2c_u}\right)\varepsilon. \tag{4.42}$$

To determine the complexity and accuracy of Algorithm 4.8, we express ε from (4.42), substitute it into (3.25) and introduce parameter R to simplify the presentation.
■

According to Theorem 4.5, Algorithm 4.8 is of pseudopolynomial complexity, which is due to multipliers T and ε entering (3.25). However, if demand is single-mode, piece-wise constant, the algorithm becomes strictly polynomial. This is due to the fact that the approximate integration which is carried out in Subroutines 1-4 to Algorithm 4.8 is reduced to solving systems of algebraic equations. Such modification is left as an exercise for the reader.

Special Cases of Zero Costs

The problem with no production cost, $1|1|u(t)\in[0,U]\,|d(t),T|\,L_x$, has been solved in Section 4.1.1. The problem with no inventory cost, $1|1|u(t)\in[0,U]\,|d(t),T|\,Q_u, L_x, c_x^+=0$ is solved by Algorithms 4.7-4.8 with no significant simplifications.

No Backlog

When backlogs are disallowed, i.e. for the $1|1|u(t)\in[0,U], X(t)\geq 0\,|d(t),T|\,Q_u, L_x$ problem, the solution algorithm can be obtained by minor modifications to Algorithms 4.7 and 4.8. Namely, the first two solvable cases proven in Lemmas 4.15 and 4.16 are realized only if negative inventories are allowed. They are thus not feasible due to the no-backlog constraint. Assuming that c_x^- is very large, the third and the only solvable case (Step 3 of Algorithm 4.7) simplifies because the optimal control is always fractional, $u(t)=d(t)$, and the conditions for this regime to realize are $d(t)\leq U$ and $2c_u\dot{d}(t)\leq c^+$.

Dual Formulation

Now problem (4.32), (4.2) and (4.3) has the additional constraint $X(t)\geq 0$ which affects neither Hamiltonian (4.33) nor optimality condition (4.34), but changes the co-state differential equation

$$d\psi(t)=c_x^+dt-d\mu(t),\quad \psi(T+0)=0,$$

where Lagrange multiplier $d\mu(t)\geq 0$ corresponds to the no-backlog constraint and satisfies the complementary slackness condition (4.16).

Algorithm 4.9

Step 1:

If $d(t)\leq U$ and $2c_u\dot{d}(t)\leq c^+$ along the entire planning horizon, then set the optimal solution as
$u(t)=d(t)$ for $0\leq t\leq T$, STOP;
otherwise go to the next step.

Step 2:

Set accuracy ξ required for computing the optimal solution. Select time intervals $\tau_k=[t_{left}^k, t_{right}^k]$ at which $d(t)\leq U$ and $2c_u\dot{d}(t)\leq c^+$,

$t^k_{right} < t^{k+1}_{left}$ and number them beginning from k=2. In addition, define τ_1 and τ_K as the intervals which are degenerated to a point, $\tau_1 = \{0\}$ and $\tau_K = \{T\}$, where K is the total number of obtained intervals. Calculate accuracy of integration $\varepsilon = \left(KTc_x^+ \left(U + \frac{Tc_x^+}{2c_u} \right) \right)^{-1} \xi$.

Step 3 and Step 4 are identical to Steps 2 and 3 of Algorithm 3.4.

Subroutine 1 to Algorithm 4.9

(i) Set time points $t_x = t^n_{left}$, $t_y = t^n_{right}$, $t_{out} = t^n_{left}$.

(ii) Integrate the following system of differential equations

$$\dot{X}(t) = \begin{cases} \frac{\psi(t)}{2c_u} - d(t), & \text{if } 0 \le \psi(t) \le 2c_u U, \\ U - d(t), & \text{if } 2c_u U \le \psi(t), \end{cases} \qquad X(t_{out}) = 0,$$

$$\dot{\psi}(t) = c_x^+, \qquad \psi(t_{out}) = 2c_u d(t_{out}),$$

over the interval $[t_{out}, t^m_{right}]$.

If $X(t) < 0$ for all $t \in \tau_m$, then no arc links nodes τ_n and τ_m; STOP. Otherwise, set $t_{out} = t^n_{right}$ and integrate the above system of differential equations.

If $X(t) > 0$ for all $t \in \tau_m$, then no arc links nodes τ_n and τ_m; STOP. Otherwise go to (iii).

(iii) Integrate the system from (ii) with $t_{out} = 0.5(t_x + t_y)$ and set time point $t_{in} \in \tau_m$ such that $X(t_{in}) = 0$.

If $\psi(t_{in}) > 2c_u d(t_{in})$, then $t_x = t_{out}$ and go to (iv);

if $\psi(t_{in}) < 2c_u d(t_{in})$, then $t_y = t_{out}$ and go to (iv).

(iv) If $t_y - t_x \le \varepsilon$, then set $arc_{nm} = (\tau_n, \tau_m) \in A$ and its three weights $a(arc_{nm}) = t_{out}$, $b(arc_{nm}) = t_{in}$ and

$$c(arc_{nm}) = \int_{t_{out}}^{t_{in}} c_x^+ X(t) + c_u u(t)^2 \, dt \text{ ; go to (v).}$$

Otherwise go to (iii).

(v) If there exist arcs arc_{jn} such that $\left|b(arc_{jn})-a(arc_{nm})\right|<\varepsilon$, then set $a(arc_{nm})=\max_j b(arc_{jn})$. For each arc arc_{jn} entering node n, update weights

$$SD^n_{jm}=\begin{cases}\displaystyle\int_{b(arc_{jn})}^{a(arc_{nm})} c_u d(t)^2\,dt, & \text{if } a(arc_{nm})\geq b(arc_{jn});\\ -1, & \text{otherwise.}\end{cases}\quad \text{STOP.}$$

Otherwise, STOP.

Discussion

Step 1 of Algorithm 4.9 provides a closed form optimal control, $u(t)=d(t)$, along the entire planning horizon when demand is weak, $d(t)\leq U$ and $2c_u\dot{d}(t)\leq c^+$. This solution is due to $X(t)=0$, $\psi(t)=2c_u d(t)$, $\dot{\mu}(t)=c_x^+-2c_u\dot{d}(t)$ for $0\leq t\leq T$ and jump $d\mu(T)=-d\psi(T)=2c_u d(T)$.

The other steps of Algorithm 4.9 are similar to the corresponding steps of Algorithm 4.8, and therefore approximate the globally optimal solution with the same accuracy and complexity when c_x^- is eliminated. In Step 2 of Algorithm 4.9, the condition for selecting time intervals τ_k is simplified, while the other two steps are identical to those of Algorithms 4.8 and 3.4. However, Step 3 invokes four subroutines which are now slightly changed. Namely, the co-state differential equation integrated iteratively in the subroutines is modified as presented in procedure (ii) of Subroutine 1 to Algorithm 4.9. Subroutines 2-4 to Algorithm 4.9 are omitted since they are modified accordingly. Note, that with respect to (4.16), $d\mu(t)=0$ at time intervals of positive inventories. However, jumps $d\mu(t)>0$ at points $t=t_{out}$ and $t=t_{in}$ are formally possible. The following lemma states that there are no jumps at t_{out} and t_{in}, and, thus, the system of the state and co-state equations used at Step 3 is correct.

LEMMA 4.20. (Correctness of Algorithm 4.9) Given two time intervals $[t_1,t_{out}]$ and $[t_{in},t_2]$ at which $X(t)=0$ and $\psi(t)=2c_u d(t)$ (the fractional regime), the following condition holds, $d\mu(t_{in})=d\mu(t_{out})=0$.

PROOF. The proof is identical to that of Lemma 3.12. ■

4.1.4 NON-LINEAR INVENTORY AND PRODUCTION COSTS

Both costs entering the objective are now non-linear. This further complicates the optimal behavior of the system; if the problem cannot be solved as though it is uncapacitated, then only a numerical approximation of the optimal solution is available.

Backlogs

The problem to be considered is one-machine, one-product-type scheduling with backlogs and limited production rate to minimize the quadratic inventory and production costs, that is, the $1|1|u(t)\in[0,U\,|d(t),T|\,Q_u,Q_x$ problem which is stated as follows

$$J=\int_0^T\left[c_x^+X^+(t)^2+c_x^-X^-(t)^2+c_uu(t)^2\right]dt\to\min, \tag{4.43}$$

subject to (4.2) and (4.3).

Dual Formulation

The Hamiltonian, H, resulting from the maximum principle is as follows

$$H(t)=-\left[c_x^+X^+(t)^2+c_x^-X^-(t)^2+c_uu(t)^2\right]+\psi(t)\left[u(t)-d(t)\right], \tag{4.44}$$

where co-state variable $\psi(t)$ satisfies the co-state differential equation (4.22). Maximization of the Hamiltonian as a function of $u(t)$ results, as in the previous section, in (4.34).

Algorithm 4.10

Step 1:

Run Step 1 of Algorithm 3.7. If $u(t)\le U$ for $0\le t\le T$, then the found solution is optimal; STOP.
Otherwise go to the next step.

Step 2:

(i) Set accuracy ξ required for computing the optimal objective value.

Calculate upper co-state variable estimate $\hat{\psi}=2c_x^-\int_0^T\int_0^t d(\tau)d\tau dt$,

parameter $g=\sqrt{\max\left\{\frac{c_x^+}{c_u},\frac{c_x^-D(T)}{c_uUT}\right\}}$, accuracy of the dichotomous search $\hat{\varepsilon}=\left(2TU(1+T^2g^2)ch(gT)\right)^{-1}\xi$ and accuracy of integration $\varepsilon=\left(2TUc_ug^2\right)^{-1}\hat{\varepsilon}$. Assign $\psi^{\max}=\psi^{\min}=\hat{\psi}$.

(ii) Integrate the following system of ordinary differential equations

$$\dot{\psi}(t) = \begin{cases} 2c_x^+ X(t), & \text{if } X(t) > 0; \\ 2c_x^- X(t), & \text{if } X(t) \le 0; \end{cases} \tag{4.45}$$

$$\dot{X}(t) = \begin{cases} \dfrac{\psi(t)}{2c_u} - d(t), & \text{if } 0 \le \psi(t) \le 2c_u U, \\ U - d(t), & \text{if } 2c_u U \le \psi(t), \end{cases} \tag{4.46}$$

over the interval $[0,T]$ with initial conditions $X(0) = 0$ and $\psi(0) = 0.5(\psi^{\min} + \psi^{\max})$.

(iii) If $\psi(T) \ge 0$, then assign $\psi^{\max} = \psi(0)$,
otherwise, assign $\psi^{\min} = \psi(0)$.

(iv) If $\psi^{\max} - \psi^{\min} > \hat{\varepsilon}$, then go to step (ii), otherwise STOP.

Discussion

Step 1 of Algorithm 4.10 gives a closed form optimal control, if the problem can be solved as though it is uncapacitated. If, however, the exact optimal solution cannot be found at Step 1, an effective approximation with a given accuracy is provided at Step 2 of the algorithm for an arbitrary demand. This step is similar to Step 2 of Algorithm 3.7, and uses the dichotomous search for locating $\psi(0)$ - the missing initial condition of the system of the state and co-state differential equations, (4.45) and (4.46).

LEMMA 4.21. Given problem (4.43), (4.2) and (4.3), the co-state variable $\psi(t)$ is bounded as $\psi(t) \in [0, \hat{\psi}]$ for $t \in [0,T]$, where $\hat{\psi} = 2c_x^- \int_0^T \int_0^t d(\tau)d\tau dt$.

PROOF. The proof is identical to that of Lemma 3.13. ■

LEMMA 4.22. Given problem (4.43), (4.2) and (4.3), slack function F_5 is non-decreasing.

PROOF. Since the proof is analogous to that for Lemma 3.7, it is omitted here.■

LEMMA 4.23. (Correctness of Algorithm 4.10 Step 2) Given a solution $(X(t), \psi(t))$ of system (4.45)-(4.46) such that $F_5(\psi(0)) = 0$, then the optimal control is

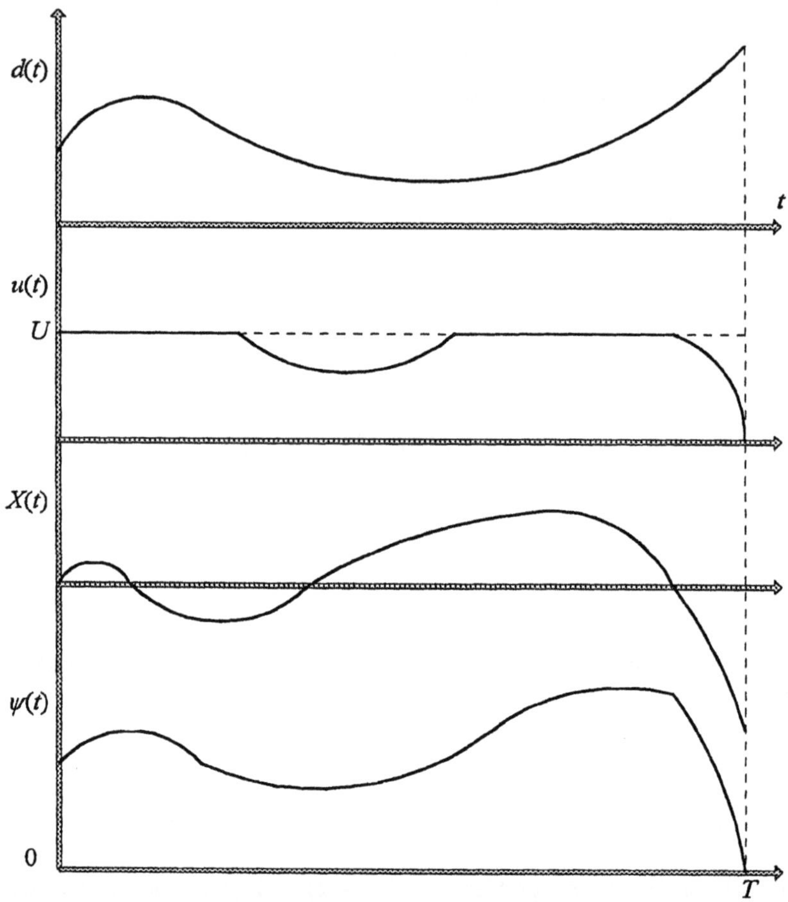

Figure 4.7. Demand $d(t)$, optimal control $u(t)$, state $X(t)$ and co-state $\psi(t)$ variables for the $1|1|u(t)\in[0,U]\,|d(t),T|\,Q_u, Q_x$ problem.

$$u(t)=\begin{cases}\dfrac{\psi(t)}{2c_u}, & \text{if } \psi(t)\le 2c_uU,\\ U, & \text{if } 2c_uU\le\psi(t).\end{cases} \tag{4.47}$$

PROOF. Any solution of system (4.45)-(4.46) necessarily meets all conditions of the maximum principle except for the co-state boundary condition. The root of the slack function, $F_5(\psi(0))=0$, implies that the boundary condition $\psi(T)=0$ is also satisfied, i.e. the optimal control is determined by (4.34), as stated in the lemma (see Figure 4.7).

THEOREM 4.6. Problem (4.43), (4.2) and (4.3), is solvable with accuracy ξ in $O\left(\frac{2T^2 g^2 c_u U}{\hat{\psi}}\frac{\hat{R}}{\xi}\log\frac{\hat{R}}{\xi}\right)$ time,

where $\hat{R} = 2TU\hat{\psi}ch(gT)\left(1+g^2T^2\right)$ and $g = \sqrt{\max\left\{\frac{c_x^+}{c_u}, \frac{c_x^- D(T)}{c_u UT}\right\}}$.

PROOF. First note, that the complexity of Algorithm 4.10 Step 2 is the same as that of Algorithm 3.7 Step 2, i.e. $O\left(\frac{T}{\varepsilon}\log\frac{\hat{\psi}}{\hat{\varepsilon}}\right)$.

To estimate the relationship between computational accuracy of the objective ξ, accuracy of integration ε and accuracy of the dichotomous search $\hat{\varepsilon}$, we vary the objective function (4.43)

$$\xi \le 2T\max\{c_x^+ \max_t X^+(t), c_x^- \max_t X^-(t)\}\max_t|\Delta X(t)| + 2Tc_u \max_t u(t)\max_t|\Delta u(t)|. \quad (4.48)$$

The value of $\max_t u(t)$ is evidently *U*. The maximum inventory and backlog are easy to estimate as

$$\max_t X^+(t) = T\max_t u(t) = TU \quad \text{and} \quad \max_t X^-(t) = \int_0^T d(t)dt = D(T). \quad (4.49)$$

The value of $\max_t|\Delta u(t)|$ is also estimated from (4.34) as $\frac{1}{2c_u}\max_t|\Delta\psi(t)|$. To estimate the error of the co-state variable, we solve the following differential equation

$$\Delta\ddot{\psi}(t) = g^2\Delta\psi(t), \quad \Delta\psi(0) = 2\hat{\varepsilon}, \quad \Delta\dot{\psi}(0) = 0,$$

where $\hat{\varepsilon}$ is determined as

$$\hat{\varepsilon} = \max_t \dot{\psi}(t)\varepsilon = 2\max\{c_x^+ \max_t X^+(t), c_x^- \max_t X^-(t)\}\varepsilon = 2UTc_u g^2\varepsilon. \quad (4.50)$$

Thus, as with Theorem 3.3, we find $\Delta\psi(t) = 2ch(gt)\hat{\varepsilon}$, and $\max_t|\Delta\psi(t)| = 2ch(gT)\hat{\varepsilon}$.

The upper estimates for $\max_t|\Delta u(t)|$ and $\max_t|\Delta X(t)|$ are $\max_t|\Delta u(t)| = \frac{1}{2c_u}\max_t|\Delta\psi(t)|$

and $\max_t |\Delta X(t)| = \frac{T}{2c_u} \max_t |\Delta \psi(t)|$ respectively. By substituting the found estimates into (4.49), we obtain the worst estimate of ξ as a function of $\hat{\varepsilon}$

$$\xi = 2TUch(gT)\left(1+g^2T^2\right)\hat{\varepsilon}\,. \tag{4.51}$$

To obtain the final estimates of the complexity and accuracy for Algorithm 3.7 Step 2, we express $\hat{\varepsilon}$ from (4.51) and ε from (4.50), substitute them into (3.39) and introduce parameter $\hat{R}$ to simplify the presentation.
■

Special Cases of Zero Costs

The problem with no production cost, $1|1|u(t) \in [0,U]\,|d(t),T|\,Q_x$, has been solved in Section 4.1.2. The problem with no inventory cost, $1|1|u(t) \in [0,U]\,|d(t),T|\,Q_u, Q_x, c_x^+ = 0$ is solved by Algorithm 4.10 with no modifications.

No Backlog

To solve the $1|1|u(t) \in [0,U], X(t) \geq 0\,|d(t),T|\,Q_u, Q_x$ problem, we reduce it to a search for the shortest path on a 3-weighted graph. Prior to defining the graph and the algorithm for the search, we formulate the dual formulation for the no-backlog problem.

Dual Formulation

Now problem (4.43), (4.2) and (4.3) has the additional constraint $X(t) \geq 0$ which affects neither Hamiltonian (4.44) nor optimality conditions (4.34), but changes the co-state differential equation

$$d\psi(t) = c_x^+ X(t)dt - d\mu(t), \quad \psi(T+0) = 0,$$

where Lagrange multiplier $d\mu(t) \geq 0$ corresponds to the no-backlog constraint and satisfies complementary slackness condition (4.16).

Algorithm 4.11

Step 1:

Run Algorithm 3.8. If $u(t) \leq U$ for $0 \leq t \leq T$, then the found solution is optimal; STOP.

Otherwise go to the next step.

Step 2:

Set accuracy ξ required for computing the optimal solution. Calculate parameters $y = \sqrt{\frac{c_x^+}{c_u}}$ and $B = \max_t \dot{d}(t)$. Select time intervals

$\tau_k = [t_{left}^k, t_{right}^k]$ at each of which the demand function decreases, $t_{right}^k < t_{left}^{k+1}$, and number them beginning from k=2. In addition, define τ_1 and τ_K as the intervals which are degenerated to a point, $\tau_1 = \{0\}$ and $\tau_K = \{T\}$, where K is the total number of obtained intervals.

Calculate accuracy of integration $\varepsilon = \left(2TUBKc_u ch(yT)\left(1+y^2T^2\right)\right)^{-1}\xi$.

Step 3 and Step 4 are identical to Steps 2 and 3 of Algorithm 3.4.

Subroutine 1 to Algorithm 4.11

(i) Set time points $t_x = t_{left}^n$, $t_y = t_{right}^n$, $t_{out} = t_{left}^n$.

(ii) Integrate the following system of differential equations

$$\dot{X}(t) = \begin{cases} \dfrac{\psi(t)}{2c_u} - d(t), & \text{if } 0 \le \psi(t) \le 2c_u U, \\ U - d(t), & \text{if } 2c_u U \le \psi(t), \end{cases} \qquad X(t_{out}) = 0,$$

$$\dot{\psi}(t) = 2c_x^+ X(t), \qquad \psi(t_{out}) = 2c_u d(t_{out}),$$

over the interval $[t_{out}, t_{right}^m]$.

If $X(t) < 0$ for any $t < t_{left}^m$, then no arc links nodes τ_n and τ_m; STOP.

Otherwise, set $t_{out} = t_{right}^n$ and integrate the above system.

If $X(t) > 0$ for all $t \in \tau_m$, then no arc links nodes τ_n and τ_m; STOP.

Otherwise go to (iii)

(iii) Integrate the system from (ii) with $t_{out} = 0.5(t_x + t_y)$;

If $X(t) < 0$ for any $t < t_{left}^m$ then $t_y = t_{out}$ and go to (iv),

otherwise set time point $t_{in} \in \tau_m$ such that $X(t_{in}) = 0$.

If $\psi(t_{in}) > 2c_u d(t_{in})$, then $t_x = t_{out}$ and go to (iv);

if $\psi(t_{in}) < 2c_u d(t_{in})$, then $t_y = t_{out}$ and go to (iv).

(iv) If $t_y - t_x \le \varepsilon$, then set $arc_{nm} = (\tau_n, \tau_m) \in A$ and its three weights

$a(arc_{nm}) = t_{out}$, $b(arc_{nm}) = t_{in}$ and

$$c(arc_{nm}) = \int_{t_{out}}^{t_{in}} c_x^+ X(t)^2 + c_u u(t)^2\, dt \text{ ; go to (v).}$$

Otherwise go to (iii).

(v) If there exist arcs arc_{jn} such that $\left|b(arc_{jn})-a(arc_{nm})\right|<\varepsilon$, then set $a(arc_{nm})=\max_j b(arc_{jn})$. For each arc arc_{jn} entering node n, update weights

$$SD_{jm}^{n}=\begin{cases}\displaystyle\int_{b(arc_{jn})}^{a(arc_{nm})} c_u d(t)^2\,dt, & \text{if } a(arc_{nm})\geq b(arc_{jn});\\ -1, & \text{otherwise.}\end{cases}\quad \text{STOP.}$$

Otherwise, STOP.

Discussion

Step 1 of Algorithm 4.11 solves the problem exactly as though it is uncapacitated and then verifies whether the capacity is exceeded or not. The other steps of Algorithm 4.11 which are similar to Algorithm 3.4, approximate the globally optimal solution if feasibility condition (4.14) is satisfied. At Step 2, the condition of selecting time intervals τ_k at which the fractional production can occur is changed. The other two steps are identical to those of Algorithm 3.4. However, Step 3 invokes four subroutines which are now slightly changed. Namely, the no-backlog condition is additionally checked and the co-state differential equation integrated iteratively in the subroutines is modified as presented in procedure (ii) of Subroutine 1 to Algorithm 4.11. Subroutines 2-4 to Algorithm 4.11 are omitted since they are modified accordingly.

Note, that with respect to (4.16), $d\mu(t)=0$ at time intervals of positive inventories. Although jumps of the μ-function, $d\mu(t)>0$, at points $t=t_{out}$ and $t=t_{in}$ are formally possible, its value is zero, which is proved analogously to the proof of Lemma 3.12.

The new condition for selecting intervals of possible fractional production $u(t)=d(t)$ is stated in the following lemma.

LEMMA 4.24. (Correctness of Algorithm 4.11 Steps 1 and 2) Given problem (4.43), (4.2) and (4.3), with backlog disallowed, if demand decreases at K intervals of time τ_k, $k=1,\ldots,K$, then the optimal solution contains at most K time intervals of zero inventory, $\chi_k=\{t\in\tau_k \mid X(t)=0\}$.

PROOF. The proof is identical to that of Lemma 3.16. ■

The upper and lower bounds of the co-state variable remain the same as for the uncapacitated problem.

LEMMA 4.25. Given problem (4.32), (4.2) and (4.3) with backlog disallowed, the co-state variable $\psi(t)$ is bounded as $\psi(t)\in[0,2c_u\max_{t'} d(t')]$ for $t\in[0,T]$.

PROOF. The proof is identical to that of Lemma 3.17. ■

THEOREM 4.7. Given the demand as a differentiable function of time which satisfies (4.14) and $\dot{d}(t) \le B$ for $t \in [0,T]$, then problem (4.43), (4.2) and (4.3), with disallowed backlog is solvable with accuracy ξ in $O\left(\max\{K^4, \frac{K^3 R}{\xi}\log\frac{KR}{\xi}\}\right)$ time, where K is the number of time intervals at which demand decreases, $R = 2UBT^2 c_u ch(yT)\left(1+y^2T^2\right)$ and $y = \sqrt{\frac{c_x^+}{c_u}}$.

PROOF. Complexity estimate (3.25) is valid for Algorithm 4.11, since only Step 1, which needs $O(K)$ operations, has been changed in comparison with Algorithm 3.4.

However, the complexity estimate involves the accuracy of the algorithm. To find the new estimate, we evaluate the relationship between required accuracy ξ of the objective and computational accuracy ε of the dichotomous search. By varying the objective function (4.43), we obtain (4.48) and estimate its terms.

The value of $\max_t u(t)$ is evidently U. Thus, the upper estimate of the maximum value for $X(t)$ is TU. The value of $\max_t |\Delta u(t)|$ is also estimated from (4.34) as $\frac{1}{2c_u}\max_t|\Delta\psi(t)|$. From (4.45)-(4.46), the error of the co-state variable $\Delta\psi(t)$ in the dichotomous search satisfies

$$\Delta\ddot{\psi}(t) = \frac{c_x^+}{c_u}\Delta\psi(t)$$

for each link starting at point t_{out} at Step 2 of the algorithm. The boundary conditions for this second-order differential equation are $\Delta\psi(t_{out}) = 2c_u\dot{d}(t_{out})\varepsilon$ and $\Delta\dot{\psi}(t_{out}) = 0$. By solving this equation, we obtain $\Delta\psi(t) = 2c_u\dot{d}(t_{out})\varepsilon ch(y(t-t_{out}))$. By taking into account that there can be at most K links along the planning horizon T, $\max_t|\Delta\psi(t)| \le 2c_u BK\varepsilon ch(yT)$. The value of $\max_t|\Delta X(t)|$ is just $T\max_t|\Delta u(t)| = TBK\varepsilon ch(yT)$. By substituting the found estimates into (4.48), the worst accuracy estimate ξ is obtained as a function of ε

$$\xi = 2TUBKc_u ch(yT)\left(1+y^2T^2\right)\varepsilon. \tag{4.52}$$

To determine the complexity and accuracy of Algorithm 4.11, we express ε from (4.52), substitute it into (3.25) and introduce parameter R to simplify the presentation.

■

4.2 DUE DATE

This section addresses the scheduling problems of meeting a demand required by a due date under a limited production capacity.

4.2.1 LINEAR PERFORMANCE MEASURE

Backlogs

The problem to be considered is one-machine, one-product-type scheduling with limited production rate and backlogs to minimize the total linear cost of inventory and production, that is, the $1|1|u(t)\in[0,U]|D,T|L_u,L_x$ problem which is stated as follows

$$J=\int_0^T\left[c_x^+Z^+(t)+c_x^-Z^-(t)+c_uu(t)\right]dt\to\min, \tag{4.53}$$

$$Z^+(t)=\max\{0,X(t)-D(t)\},\, Z^-(t)=\max\{0,D(t)-X(t)\},$$

subject to

$$\dot{X}(t)=u(t),\qquad X(0)=X_0, \tag{4.54}$$

$$0\le u(t)\le U, \tag{4.55}$$

where

$$D(t)=\begin{cases}D\ , & \text{if } t\ge t_d;\\ 0, & \text{otherwise;}\end{cases} \tag{4.56}$$

and D is the amount of the product-type to be produced by the due date t_d .

Dual Formulation

The Hamiltonian, H, resulting from the maximum principle is

$$H(t)=-\left[c_x^+Z^+(t)+c_x^-Z^-(t)+c_uu(t)\right]+\psi(t)u(t). \tag{4.57}$$

The co-state variable, $\psi(t)$ in (4.57), satisfies the following co-state differential equation with its transversality condition

$$\dot{\psi}(t) = \begin{cases} c_x^+, \text{ if } X(t) > D(t); \\ -c_x^-, \text{ if } X(t) < D(t); \\ a,\ a \in [-c_x^-, c_x^+], \text{ if } X(t) = D(t); \end{cases} \qquad \psi(T) = 0. \tag{4.58}$$

Maximization of the Hamiltonian implies inequality (4.6), that is, if $\psi(t) = c_u$ holds at an interval of time, then the optimal control $u(t)$ can be arbitrary within the control limits (4.55) at this interval (fractional production regime). Otherwise, if $\psi(t) < c_u$, no control is undertaken, $u(t)$=0; if $\psi(t) > c_u$, production at the maximum rate is optimal, $u(t)$=U.

Algorithm 4.12

Step 1:

Calculate time points $t_1 = t_d - \dfrac{c_x^- D}{(c_x^+ + c_x^-)U}$ and $t_2 = t_d + \dfrac{c_x^+ D}{(c_x^+ + c_x^-)U}$.

If $t_2 + \dfrac{c_u}{c_x^-} \le T$ and $t_1 \ge 0$, then set the optimal solution as

$u(t) = 0$ for $0 \le t < t_1$ and $t_2 < t \le T$,

$u(t) = U$ for $t_1 \le t \le t_2$; STOP.

Otherwise go to the next step.

Step 2:

Assign time point $t_2 = \dfrac{D}{U}$.

If $t_2 + \dfrac{c_u}{c_x^-} \le T$ and $t_1 \le 0$, then assign $t_1 = 0$ and set the optimal solution as

$u(t) = U$ for $t_1 \le t \le t_2$,

$u(t) = 0$ for $t_2 < t \le T$; STOP.

Otherwise, solution $u(t)$=0 over the entire planning horizon is optimal, STOP.

Example 4.4

Consider the $1|1|u(t) \in [0,U]\,|D,T|\,L_u, L_x$ problem with parameters T=10hr, U=3unit/hr, c_u=1\$/unit, c_x^+=0.2\$/hr.unit, c_x^-=1\$/hr.unit and demand D=15units and $t_d = 7$ hr.

Step 1 of Algorithm 4.12 calculates time points $t_1 = t_d - \frac{c_x^- D}{(c_x^+ + c_x^-)U} = 2.83$ and $t_2 = t_d + \frac{c_x^+ D}{(c_x^+ + c_x^-)U} = 7.83$. Since $t_2 + \frac{c_u}{c_x^-} \le T$ and $t_1 \ge 0$, this step sets the optimal solution as

$u(t) = 0$ for $0 \le t < 2.83$ and $7.83 < t \le 10$,

$u(t) = 3$ unit/hr for $2.83 \le t \le 7.83$.

■

Discussion

Algorithm 4.12 solves the problem in $O(1)$ time. Note that in contrast to the problems dealing with general demands, no fractional production regime occurs if the demand is concentrated at a due date. Furthermore, the full production regime, if started, is never preempted until the demand is completed.

Next, no production at all may become profitable if either the due date is close to the end of the planning horizon, so that $t_d + \frac{c_x^+ D}{(c_x^+ + c_x^-)U} + \frac{c_u}{c_x^-} > T$, or the planning horizon and, thus, the due date, are quite short, so that $t_d < \frac{c_x^- D}{(c_x^+ + c_x^-)U}$ and $T < \frac{D}{U} + \frac{c_u}{c_x^-}$.

LEMMA 4.26. (Correctness of Algorithm 4.12) Given problem (4.53)-(4.55), $t_1 = t_d - \frac{c_x^- D}{(c_x^+ + c_x^-)U}$, $t_2 = t_d + \frac{c_x^+ D}{(c_x^+ + c_x^-)U}$ and $t_3 = \frac{D}{U}$, the optimal control is determined as follows

- if $t_2 + \frac{c_u}{c_x^-} \le T$ and $t_1 \ge 0$, then

$u(t) = 0$ for $0 \le t < t_1$ and $t_2 < t \le T$; $u(t) = U$ for $t_1 \le t \le t_2$;

- if $t_3 + \frac{c_u}{c_x^-} \le T$ and $t_1 \le 0$, then

$u(t) = U$ for $0 \le t < t_3$, $u(t) = 0$ for $t_3 \le t \le T$;

- otherwise, $u(t)=0$ for $0 \le t \le T$.

PROOF. First note that the fractional production regime condition $\psi(t) = c_u$, at an interval of time, implies $\dot{\psi}(t) = 0$, which can be met only if $X(t) = D(t)$. Thus, due to the special form of the demand (4.56), the control effort over the fractional regime is necessarily equal to zero before and after the due date.

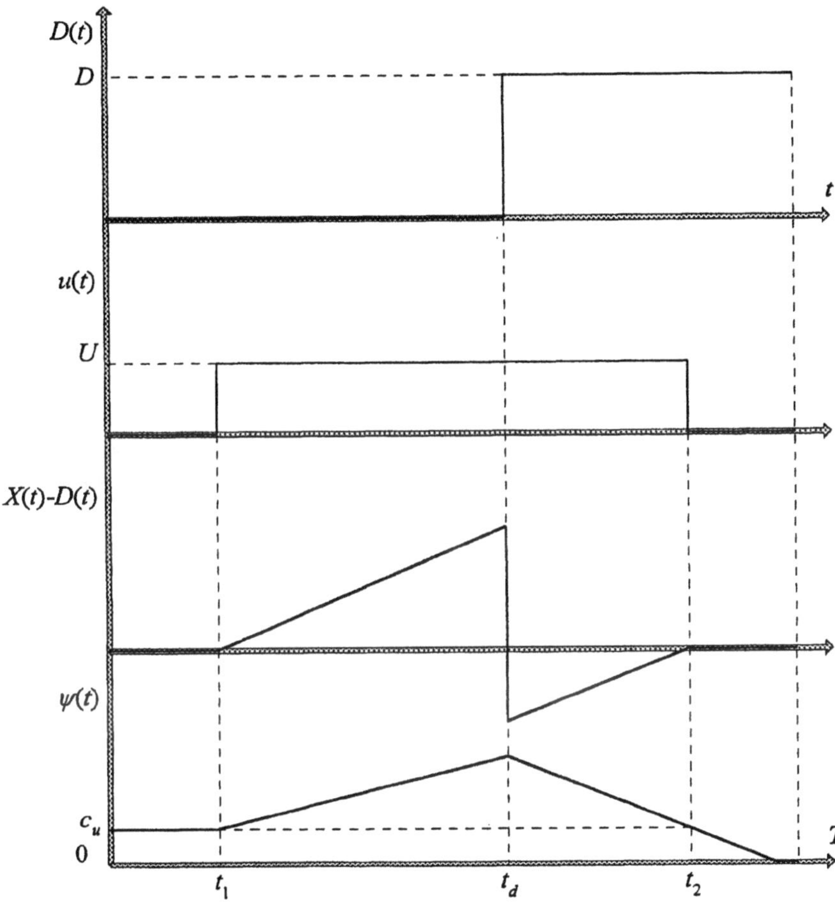

Figure 4.8. Demand $D(t)$, optimal control $u(t)$, state $X(t)$ and co-state $\psi(t)$ variables for a special case of the $1|1|u(t)\in[0,U]|D,T|\,L_u, L_x$ problem.

Consider the solution which includes a sequence of regimes depicted in Figure 4.8: fractional production, full production and no-production,

$$\psi(t)=c_u,\ \ 0\le t\le t_1;\ \ \psi(t)=c_u+c_x^+(t-t_1),\ t_1\le t\le t_d,$$
$$\psi(t)=c_x^-(\hat{T}-t),\ \ t_d\le t\le \hat{T},\ \psi(t)=0,\ \ \hat{T}\le t\le T; \tag{4.59}$$

$$X(t)=0,\ \ u(t)=0,\ \ 0\le t\le t_1;\ \ X(t)=U(t-t_1),\ u(t)=U, t_1\le t\le t_2,$$
$$X(t)=D,\ \ u(t)=0,\ \ t_2\le t\le T. \tag{4.60}$$

From (4.59) and (4.60), we conclude that the time points where the regimes change over satisfy the following conditions

$$X(t_2)=U(t_2-t_1)=D,\ \ \psi(t_d)=c_u+c_x^+(t_d-t_1)=c_x^-(\hat{T}-t_d),$$
$$\psi(t_2)=c_x^-(\hat{T}-t_2)=c_u. \tag{4.61}$$

Equations (4.61) constitute a system of three linear equations in three unknowns t_1, t_2 and $\hat{T}$.

Finally, the considered sequence of regimes is optimal, i.e. it satisfies the primal and dual formulations (4.54), (4.55), (4.58) and (4.6), if the points where the regimes change over are feasible, i.e. $t_1\ge 0$, $t_1<t_2<\hat{T}$ and $\hat{T}\le T$. By substituting the solution of (4.61) into the last inequalities, we prove the first condition of the lemma.

The other two conditions of the lemma are similarly proved by considering the sequence of the full production and no-production regimes or no-production at all.
■

Special Cases of Zero Costs

The no production cost problem, $1|1|u(t)\in[0,U]|D,T|L_x$, and the no inventory cost problem, $1|1|u(t)\in[0,U]|D,T|L_u, L_x, c_x^+=0$, are straightforwardly solved by Algorithm 4.12.

No Backlog

For the case when backlogs are disallowed, i.e. for the $1|1|u(t)\in[0,U], X(t)\ge 0|D,T|L_u, L_x$ problem, the solution algorithm can be obtained by an easy modification to Algorithm 3.9. Indeed, for the problem to have a feasible solution, it is necessary that $t_d\ge\frac{D}{U}$. Therefore, Step 2 of Algorithm 4.12 will never be executed, while when c_x^- is very large, the first step of Algorithm 4.12 results in $u(t)=U$ for $t_1\le t\le t_2$, where $t_1=t_d-\frac{D}{U}$ and $t_2=t_d$.

4.2.2 NON-LINEAR INVENTORY COST

It will be shown below that introducing non-linear inventory cost does not influence the optimal control regimes. However, this does affect the optimal control value. Unlike the case with the linear costs, a trade-off between the production and the backlog costs is provided by partial production. When, however, the backlog is disallowed, the optimal solution is very similar to that derived for the linear cost case.

Backlogs

The problem to be considered is one-machine, one-product-type scheduling, with limited production rate and backlogs to minimize the quadratic inventory cost and linear production cost, that is, the $1|1|u(t)\in[0,U]|D,T|\,L_u,Q_x$ problem which is stated as follows

$$J=\int_0^T\left[c_x^+Z^+(t)^2+c_x^-Z^-(t)^2+c_u u(t)\right]dt\to\min, \tag{4.62}$$

subject to (4.54) and (4.55).

Dual Formulation

The Hamiltonian, H, resulting from the maximum principle is

$$H(t)=-\left[c_x^+Z^+(t)^2+c_x^-Z^-(t)^2+c_u u(t)\right]+\psi(t)u(t). \tag{4.63}$$

The co-state variable, $\psi(t)$ in (4.63), satisfies the following co-state differential equation with its transversality condition

$$\dot{\psi}(t)=\begin{cases}2c_x^+(X(t)-D(t)), & \text{if } X(t)>D(t);\\ 2c_x^-(X(t)-D(t)), & \text{if } X(t)\le D(t);\end{cases}\qquad \psi(T)=0. \tag{4.64}$$

Maximization of the Hamiltonian implies that the control function $u(t)$ must satisfy (4.6).

Algorithm 4.13

Step 1:

If $c_u\ge 2c_x^-D(T-t_d)$, then solution $u(t)=0$, $0\le t\le T$, is optimal, STOP; otherwise go to the next step.

Step 2:

Calculate $t_1=\dfrac{1}{2}\left(\dfrac{D}{U}+T-\sqrt{\left(\dfrac{D}{U}-T\right)^2+\dfrac{2c_u}{c_x^-U}}\right)$.

If $\left(c_x^--c_x^+\right)Ut_d^2-c_x^-Ut_1^2+2c_x^-D(t_1-t_d)\ge 0$ and $t_d<t_1$, set the optimal solution as

$u(t)=U$ for $0\le t<t_1$,

$u(t)=0$ for $t_1\le t\le T$, STOP.

Otherwise go to the next step.

Step 3:

(i) Set accuracy ξ required for approximating the objective of the optimal solution. Calculate parameter $g = \dfrac{\max\{c_x^+ UT, c_x^- D\}}{c_u}$ and accuracy of the dichotomous search $\varepsilon = \left(2Uc_u\left(1+2gT\right)\right)^{-1}\xi$.

(ii) Define function $t_2(t_1) = \dfrac{1}{2}\left(t_1 + \dfrac{D}{U} + T - \sqrt{\left(t_1 + \dfrac{D}{U} - T\right)^2 + \dfrac{2c_u}{c_x^- U}}\right)$.

Set time points $t_x = t_d$, $t_y = T$, $\hat{t}_x = t_2(t_x)$ and $\hat{t}_y = t_2(t_y)$.

(iii) Calculate $t_1 = 0.5\left(t_x + t_y\right)$ and

$f = U\left(c_x^+ - c_x^-\right)\left(t_d - t_1\right)^2 + Uc_x^-\left(t_2(t_1) - t_1\right)^2 - 2c_x^- D\left(t_2(t_1) - t_d\right)$.

If $f > 0$, then $t_x = t_1$, $\hat{t}_x = t_2(t_x)$ and go to (iv);
if $f < 0$, then $t_y = t_1$, $\hat{t}_y = t_2(t_y)$ and go to (iv).

(iv) If $t_y - t_x \le \varepsilon$ and $\hat{t}_y - \hat{t}_x \le \varepsilon$, then set the optimal solution as
$u(t) = 0$ for $0 \le t < t_1$ and $t_2(t_1) \le t < T$,
$u(t) = U$ for $t_1 \le t < t_2(t_1)$, STOP.
Otherwise go to (iii).

Example 4.5

Consider the $1|1|u(t) \in [0,U]\,|D,T|\,L_u, L_x$ problem with parameters T=10hr, U=3unit/hr, c_u=1\$/unit, c_x^+=0.02\$/hr.unit^2, c_x^-=1\$/hr.unit^2 and demand D=27units and $t_d = 7$ hr.

Since $c_u \ge 2c_x^- D(T - t_d)$ is not true, Algorithm 4.13 proceeds from Step 1 to Step 2. Step 2 then calculates $t_1 = \dfrac{1}{2}\left(\dfrac{D}{U} + T - \sqrt{\left(\dfrac{D}{U} - T\right)^2 + \dfrac{2c_u}{c_x^- U}}\right) = 8.85$ hr.

Since conditions $\left(c_x^- - c_x^+\right)Ut_d^2 - c_x^- Ut_1^2 + 2c_x^- D(t_1 - t_d) \ge 0$ and $t_d < t_1$ are true, the optimal solution is set as
$u(t)$= 3unit/hr for $0 \le t < 8.85$,
$u(t)$=0 for $8.85 \le t \le 10$.
■

Discussion

Steps 1 and 2 of Algorithm 4.13 present two exactly solvable cases of $O(1)$ complexity. The first solvable case is no-production at all, which becomes profitable when the due date is close to the end of the planning horizon, so that $\frac{c_u}{2c_x^- D} \geq T - t_d$.

The other solvable case involves only two optimal control regimes: full production and no-production. Step 3 of the algorithm complements the cases solved in Steps 1 and 2. It constructs the solution which involves three regimes: fractional, full and no-production. The regimes are determined analytically, so that there is no need for numerical integration of the state and co-state differential equations. Consequently, calculation of the two time points, where the regimes change over, is reduced to finding a root of a non-linear analytic equation. Since there exists exactly one root of the equation, the dichotomous search provides weakly polynomial, rather than pseudopolynomial complexity, as was the case in Section 4.1.2 (Algorithm 4.6) where the equation was not expressed analytically.

LEMMA 4.27. (Correctness of Algorithm 4.13 Steps 1-2) Given problem (4.62), (4.54) and (4.55) and $t_1 = \frac{1}{2}\left(\frac{D}{U} + T - \sqrt{\left(\frac{D}{U} - T\right)^2 + \frac{2c_u}{c_x^- U}}\right)$, the optimal control is determined as follows

- if $c_u \geq 2c_x^- D(T - t_d)$, then $u(t)$=0 for $0 \leq t \leq T$;
- if $\left(c_x^- - c_x^+\right)Ut_d^2 - c_x^- Ut_1^2 + 2c_x^- D(t_1 - t_d) \geq 0$ and $t_d < t_1$, then

$u(t)=U$ for $0 \leq t < t_1$ and $u(t)$=0 for $t_1 \leq t \leq T$.

PROOF. Consider the no-production solution

$$u(t) = 0,\ X(t) = 0,\ \psi(t) \leq c_u \ \text{ for } 0 \leq t \leq T. \tag{4.65}$$

According to (4.64) and (4.65), the co-state variable behaves as

$$\psi(t) = 2c_x^- D(T - t_d) \ \text{ for } 0 \leq t \leq t_d \text{ and } \psi(t) = 2c_x^- D(T - t) \ \text{ for } t_d \leq t \leq T.$$

The solution is feasible if $\psi(t)$ does not exceed the value of c_u, i.e. $c_u \geq 2c_x^- D(T - t_d)$. This proves the first condition of the lemma.

To prove the second condition of the lemma, we consider a solution which changes from the full production to the no-production regime at point t_1

$$u(t) = U,\ X(t) = Ut,\ \psi(t) \geq c_u \ \text{ for } 0 \leq t \leq t_1. \tag{4.66}$$

$$u(t)=0\,,\ X(t)=Ut_1\,,\ \psi(t)<c_u \text{ for } t_1\le t\le T\,. \tag{4.67}$$

Conditions (4.66) and (4.67) imply that $\psi(t_1)=c_u$ which, together with the transversality condition $\psi(T)=0$, forms two equations in two unknowns $\psi(0)$ and t_1

$$2c_x^-(T-t_1)(D-Ut_1)=c_u,$$

$$\psi(0)+c_x^+Ut_d^2+\int_{t_d}^{t_1}2c_x^-(Ut-D)dt=c_u\,.$$

By taking the left root of the first equation, we ensure that $t_1\le T$ and $t_1\le\frac{D}{U}$. The requirements $\psi(0)\ge c_u$ (see (4.66)) and $t_1>t_d$ ensure the considered solution is feasible that proves the second part of the lemma.

■

Definition 4.5. Given the following system of differential equations

$$\dot X(t)=\begin{cases}U, \text{ if } \psi(t)\ge c_u;\\ 0, \text{ if } \psi(t)<c_u;\end{cases}\quad \dot\psi(t)=\begin{cases}2c_x^+(X(t)-D(t)), \text{ if} X(t)>D(t);\\ 2c_x^-(X(t)-D(t)), \text{ if} X(t)\le D(t);\end{cases}$$

with initial conditions $X(t_1)=0$ and $\psi(t_1)=c_u$, F_{10} is a slack function which maps $t_1\to\psi(T)$.

LEMMA 4.28. Slack function F_{10} is a non-increasing function of its argument.

PROOF. The proof is similar to that of Lemma 3.7. ■

LEMMA 4.29. (Correctness of Algorithm 4.13 Step 3) Given problem (4.62), (4.54), (4.55) and $\bar t=\frac{1}{2}\left(\frac{D}{U}+T-\sqrt{\left(\frac{D}{U}-T\right)^2+\frac{2c_u}{c_x^-U}}\right)$, if $c_u<2c_x^-D(T-t_d)$ and either

$$\left(c_x^- -c_x^+\right)Ut_d^2-c_x^-U\bar t^{\,2}+2c_x^-D(\bar t-t_d)<0 \text{ or } t_d\ge\bar t\,,$$

then the optimal control is determined as follows
$u(t)$=0 for $0\le t\le t_1$ and $t_2\le t\le T$;
$u(t)$=U for $t_1\le t<t_2$, where

$t_2 = \frac{1}{2}\left(t_1 + \frac{D}{U} + T - \sqrt{\left(t_1 + \frac{D}{U} - T\right)^2 + \frac{2c_u}{c_x^- U}}\right)$ and t_1 is the root of the equation

$f(t_1) = U\left(c_x^+ - c_x^-\right)\left(t_d - t_1\right)^2 + Uc_x^-\left(t_2 - t_1\right)^2 - 2c_x^- D\left(t_2 - t_d\right) = 0$.

PROOF. From Lemma 4.27, it follows that the conditions of this lemma ensure the optimal solution to include three regimes: fractional, full and no-production as depicted in Figure 4.9,

$$u(t) = 0,\ X(t) = 0,\ \psi(t) = c_u \text{ for } 0 \le t \le t_1,$$

$$u(t) = U,\ X(t) = Ut,\ \psi(t) \ge c_u \text{ for } t_1 \le t \le t_2, \tag{4.68}$$

$$u(t) = 0,\ X(t) = U(t_2 - t_1),\ \psi(t) < c_u \text{ for } t_2 \le t \le T.$$

The last condition in (4.68) implies that $\psi(t_2) = c_u$, i.e.

$$\psi(t_2) = -2c_x^-(T - t_2)(U(t_2 - t_1) - D) = c_u. \tag{4.69}$$

From the second condition in (4.68), we conclude that

$$\int_{t_1}^{t_2} \dot{\psi}(t)dt = 0. \tag{4.70}$$

By integrating (4.70) with respect to the co-state equation (4.64), we obtain the following algebraic equation in unknowns t_1 and t_2

$$c_u + U\left(c_x^+ - c_x^-\right)\left(t_d - t_1\right)^2 + Uc_x^-\left(t_2 - t_1\right)^2 - 2c_x^- D\left(t_2 - t_d\right) = c_u. \tag{4.71}$$

By taking into account (4.69), t_2 depends on t_1 and thus, the left-hand side of (4.71) can be denoted as function $f(t_1)$. According to (4.70) and Definition 4.5, this function coincides with slack function F_{10} at interval $[t_1, t_2]$. Finally, since F_{10} is proved to be monotone, $f(t_1)$ is also monotone at interval $[t_1, t_2]$ and, hence, equation (4.71) has a unique root that proves the lemma.

■

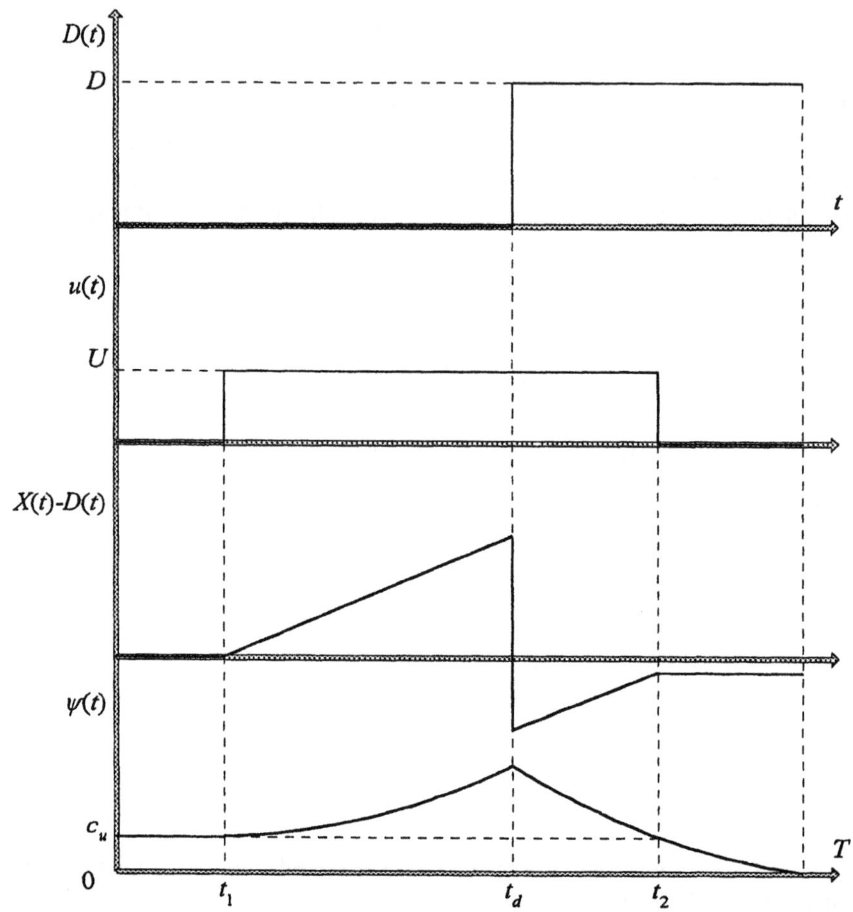

Figure 4.9. Demand $D(t)$, optimal control $u(t)$, state $X(t)$ and co-state $\psi(t)$ variables for a special case of the $1|1|u(t)\in[0,U]\,|D,T|\,L_u,Q_x$ problem.

THEOREM 4.8. Problem (4.62), (4.54) and (4.55) is solvable with accuracy ξ in $O\left(\log\frac{R}{\xi}\right)$ time, where $R=2TUc_u(1+2gT)$ and $g=\dfrac{\max\{c_x^+UT,c_x^-D\}}{c_u}$.

PROOF. First note that Steps 1 and 2 of Algorithm 4.13 present the cases when the problem is exactly solvable in $O(1)$ time. If, however, the conditions of these two steps are not met, the third step utilizes the dichotomous search for the root of slack function F_{10} with a given accuracy ε in $O\left(\log\frac{T}{\varepsilon}\right)$ operations.

Finally, we estimate the relationship between computational accuracy of the objective ξ and accuracy of the dichotomous search ε. By varying objective function (4.62), we determine

$$\xi \le 2T\max\{c_x^+ \max_t Z^+(t), c_x^- \max_t Z^-(t)\} \max_t |\Delta Z(t)| + c_u \int_0^T |\Delta u(t)| dt\,. \tag{4.72}$$

Since the dichotomous search in Step 3 ensures that the switching points t_1 and t_2 are calculated with accuracy ε, the cumulative error in control $\int_0^T |\Delta u(t)| dt$ is at most εU for each of these points. Evidently, no other time point results in the control error. The maximum inventory and backlog are easy to estimate as

$$\max_t Z^+(t) = T\max_t u(t) = TU \text{ and } \max_t Z^-(t) = D\,.$$

By using again the fact that the control error is due to the two switching points only, the error in the state variable is accumulated as follows

$$\max_t |\Delta Z(t)| = \max_t |\Delta X(t)| = \max_t \int_0^t |\Delta u(\tau)| d\tau = 2U\varepsilon\,.$$

By substituting the found estimates into (4.72) , we obtain maximal error ξ as a function of ε

$$\xi = 2Uc_u \max\{c_x^+ UT, c_x^- D\}\left(1+2gT\right)\varepsilon\,. \tag{4.73}$$

To obtain the complexity and accuracy of Algorithm 4.13, we express ε from (4.73), substitute it into the complexity of the dichotomous search and introduce parameter R to simplify the presentation.
■

Special Cases of Zero Costs

First note that the $1|1|u(t)\in[0,U]\,|D,T|\,L_u, Q_x, c_x = 0$ problem is solvable by Algorithm 4.13 with no significant simplification. However, the $1|1|u(t)\in[0,U]\,|D,T|\,Q_x$ problem is exactly solvable in $O(1)$ time. Indeed, by setting $c_u = 0$ in Algorithm 4.13, we find that the condition of Step 1 never holds, while Steps 2 and 3 simplify. Specifically, the dichotomous search at Step 3 is no longer needed because the root of the slack function, which now also simplifies, is found analytically. Since such an algorithm is a special case of Algorithm 4.13, we present it below without further discussion.

Algorithm 4.14
Step 1:

Calculate $t_1 = \min\left\{T, \frac{D}{U}\right\}$.

If $\left(c_x^- - c_x^+\right)Ut_d^2 - c_x^- Ut_1^2 + 2c_x^- D(t_1 - t_d) \geq 0$, set the optimal solution as

$u(t)=U$ for $0 \leq t < t_1$,

$u(t)=0$ for $t_1 \leq t \leq T$, STOP.

Otherwise go to the next step.

Step 2:

Find root t_1 of the equation

$$U\left(c_x^+ - c_x^-\right)\left(t_d - t_1\right)^2 + Uc_x^-\left(T - t_1\right)^2 - 2c_x^- D\left(T - t_d\right) = 0.$$

If $T - \frac{D}{U} \leq t_1 \leq t_d$, then set the optimal solution as

$u(t) = 0$ for $0 \leq t < t_1$,

$u(t) = U$ for $t_1 \leq t \leq T$, STOP.

Otherwise, find root t_1 of the equation

$$U\left(c_x^+ - c_x^-\right)\left(t_d - t_1\right)^2 + c_x^- \frac{D^2}{U} - 2c_x^- D\left(t_1 + \frac{D}{U} - t_d\right) = 0.$$

Set the optimal solution as

$u(t) = 0$ for $0 \leq t < t_1$ and $t_1 + \frac{D}{U} \leq t \leq T$.

$u(t) = U$ for $t_1 \leq t < t_1 + \frac{D}{U}$, STOP.

No Backlog

For the problem with disallowed backlogs, i.e. for the $1|1|u(t) \in [0,U], X(t) \geq 0|D,T|L_u, Q_x$ problem, no feasible solution exists if the due date is too short, $t_d < \frac{D}{U}$. Otherwise, if $t_d \geq \frac{D}{U}$, the problem becomes exactly solvable in $O(1)$ time and the solution algorithm is obtained by a modification to Algorithm 4.13. Indeed, when c_x^- is very large, Steps 1 and 2 of Algorithm 4.13 will never be executed, while Step 3 always results in $t_1 = t_d - \frac{D}{U}$ and $t_2 = t_d$.

4.2.3 NON-LINEAR PRODUCTION COST

In this section we consider the combination of a piece-wise linear inventory and backlog cost function and a non-linear production cost function. This change results in a large variety of closed form solutions, depending on the relations between system parameters.

Backlogs

The problem to be considered is one-machine, one-product-type scheduling with backlogs and limited production rate to minimize the piece-wise inventory cost and quadratic production cost, that is, the $1|1|u(t)\in[0,U]|D,T|Q_u, L_x$ problem which is stated as follows

$$J=\int_0^T\left[c_x^+Z^+(t)+c_x^-Z^-(t)+c_u u(t)^2\right]dt\to\min, \tag{4.74}$$

subject to (4.54) and (4.55).

Dual Formulation

The Hamiltonian, H, resulting from the maximum principle is

$$H(t)=-\left[c_x^+Z^+(t)+c_x^-Z^-(t)+c_u u(t)^2\right]+\psi(t)u(t). \tag{4.75}$$

The co-state variable, $\psi(t)$ in (4.75), satisfies the co-state differential equation (4.58). Maximization of the Hamiltonian implies that

$$u(t)=\begin{cases}U, & \text{if } \psi(t)>2c_uU,\\ 0, & \text{if } \psi(t)<0,\\ \dfrac{\psi(t)}{2c_u}, & \text{if } 0\le\psi(t)\le 2c_uU,\end{cases} \tag{4.76}$$

The first two conditions in (4.76), which are similar to those found for the case of linear production cost, determine full production and no-production regimes respectively. The last condition presents the partial production regime which allows production at an intermediate rate less than the capacity U.

Algorithm 4.15

Step 1:

Run Algorithm 3.11. If $u(t)\le U$ for $0\le t\le T$, then the optimal solution found by Algorithm 3.11 is valid. STOP.
Otherwise, go to the next step.

Step 2:

Calculate $t_1 = t_d - \frac{c_x^-}{c_x^+ + c_x^-}\frac{D}{U} - \frac{c_u U}{c_x^+}$.

If $t_1 \geq 0$, then calculate the time required to complete the demand $t_2 = t_d + \frac{c_x^+}{c_x^+ + c_x^-}\frac{D}{U} + \frac{c_u U}{c_x^-}$ and proceed to the next step, otherwise go to Step 4.

Step 3:

If $t_2 \leq T$, then calculate the time points where the full production starts and terminates

$t_x = t_d - \frac{c_x^-}{c_x^+ + c_x^-}\frac{D}{U} + \frac{c_u U}{c_x^+}$ and $t_y = t_d + \frac{c_x^+}{c_x^+ + c_x^-}\frac{D}{U} - \frac{c_u U}{c_x^-}$.

Set the optimal solution as

$u(t) = \frac{c_x^+}{2c_u}(t - t_1), \ t_1 \leq t \leq t_x,$

$u(t) = U, \ t_x \leq t \leq t_y,$

$u(t) = \frac{c_x^-}{2c_u}(t_2 - t), \ t_y \leq t \leq t_2,$

$u(t) = 0, \ 0 \leq t \leq t_1$ and $t_2 \leq t \leq T$; STOP.

Otherwise, $t_2 > T$ and if $t_d - \frac{c_x^-}{c_x^+}(T - t_d) \geq 0$, then calculate the time point where the singular production starts $t_1 = t_d - \frac{c_x^-}{c_x^+}(T - t_d)$ and time points where the full production starts and terminates, $t_x = t_d - \frac{c_x^-}{c_x^+}(T - t_d) + \frac{2c_u U}{c_x^+}$ and $t_y = T - \frac{2c_u U}{c_x^-}$.

Set the optimal solution as

$u(t) = \frac{c_x^+}{2c_u}(t - t_1), \ t_1 \leq t \leq t_x, \ \ u(t) = U, \ t_x \leq t \leq t_y,$

$u(t) = \frac{1}{2c_u}c_x^-(T - t), \ t_y \leq t \leq T$, and $u(t) = 0, \ 0 \leq t \leq t_1$, STOP.

Otherwise go to the next step.

Step 4:

Calculate t_x as a root of the following equation

$$t_x^2+\frac{4c_uU}{c_x^-}t_x-\frac{4c_u^2U^2}{c_x^+c_x^-}+\frac{4c_uD}{c_x^+}-\frac{4c_uUt_d(c_x^++c_x^-)}{c_x^+c_x^-}=0\,.$$

If $0\le t_x\le t_d$ and $t_d+\frac{2c_uU}{c_x^-}-\frac{c_x^+}{c_x^-}(t_x-t_d)\le T$, then calculate

$$t_2=t_d+\frac{2c_uU}{c_x^-}-\frac{c_x^+}{c_x^-}(t_x-t_d) \text{ and } t_y=t_d-\frac{c_x^+}{c_x^-}(t_x-t_d)\,.$$

Set the optimal solution as

$$u(t)=U+\frac{c_x^+}{2c_u}(t-t_x),\ 0\le t\le t_x,$$

$$u(t)=U\,,\ t_x\le t\le t_y,$$

$$u(t)=\frac{c_x^-}{2c_u}(t_2-t),\ t_y\le t\le t_2,$$

$u(t)=0$, $t_2\le t\le T$; STOP.

Otherwise go to the next step.

Step 5:

Calculate $t_x=t_d+\frac{2c_uU}{c_x^+}-\frac{c_x^-}{c_x^+}(T-t_d)$ and $t_y=T-\frac{2c_uU}{c_x^-}$.

If $0\le t_x\le t_d$ then set the optimal solution as

$$u(t)=U+\frac{c_x^+}{2c_u}(t-t_x),\ 0\le t\le t_x,$$

$$u(t)=U\,,\ t_x\le t\le t_y,$$

$u(t)=\frac{c_x^-}{2c_u}(T-t),\ t_y\le t\le T$, STOP.

Otherwise, go to the next step.

Step 6:

Calculate $t_2=\frac{c_uU}{c_x^-}+\frac{D}{U}$ and $t_y=-\frac{c_uU}{c_x^-}+\frac{D}{U}$.

If $t_2\le T$, set the optimal solution as

$$u(t)=U\,,\ 0\le t\le t_y,$$

$$u(t)=\frac{c_x^-}{2c_u}(t_2-t),\ t_y\le t\le t_2,$$

$u(t)=0$, $t_2\le t\le T$; STOP.

Otherwise, calculate $t_y=T-\frac{2c_uU}{c_x^-}$ and set the optimal solution as

$$u(t) = U\,,\quad 0 \le t \le t_y\,,$$

$$u(t) = \frac{c_x^-}{2c_u}(T-t),\quad t_y \le t \le T\,,\ \text{STOP.}$$

Example 4.6

Consider the $1|1|u(t) \in [0,U]\,|D,T|\,Q_u, L_x$ problem with parameters T=10hr, U=3unit/hr, c_u=0.1\$hr/unit², c_x^+=0.2\$/hr.unit, c_x^-=1\$/hr.unit and demand D=15units and t_d = 7 hr.

Step 1 of Algorithm 4.15 runs Algorithm 3.11 and finds that the problem cannot be solved as though it is uncapacitated. Step 2 calculates time points

$$t_1 = t_d - \frac{c_x^-}{c_x^+ + c_x^-}\frac{D}{U} - \frac{c_u U}{c_x^+} = 1.33\text{hr and } t_2 = t_d + \frac{c_x^+}{c_x^+ + c_x^-}\frac{D}{U} + \frac{c_u U}{c_x^-} = 8.13\text{hr}.$$

Since $t_2 \le T$ and $t_1 \ge 0$, Step 3 calculates $t_x = t_d - \dfrac{c_x^-}{c_x^+ + c_x^-}\dfrac{D}{U} + \dfrac{c_u U}{c_x^+}$ =4.33hr

and $t_y = t_d + \dfrac{c_x^+}{c_x^+ + c_x^-}\dfrac{D}{U} - \dfrac{c_u U}{c_x^-}$ =7.53hr and sets the optimal solution as

$u(t) = (t - 1.33)\,\text{unit/hr},\ 1.33 \le t \le 4.33\,,$

$u(t) = 3$ unit/hr, $4.33 \le t \le 7.53\,,$

$u(t) = 5(8.13 - t)\,\text{unit/hr},\ 7.53 \le t \le 8.13\,,$

$u(t) = 0,\ 0 \le t \le 1.33$ and $8.13 \le t \le 10\,.$

■

Discussion

According to Algorithm 4.15, problem (4.74), (4.54) and (4.55) is solvable in $O(1)$. The solution contains three types of optimal production regimes: no-production, partial production and full production. The most general sequence of the regimes starts with no-production, reaches full production through a gradual increase of the partial production, then decreases to the no-production via the partial production regime. Besides this general sequence, Algorithm 4.15 solves all the special cases which occur when some regimes vanish from the sequence. Specifically, all cases which do not involve the full production at all are computed by Algorithm 3.11 developed for the corresponding uncapacitated problem.

Note that the optimal control found by Algorithm 4.15 is always a continuous function of time. Therefore, the partial production regime never vanishes.

THEOREM 4.9. Algorithm 4.15 solves problem (4.74), (4.54) and (4.55) to optimality in $O(1)$.

PROOF. First of all, Step 1 of Algorithm 4.15 explicitly verifies whether the problem can be solved as uncapacitated. If this is the case, then Algorithm 3.11 is applied. Thus, Steps 2-6 of Algorithm 4.15 solve the remaining cases where the full production regime is necessarily involved. Let us consider the most general case which includes:

- the no-production regime at the beginning and at the end of the planning horizon

$$u(t)=0,\ \ X(t)=0 \text{ and } \psi(t)=0 \text{ for } 0\le t\le t_1 \text{ and}$$

$$u(t)=0,\ \ X(t)=D \text{ and } \psi(t)=0 \text{ for } t_2\le t\le T;$$

- the partial production regime

$$u(t)=\frac{\psi(t)}{2c_u},\ \ X(t)=U(t-t_1) \text{ and } 0\le\psi(t)\le 2c_uU \text{ for } t_1\le t\le t_x \text{ and } t_y\le t\le t_2;$$

- the full production regime

$$u(t)=U\,,\ \ X(t)=U(t-t_1) \text{ and } \psi(t)\ge 2c_uU \text{ for } t_x\le t\le t_y.$$

By integrating the state and co-state equations (4.54) and (4.58), we find

$$X(t_2)=X(t_1)+\frac{1}{2}U(t_x-t_1)+U(t_y-t_x)+\frac{1}{2}U(t_2-t_y),$$

$$\psi(t_x)=\psi(t_1)+c_x^+(t_x-t_1),$$

$$\psi(t_y)=\psi(t_2)+c_x^-(t_2-t_y), \tag{4.77}$$

$$\psi(t_d)=\psi(t_1)+c_x^+(t_d-t_1)=\psi(t_2)+c_x^-(t_2-t_d).$$

By substituting the boundary conditions of the regimes into equations (4.77), we determine the following system of four linear equations in four unknowns t_1, t_x, t_y, t_2

$$D=\frac{1}{2}U(t_x-t_1)+U(t_y-t_x)+\frac{1}{2}U(t_2-t_y),$$
$$2c_uU=c_x^+(t_x-t_1),$$
$$2c_uU=c_x^-(t_2-t_y), \tag{4.78}$$
$$c_x^+(t_d-t_1)=c_x^-(t_2-t_d).$$

Solution of system (4.78) proves Steps 2 and 3 of Algorithm 4.15 if $t_1 \geq 0$ and $t_2 \leq T$. If, however, $t_2 > T$, the terminal no-production regime vanishes. Consequently, demand is not fully completed and there remain only three switching points t_1, t_x, t_y, which are found from system (4.78) by eliminating the first equation and setting $t_2 = T$. The other steps of Algorithm 4.5 deal with special cases of the sequence of regimes considered above. The proof of their correctness is similar and left as an exercise for the reader.

■

Special Cases of Zero Costs

First, the zero production cost problem, $1|1|u(t)\in[0,U]|D,T|L_x$ has been solved in Section 4.2.1. Secondly, consider the zero inventory cost problem, $1|1|u(t)\in[0,U]|D,T|L_u,Q_x,c_x^+=0$. By allowing very small values of c_x^+, one can observe that Steps 2-5 are no longer needed and the optimal solution is always set either at Step 1 (which calls in this case for only Algorithm 3.11 Step 3) or at Step 6.

No Backlog

When backlogs are disallowed, i.e. for the $1|1|u(t)\in[0,U], X(t)\geq 0|D,T|Q_u,L_x$ problem, the solution algorithm is simplified, because the demand must be completed on the due date.

Dual Formulation

Now problem (4.74), (4.54) and (4.55) has the additional constraint $X(t)\geq D(t)$ which affects neither Hamiltonian (4.75) nor optimality condition (4.76), but changes the co-state differential equation

$$d\psi(t)=c_x^+dt-d\mu(t),\quad \psi(T+0)=0,$$

where Lagrange multiplier $d\mu(t)\geq 0$ corresponds to the no-backlog constraint and satisfies the complementary slackness condition

$$\int_0^T (X(t)-D(t))d\mu(t)=0. \tag{4.79}$$

Algorithm 4.16

Step 1:

Run Algorithm 3.12. If $u(t) \le U$ for $0 \le t \le T$, then the optimal solution found by Algorithm 3.12 is valid. STOP.
Otherwise go to the next step.

Step 2:

Calculate $t_1 = t_d - \frac{D}{U} - \frac{c_u U}{c_x^+}$.

If $t_1 \ge 0$, then proceed to the next step, otherwise go to Step 4.

Step 3:

Calculate the time points where the full production starts

$$t_x = t_d - \frac{D}{U} + \frac{c_u U}{c_x^+}.$$

Set the optimal solution as

$$u(t) = \frac{c_x^+}{2c_u}(t - t_1), \quad t_1 \le t \le t_x,$$

$$u(t) = U, \quad t_x \le t \le t_d,$$

$$u(t) = 0, \quad 0 \le t \le t_1 \text{ and } t_d \le t \le T; \text{ STOP.}$$

Step 4:

Calculate $t_x = \sqrt{\frac{4c_u}{c_x^+}(D - Ut_d)}$ and set the optimal solution as

$$u(t) = U + \frac{c_x^+}{2c_u}(t - t_x), \quad 0 \le t \le t_x,$$

$$u(t) = U, \quad t_x \le t \le t_d,$$

$$u(t) = 0, \quad t_d \le t \le T; \text{ STOP.}$$

Discussion

Algorithm 4.16 is of $O(1)$ complexity. It presents a few analytical forms of the optimal solution possible for the problem with backlog disallowed. In these cases, the demand is fully completed by the due date, while the production is either started from the beginning of the planning horizon or from a switching point before which no-production regime is beneficial. Note that the problem has a feasible solution, i.e. the demand can be completed by the due date, only if $t_d \ge \frac{D}{U}$.

LEMMA 4.30. (Correctness of Algorithm 4.16 Step 2-4) Given problem (4.74), (4.54) and (4.55) with backlog disallowed,

- If $t_d \geq \frac{D}{U} + \frac{c_u U}{c_x^+}$ and $\frac{D}{U} \geq \frac{c_u U}{c_x^+}$, then $u(t) = \frac{c_x^+}{2c_u}(t - t_1)$, $t_1 \leq t \leq t_x$;

  $u(t) = U$, $t_x \leq t \leq t_d$; $u(t) = 0$, $0 \leq t \leq t_1$ and $t_d \leq t \leq T$,

  where $t_1 = t_d - \frac{D}{U} - \frac{c_u U}{c_x^+}$ and $t_x = t_d - \frac{D}{U} + \frac{c_u U}{c_x^+}$.

- If $\sqrt{\frac{4c_u}{c_x^+}(D - Ut_d)} \leq t_d$ and $t_d \leq \frac{D}{U} + \frac{c_u U}{c_x^+}$, then

  $u(t) = U + \frac{c_x^+}{2c_u}(t - t_x)$, $0 \leq t \leq t_x$; $u(t) = U$, $t_x \leq t \leq t_d$; $u(t) = 0$, $t_d \leq t \leq T$,

  where $t_x = \sqrt{\frac{4c_u}{c_x^+}(D - Ut_d)}$.

PROOF. To prove the lemma, we construct a solution for the primal and dual problems which is feasible and thus optimal. There can be two types of such solutions. The first one assumes a no-production regime in the beginning of the planning horizon (see Figure 4.10). The solution, which satisfies the state and co-state equations in this case, is

$$X(t) = 0,\ u(t) = 0 \text{ and } \dot{\mu}(t) = 0,\ \psi(t) = 0,\ 0 \leq t \leq t_1;$$

$$u(t) = \frac{\psi(t)}{2c_u},\ t_1 \leq t \leq t_x \text{ and } u(t) = U,\ t_x \leq t < t_d;$$

$$X(t) = \int_{t_1}^{t} u(\tau)d\tau,\ \psi(t) = c_x^+(t - t_1),\ \dot{\mu}(t) = 0,\ t_1 \leq t < t_d; \tag{4.80}$$

$$d\psi(t_d) = -d\mu(t_d),\ d\mu(t_d) > 0;$$

$$X(t) = D,\ u(t) = 0 \text{ and } \psi(t) = 0,\ \dot{\mu}(t) = 0,\ t_d < t \leq T.$$

To find the two unknown switching points, we integrate the state and co-state equations

$$X(t_d) = X(t_1) + \int_{t_1}^{t_d} u(t)dt.$$

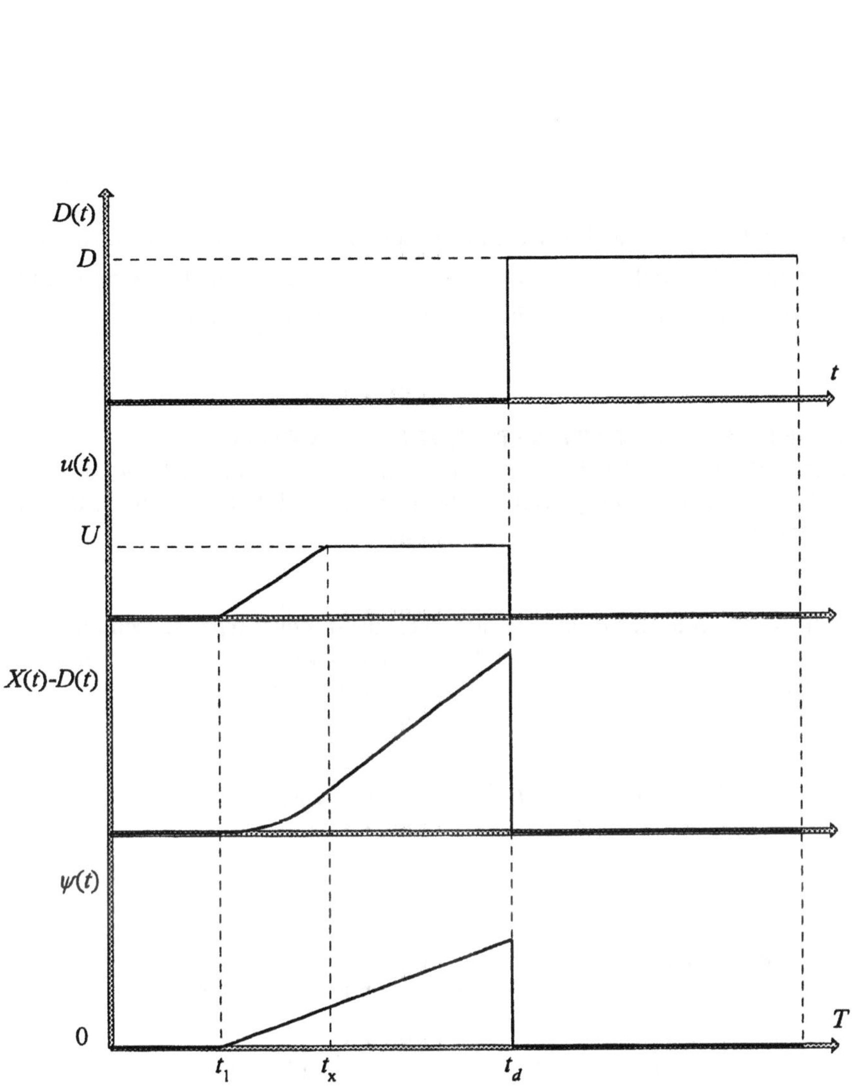

Figure 4.10. Demand $D(t)$, optimal control $u(t)$, state $X(t)$ and co-state $\psi(t)$ variables for a special case of the $1|1|u(t)\in[0,U], X(t)\geq 0\,|D,T|\,Q_u, L_x$ problem.

$$\psi(t_x)=\psi(t_1)+c_x^+(t_x-t_1)$$

and use the boundary conditions determined by (4.80)

$$D=\frac{1}{2}U(t_x-t_1)+U(t_d-t_x), \tag{4.81}$$

$$2c_u U = c_x^+(t_x - t_1).$$

Solution of the system (4.81) with respect to two unknowns t_x and t_1 proves the first case of the lemma, if $t_1 \geq 0$ and $t_x \leq t_d$. The second case, where the no-production regime at the beginning of the planning horizon vanishes, is proved similarly.
■

4.2.4 NON-LINEAR INVENTORY AND PRODUCTION COSTS

Although both costs entering the objective are now non-linear, the optimal solution of the problem, with and without backlogs, can still be approximated with a given accuracy in weakly polynomial time.

Backlogs

The problem to be considered is one-machine, one-product-type scheduling with backlogs and limited capacity to minimize the quadratic inventory and production costs, that is, the $1|1|u(t) \in [0, U]\,|D,T|\,Q_u, Q_x$ problem which is stated as follows

$$J = \int_0^T \left[c_x^+ Z^+(t)^2 + c_x^- Z^-(t)^2 + c_u u(t)^2\right] dt \to \min, \tag{4.82}$$

subject to (4.54) and (4.55).

Dual Formulation

The Hamiltonian, H, resulting from the maximum principle is

$$H(t) = -\left[c_x^+ Z^+(t)^2 + c_x^- Z^-(t)^2 + c_u u(t)^2\right] + \psi(t) u(t). \tag{4.83}$$

The co-state variable $\psi(t)$ satisfies the co-state differential equation (4.64). Maximization of the Hamiltonian as a function of $u(t)$ results, as in the previous section, in

$$u(t) = \begin{cases} U, & \text{if } \psi(t) > 2c_u U, \\ 0, & \text{if } \psi(t) < 0, \\ \dfrac{\psi(t)}{2c_u}, & \text{if } 0 \leq \psi(t) \leq 2c_u U. \end{cases} \tag{4.84}$$

Algorithm 4.17

Step 1:

Run Algorithm 3.13. If $u(t) \le U$ for $0 \le t \le T$, then the optimal solution found by Algorithm 3.13 is valid, STOP.
Otherwise go to the next step.

Step 2:

(i) Set accuracy ξ required for approximating the objective of the optimal solution. Calculate parameters $y_1 = \sqrt{\frac{c_x^+}{c_u}}$, $y_2 = \sqrt{\frac{c_x^-}{c_u}}$,

$$g = \sqrt{\frac{\max\{c_x^+, c_x^-, c_x^- \frac{D}{UT}\}}{c_u}} \text{ and } \varepsilon = \left(4c_u T^2 U^2 g^2 \left(1 + g^2 T^2\right) ch(gT)\right)^{-1} \xi .$$

Set time points $t_1 = t_d$, $t_2 = T$.

(ii) Calculate $t_y = 0.5(t_1 + t_2)$, and

$f = c_x^- (D - Ut_y) - c_u U y_2 cth(y_2 (T - t_y))$.

If $f > 0$, then $t_1 = t_y$ and go to (iii);

if $f < 0$, then $t_2 = t_y$ and go to (iii).

(iii) If $t_1 - t_2 > \varepsilon$, then go to (ii). Otherwise, if

$2c_x^- (t_y - t_d)(D - Ut_y) + c_x^- U (t_y - t_d)^2 - c_x^+ U t_d^2 \ge 0$,

then set the optimal solution as

$u(t) = U$ for $0 \le t < t_y$,

$u(t) = U \dfrac{sh(y_2 (T - t))}{sh(y_2 (T - t_y))}$ for $t_y \le t \le T$, STOP.

Otherwise go to the next step.

Step 3:

(i) Set time points $\hat{t}_1 = 0$, $\hat{t}_2 = t_d$.

(ii) Set time points $t_1 = t_d$, $t_2 = T$.

(iii) Calculate $t_y = 0.5(t_1 + t_2)$ and

$$f_1 = -U(t_y - t_x) + D - \frac{U}{y_1} th(y_1 t_x) - \frac{U}{y_2} cth(y_2 (T - t_y)) .$$

If $f_1 > 0$, then $t_1 = t_y$ and go to (iv);

if $f_1 < 0$, then $t_2 = t_y$ and go to (iv).

(iv) If $t_2 - t_1 > \varepsilon$, then go to (iii); Otherwise go to (v).

(v) Calculate $t_x = 0.5(\hat{t}_1 + \hat{t}_2)$ and

$$f_2 = 2c_u y_1 (t_d - t_x) th(y_1 t_x) + c_x^+ (t_d - t_x)^2 -$$
$$- 2c_u y_2 (t_y - t_d) cth(y_2 (T - t_y)) - c_x^- (t_y - t_d)^2 .$$

If $f_2 > 0$, then $\hat{t}_1 = t_x$ and go to (vi);

if $f_2 < 0$, then $\hat{t}_2 = t_x$ and go to (vi).

(vi) If $\hat{t}_2 - \hat{t}_1 > \varepsilon$, then go to (ii), otherwise set the optimal solution as

$$u(t) = U \frac{ch(y_1 t)}{ch(y_1 t_x)} \quad \text{for } 0 \le t < t_x,$$

$$u(t) = U \quad \text{for } t_x \le t < t_y,$$

$$u(t) = U \frac{sh(y_2 (T - t))}{sh(y_2 (T - t_y))} \quad \text{for } t_y \le t \le T, \text{ STOP.}$$

Discussion

According to Algorithm 4.17, the solution of problem (4.82), (4.54) and (4.55) includes three production regimes - partial production, full production and again partial production. The general case of the problem is approximated by Step 3 in $O\left(\left(\log \frac{R}{\xi}\right)^2\right)$ time. In addition, Algorithm 4.17 solves two special cases which occur when some regimes vanish from the sequence. Specifically, the case when the full production is not involved at all is computed in $O(1)$ time by Step 1 which runs Algorithm 3.13 developed for the uncapacitated problem. The case when the initial partial production vanishes, is solved by Step 2 in $O\left(\log \frac{R}{\xi}\right)$ time.

LEMMA 4.31. Given problem (4.82), (4.54) and (4.55), the co-state variable $\psi(t)$ is non-negative, $\psi(t) \ge 0$ for $t \in [0, T]$.

PROOF. The proof of this lemma is similar to the proof of the lower bound in Lemma 3.8 and therefore is omitted.

■

LEMMA 4.32. (Correctness of Algorithm 4.17) Given problem (4.82), (4.54) and (4.55), and parameters $y_1 = \sqrt{\frac{c_x^+}{c_u}}$, $y_2 = \sqrt{\frac{c_x^-}{c_u}}$,

- if $2c_x^-(t_y - t_d)(D - Ut_y) + c_x^- U(t_y - t_d)^2 - c_x^+ Ut_d^2 \geq 0$, then the optimal solution is

$u(t) = U$ for $0 \leq t < t_y$ and $u(t) = U\frac{sh(y_2(T-t))}{sh(y_2(T-t_y))}$ for $t_y \leq t \leq T$,

where point t_y satisfies equation $c_x^-(D - Ut_y) - c_u U y_2 cth(y_2(T - t_y)) = 0$;

- otherwise, the optimal solution is

$u(t) = U\frac{ch(y_1 t)}{ch(y_1 t_x)}$ for $0 \leq t < t_x$, $u(t) = U$ for $t_x \leq t < t_y$ and

$u(t) = U\frac{sh(y_2(T-t))}{sh(y_2(T-t_y))}$ for $t_y \leq t \leq T$,

where points t_x and t_y satisfy the system of two algebraic equations

$$-U(t_y - t_x) + D - \frac{U}{y_1} th(y_1 t_x) - \frac{U}{y_2} cth(y_2(T - t_y)) = 0.$$

$$2c_u y_1(t_d - t_x) th(y_1 t_x) + c_x^+(t_d - t_x)^2 - 2c_u y_2(t_y - t_d) cth(y_2(T - t_y)) - c_x^-(t_y - t_d)^2 = 0,$$

and $th(x) = \frac{sh(x)}{ch(x)}$, $cth(x) = \frac{ch(x)}{sh(x)}$.

PROOF. Since $D(t) = 0$ for $0 \leq t < t_d$, the inventory is non-negative at this time interval and the co-state differential equation (4.64) simplifies to

$$\dot{\psi}(t) = 2c_x^+ X(t). \tag{4.85}$$

Next, as proven in Theorem 4.9, at the remaining part of the planning horizon $t_d < t \leq T$, the optimal solution of the problem is such that $X(t) \leq D$.

Given $X(t) \leq D$ for $t_d < t \leq T$, the co-state differential equation (4.64) simplifies to

$$\dot{\psi}(t) = 2c_x^-(X(t) - D). \tag{4.86}$$

By differentiating (4.85) and (4.86) and taking into account (4.76) and Lemma 4.31, we obtain

$$\ddot{\psi}(t) = \begin{cases} 2c_x^+ U, & \text{if } 2c_u U \leq \psi(t) \\ \frac{c_x^+}{c_u}\psi(t), & \text{if } 0 \leq \psi(t) \leq 2c_u U \end{cases} \qquad 0 \leq t < t_d, \tag{4.87}$$

$$\ddot{\psi}(t)=\begin{cases}2c_x^-U, & \text{if } 2c_uU\le\psi(t)\\ \dfrac{c_x^-}{c_u}\psi(t), & \text{if } 0\le\psi(t)\le 2c_uU\end{cases}\qquad t_d<t\le T. \tag{4.88}$$

Thus, the system of the state and co-state differential equations has been presented as a second-order, ordinary differential equation. The boundary conditions for (4.87)-(4.88) are the transversality conditions

$$\dot{\psi}(0)=0\,,\qquad \psi(T)=0\,. \tag{4.89}$$

In addition, due to the jump of the demand function $D(t)$ at point t_d, the co-state equation (4.64) results in a jump in the derivative of the co-state variable, so that

$$\psi(t_d-0)=\psi(t_d+0)\,, \tag{4.90}$$

$$\frac{\dot{\psi}(t_d-0)}{2c_x^+}-D=\frac{\dot{\psi}(t_d+0)}{2c_x^-}\,. \tag{4.91}$$

Consider now the first case stated in the lemma, that is, the optimal control is characterized by the full production and partial production regimes with a single switching point t_y. In this case, equations (4.87) and (4.88) take the following form

$$\ddot{\psi}(t)=2c_x^+U\,,\qquad 0\le t<t_y, \tag{4.92}$$

$$\ddot{\psi}(t)=\frac{c_x^-}{c_u}\psi(t)\,,\qquad t_y<t\le T. \tag{4.93}$$

By analytically solving equations (4.92) and (4.93) with boundary conditions (4.89) and (4.90), we obtain

$$\psi(t)=2c_uU+2c_x^-(t_y-t_d)(D-Ut_y)+c_x^-U(t_y-t_d)^2-c_x^+Ut_d^2\,,\ \text{for } 0\le t\le t_d\,.$$

$$\psi(t)=2c_uU+2c_uUy_2(t_y-t)cth(y_2(T-t_y))+c_x^-U(t_y-t)^2\,,\ \text{for } t_d\le t\le t_y,$$

$$\psi(t)=2c_uU\frac{sh(y_2(T-t))}{sh(y_2(T-t_y))}\ \text{for } t_y\le t\le T\,.$$

Finally, by taking into account boundary condition (4.91), we find that t_y is determined by the following equation

$$c_x^-(D - Ut_y) - c_u U y_2 cth(y_2(T - t_y)) = 0,$$

if $\psi(0) \geq 2c_u U$, as stated in the first condition of the lemma. The second condition of the lemma is proved similarly and the optimal behavior of the system for this case is illustrated in Figure 4.11.

■

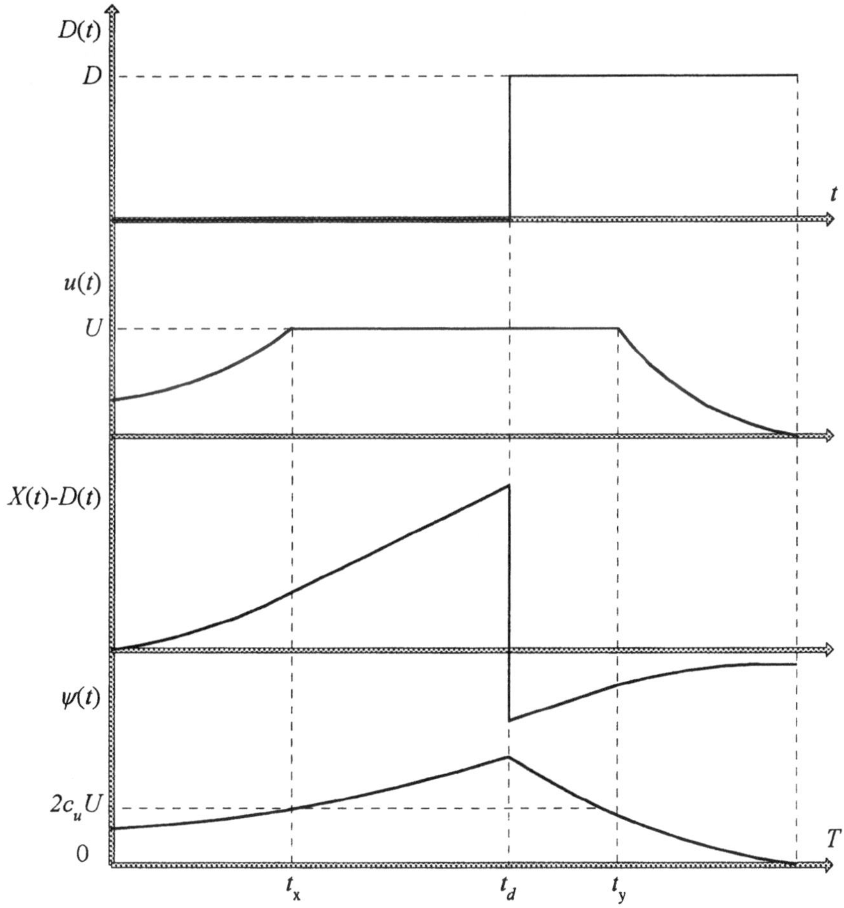

Figure 4.11. Demand $D(t)$, optimal control $u(t)$, state $X(t)$ and co-state $\psi(t)$ variables for a special case of the $1|1|u(t) \in [0, U]|D,T|Q_u, Q_x$ problem.

THEOREM 4.10. Problem (4.82), (4.54) and (4.55) is solvable with accuracy ξ in $O\left(\left(\log\frac{R}{\xi}\right)^2\right)$ time, where $R = 4c_u T^3 U^2 g^2 \left(1+g^2T^2\right)ch(gT)$ and

$$g = \sqrt{\frac{\max\{c_x^+, c_x^-, c_x^- \frac{D}{UT}\}}{c_u}}.$$

PROOF. First of all, Step 1 of Algorithm 4.17 explicitly verifies whether the problem can be solved in $O(1)$ time as uncapacitated. If this is the case, then Algorithm 3.13 is applied. Furthermore, Steps 2 and 3 of Algorithm 4.17 solve approximately the remaining cases where the full production regime is necessarily involved. At Step 2, the dichotomous search for the root of a non-increasing function $f(t_y) = c_x^-(D - Ut_y) - c_u U y_2 cth(y_2(T - t_y))$ is applied. Given accuracy ε of the search, this step requires $O\left(\log\frac{T}{\varepsilon}\right)$ operations. The last step of the algorithm utilizes the dichotomous search for t_x and t_y as the roots of the following functions

$$f_1 = -U(t_y - t_x) + D - \frac{U}{y_1} th(y_1 t_x) - \frac{U}{y_2} cth(y_2(T - t_y)) = 0,$$

$$f_2 = 2c_u y_1 (t_d - t_x) th(y_1 t_x) + c_x^+ (t_d - t_x)^2 - 2c_u y_2 (t_y - t_d) cth(y_2(T - t_y)) - c_x^- (t_y - t_d)^2 = 0.$$

Since f_1 is non-increasing in t_y for all t_x and f_2 is non-increasing in t_x for all t_y, the complexity of this double search is $O\left(\left(\log\frac{T}{\varepsilon}\right)^2\right)$.

Finally, we estimate the relationship between computational accuracy of the objective ξ and accuracy of the dichotomous search ε. By varying the objective function (4.82), we determine

$$\xi \le 2T \max\{c_x^+ \max_t Z^+(t), c_x^- \max_t Z^-(t)\} \max_t |\Delta Z(t)| + 2c_u UT \max_t |\Delta u(t)|. \quad (4.94)$$

The maximum inventory and backlog are easy to estimate as

$$\max_t Z^+(t) = T \max_t u(t) = TU \text{ and } \max_t Z^-(t) = D.$$

The value of $\max_t |\Delta u(t)|$ is estimated from (4.84) as $\frac{1}{2c_u}\max_t |\Delta\psi(t)|$. To estimate the error of the co-state variable, we solve the following differential equation

$$\Delta\ddot{\psi}(t) = g^2\Delta\psi(t),\ \ \Delta\psi(t_y) = 2\hat{\varepsilon}\ ,\ \ \Delta\dot{\psi}(0) = 0, \tag{4.95}$$

where $\hat{\varepsilon}$ is determined by

$$\hat{\varepsilon} = \max_t \dot{\psi}(t)\varepsilon = 2\max\{c_x^+ \max_t X^+(t), c_x^- \max_t X^-(t)\}\varepsilon = 2UTc_u g^2\varepsilon. \tag{4.96}$$

Since $\max_t |\Delta\psi(t)|$ found as a solution of (4.95) is less than $2ch(gT)\hat{\varepsilon}$ (see proof of Theorem 3.3), the upper estimates for $\max_t |\Delta u(t)|$ and $\max_t |\Delta X(t)|$ are $\max_t |\Delta u(t)| = \frac{1}{2c_u}\max_t |\Delta\psi(t)| = \frac{\hat{\varepsilon}}{c_u}ch(yT)$ and $\max_t |\Delta X(t)| = \frac{T}{2c_u}\max_t |\Delta\psi(t)| = \frac{T\hat{\varepsilon}}{c_u}ch(yT)$ respectively. By substituting the found estimates into (4.94), we obtain the maximum error ξ as a function of $\hat{\varepsilon}$

$$\xi = 2TUch(gT)\left(1+g^2T^2\right)\hat{\varepsilon}. \tag{4.97}$$

To obtain the final estimates of the complexity and accuracy for Algorithm 4.17 Steps 2 and 3, we express $\hat{\varepsilon}$ from (4.97) and ε from (4.96), substitute them into the complexity of Step 3 and introduce parameter R to simplify the presentation.
■

Special Cases of Zero Costs

The problem with no production cost, $1|1|u(t)\in[0,U]\,|D,T|\,Q_x$, has been solved in Section 4.2.2. The problem with no inventory cost, $1|1|u(t)\in[0,U]\,|D,T|\,Q_u, Q_x, c_x^+ = 0$ is solved by either Step 1 or Step 2 of Algorithm 3.17. The most complex Step 3 of the algorithm is no longer needed.

No Backlog

Although the algorithm for the $1|1|u(t)\in[0,U], X(t)\geq 0\,|D,T|\,Q_u, Q_x$ problem is simpler than that when backlog is allowed, the solution can still be only approximated with a given accuracy.

Dual Formulation

Now problem (4.82), (4.54) and (4.55) has the additional constraint $X(t) \geq D(t)$ which affects neither Hamiltonian (4.83) nor optimality condition (4.84), but changes the co-state differential equation

$$d\psi(t) = c_x^+\left(X(t) - D(t)\right)dt - d\mu(t), \quad \psi(T+0) = 0, \tag{4.98}$$

where Lagrange multiplier $d\mu(t) \geq 0$ corresponds to the no-backlog constraint and satisfies complementary slackness condition (4.79).

Algorithm 4.18

Step 1:

Run Algorithm 3.14. If $u(t) \leq U$ for $0 \leq t \leq T$, then the optimal solution found by Algorithm 3.14 is valid. STOP.
Otherwise go to the next step.

Step 2:

(i) Set accuracy ξ required for approximating the objective of the optimal solution. Calculate parameters $y_1 = \sqrt{\dfrac{c_x^+}{c_u}}$,

and $\varepsilon = \left(4c_u T^2 U^2 g^2\left(1 + g^2 T^2\right)ch(gT)\right)^{-1}\xi$.
Set time points $t_1 = 0$, $t_2 = t_d$.

(ii) Calculate $t_x = 0.5\left(t_1 + t_2\right)$, and

$f = \dfrac{U}{y_1} th(y_1 t_x) + U(t_d - t_x) - D$.
If $f > 0$, then $t_1 = t_x$ and go to (iii);
if $f < 0$, then $t_2 = t_x$ and go to (iii).
If $t_1 - t_2 > \varepsilon$, then go to (ii). Otherwise set the optimal solution as

$u(t) = U\dfrac{ch(y_1 t)}{ch(y_1 t_x)}$ for $0 \leq t < t_x$ and,

$u(t) = U$ for $t_x \leq t \leq T$, STOP.

Discussion

When backlog is disallowed, the demand must be fully completed by the due date, i.e. $t_d \geq \dfrac{D}{U}$. Similar to the previous problem, Algorithm 4.18 solves exactly the uncapacitated case at Step 1 in $O(1)$ time. Step 2 effectively approximates the optimal solution on which the full production is either started from the beginning of the planning horizon or from a switching point before which the partial production regime is beneficial.

LEMMA 4.33. (Correctness of Algorithm 4.18) Given problem (4.82), (4.54) and (4.55) with backlog disallowed and parameter $y_1 = \sqrt{\frac{c_x^+}{c_u}}$, the optimal solution is

$$u(t) = U\frac{ch(y_1 t)}{ch(y_1 t_x)} \text{ for } 0 \le t < t_x \text{ and } u(t) = U \text{ for } t_x \le t \le T,$$

where point t_x is the root of equation $\frac{U}{y_1} th(y_1 t_x) + U(t_d - t_x) - D = 0$.

PROOF. Similar to the proof of Lemma 4.32, by considering the optimal control characterized by the partial and full production regimes with switching point t_x, we obtain

$$\ddot{\psi}(t) = \frac{c_x^-}{c_u}\psi(t), \quad \dot{\psi}(0) = 0 \quad 0 \le t < t_x, \tag{4.99}$$

$$\ddot{\psi}(t) = 2c_x^+ U, \quad \dot{\psi}(t_d) = 2c_x^+ D \quad t_x < t \le t_d. \tag{4.100}$$

The solution of these equations proves the lemma.
■

THEOREM 4.11. Problem (4.82), (4.54) and (4.55) with backlog disallowed is solvable with accuracy ξ in $O\left(\log\frac{R}{\xi}\right)$ time, where $R = 4c_u T^3 U^2 y_1^2 \left(1 + y_1^2 T^2\right) ch(y_1 T)$.

PROOF. Step 1 of Algorithm 4.18 explicitly verifies whether the problem can be solved in $O(1)$ time as uncapacitated. If this is the case, then Algorithm 3.13 is applied. Step 2 of Algorithm 4.18 solves approximately the remaining cases by the dichotomous search in $O\left(\log\frac{T}{\varepsilon}\right)$ time. The accuracy of this step is evidently the same as that of Algorithm 4.17, where the possibility of backlogs is eliminated in varying the objective.
■

4.3 DEMAND OVER INFINITE PLANNING HORIZON

This section addresses the scheduling problems characterized by periodic demand as a function of time specified over an infinite planning horizon. The periodic form of the demand causes the optimal solution to be also periodic unless the initial inventory is fixed. To provide the periodicity of the solution, we consider the problem over the finite planning horizon T, which is equal to the demand period, and impose boundary condition $X(0)=X(T)$. Again, the problems we consider differ by inventory related constraints allowing or disallowing backlogs and by the form of the objective function.

4.3.1 LINEAR PERFORMANCE MEASURE

Backlogs

The problem to be considered is one-machine, one-product-type scheduling with backlogs and limited production rate to minimize the linear cost of inventory and production over period T, that is, the $1|1|u(t)\in[0,U], X(0)=X(T)\;|d(t)|\;L_u, L_x$ problem which is stated as follows

$$J=\int_0^T\left[c_x^+X^+(t)+c_x^-X^-(t)+c_u u(t)\right]dt\to\min, \tag{4.101}$$

$$X^+(t)=\max\{X(t),0\},\; X^-(t)=\max\{-X(t),0\}$$

subject to

$$\dot X(t)=u(t)-d(t),\qquad X(0)=X(T), \tag{4.102}$$

$$0\le u(t)\le U. \tag{4.103}$$

Dual Formulation

The Hamiltonian, H, resulting from the maximum principle is

$$H(t)=-\left[c_x^+X^+(t)+c_x^-X^-(t)+c_u u(t)\right]+\psi(t)\left[u(t)-d(t)\right]. \tag{4.104}$$

The co-state variable, $\psi(t)$ in (4.104), satisfies the following co-state differential equation with corresponding transversality condition

$$\dot\psi(t)=\begin{cases}c_x^+, \text{ if } X(t)>0;\\ -c_x^-, \text{ if } X(t)<0;\\ a,\ a\in[-c_x^-,c_x^+], \text{ if } X(t)=0;\end{cases}\qquad \psi(0)=\psi(T). \tag{4.105}$$

Maximization of the Hamiltonian implies that

$$u(t)=\begin{cases}U, & \text{if } \psi(t)>c_u,\\ 0, & \text{if } \psi(t)<c_u,\\ b,\ b\in[0,U], & \text{if } \psi(t)=c_u.\end{cases} \tag{4.106}$$

The first condition in equation (4.106) defines production at the maximal rate (full production regime). The second condition is similar to that found for the corresponding unlimited-capacity, scheduling problem and defines the no-production regime. Finally, the last condition in (4.106) presents the fractional production which allows production at an intermediate rate to track the demand.

Algorithm 4.19

Step 1:

If $d(t)\le U$ for $0\le t\le T$, then
set the optimal solution as
$u(t)=d(t)$ for $0\le t\le T$, STOP.
Otherwise go to the next step.

Step 2:

If $d(t)=\begin{cases}d_1, & \text{if } 0\le t\le \bar{t},\\ d_2, & \text{if } \bar{t}<t\le T,\end{cases}$ $d_1<U$, $d_2>U$ and $U\ge \frac{1}{T}\int_0^T d(t)dt$, then

calculate

$t_2=\bar{t}-c_x^-\dfrac{(T-\bar{t})(d_2-U)}{(c_x^+ + c_x^-)(U-d_1)}$ and $t_1=t_2-T+\dfrac{(T-\bar{t})(d_2-d_1)}{U-d_1}$.

Set the optimal solution as
$u(t)=U$ for $0\le t<t_1$,
$u(t)=d(t)$ for $t_1\le t<t_2$,
$u(t)=U$ for $t_2\le t\le T$; STOP.
Otherwise, STOP.

Example 4.7

Consider the $1|1|u(t)\in[0,U], X(0)=X(T)\,|d(t)|\,L_u, L_x$ problem with parameters T=10hr, U=3unit/hr, c_u=1\$/unit, c_x^+=0.2\$/hr.unit, c_x^-=1\$/hr.unit and two-level, piece-wise constant demand $d(t)=\begin{cases}2, & \text{if } 0\le t\le 6\\ 4, & \text{if } 6<t\le 10\end{cases}$ unit/hr.

Since the second demand level is greater than the capacity, $d_2>U$, Algorithm 4.19 proceeds from Step 1 to Step 2. Step 2 calculates

$$t_2 = \bar{t} - c_x^- \frac{(T-\bar{t})(d_2 - U)}{(c_x^+ + c_x^-)(U - d_1)} = 2.67 \text{ and } t_1 = t_2 - T + \frac{(T-\bar{t})(d_2 - d_1)}{U - d_1} = 0.67$$

and sets the optimal solution as
$u(t) = 3$ unit/hr for $0 \le t < 0.67$ and for $2.67 \le t \le 10$,
$u(t) = 2$ unit/hr for $0.67 \le t < 2.67$.
■

Discussion

Algorithm 4.19 Step 1 provides a closed form optimal solution for the problem when the demand is weak, that is, it solves the problem as though it is uncapacitated. In such a case, there is only one type of optimal control regime. This is the fractional production $u(t)=d(t)$. Algorithm 4.19 Step 2 solves the problem for a two-level, piece-wise constant demand. This case is characterized by the full production regimes at the beginning and at the end of the demand period and the fractional production regime between them. Note that the production never stops, no matter what the production cost c_u is. Furthermore, due to boundary inventory constraint $X(0) = X(T)$, no feasible solution exists unless the maximum production rate is greater than the average demand $U \ge \frac{1}{T}\int_0^T d(t)dt$.

The complexity of Algorithm 4.19 is $O(1)$ and its correctness is easily verified by solving the dual formulation in the following lemma.

LEMMA 4.34. (Correctness of Algorithm 4.19) Given problem (4.101)-(4.103), the optimal control is determined as follows

- if $d(t) \le U$ for $0 \le t \le T$, then

$u(t) = d(t)$ for $0 \le t \le T$,

- if $d(t) = \begin{cases} d_1, & \text{if } 0 \le t \le \bar{t}, \\ d_2, & \text{if } \bar{t} < t \le T, \end{cases}$ $d_1 < U$, $d_2 > U$ and $U \ge \frac{1}{T}\int_0^T d(t)dt$, then

$u(t) = U$ for $0 \le t < t_1$, $u(t) = d(t)$ for $t_1 \le t < t_2$, $u(t) = U$ for $t_2 \le t \le T$,

where $t_2 = \bar{t} - c_x^- \dfrac{(T-\bar{t})(d_2 - U)}{(c_x^+ + c_x^-)(U - d_1)}$ and $t_1 = t_2 - T + \dfrac{(T-\bar{t})(d_2 - d_1)}{U - d_1}$.

PROOF. First note, that if $d(t) \le U$ for $0 \le t \le T$, the solution

$$X(t) \equiv 0, \quad \psi(t) \equiv c_u, \quad u(t) \equiv d(t)$$

is always feasible for the system of state and co-state equations and meets the optimality condition (4.106).

Next, to prove the second condition of the lemma, we consider a solution with a sequence consisting of full, fractional and once again, full production regimes. The regimes change over at time points t_1 and t_2 (see Figure 4.12)

$$X(t) > 0 \text{ for } t_2 < t < t_x \quad \text{and} \quad X(t) < 0 \text{ for } 0 < t < t_1 \text{ and } t_x < t \leq T,$$

$$\psi(t) > c_u, \quad u(t) = U \text{ for } 0 \leq t < t_1 \text{ and } t_2 \leq t \leq T,$$

$$X(t) = 0, \quad \psi(t) = c_u, \quad u(t) = d(t) \text{ for } t_1 \leq t < t_2.$$

By considering this solution at three unknown time points t_1, t_x and t_2, we find the following three boundary conditions

$$X(t_x) = 0, \quad \psi(t_2) = \psi(t_1) = \psi(T + t_1) \text{ and } X(0) = X(T). \tag{4.107}$$

By integrating the state and co-state differential equations, conditions (4.107) transform into the following system of three algebraic equations in the three unknowns

$$U(t_x - t_2) = d_1(\bar{t} - t_2) + d_2(t_x - \bar{t}), \quad c_x^+(t_x - t_2) = c_x^-(T + t_1 - t_x),$$

$$d_1\bar{t} + d_2(T - \bar{t}) = UT - (U - d_1)(t_2 - t_1). \tag{4.108}$$

Since time points t_1, t_x and t_2 satisfying (4.108) are such that $0 \leq t_1 \leq t_2 < t_x \leq T$ always holds, the constructed solution is feasible, proving the second part of the lemma.
■

Special Cases of Zero Costs

Solution Algorithm 4.19 does not include production cost coefficient c_u and thus does not change for the special cases of the zero production cost, $1|1|u(t) \in [0, U], X(0)=X(T)|d(t)|L_x$.

The no inventory cost $1|1|u(t) \in [0, U], X(0)=X(T)|d(t)|L_u, L_x, c_x^+ = 0$ problem is solved by Algorithm 4.19 by substituting $c_x^+ = 0$ into the equations for switching points t_1 and t_2

$$t_1 = 0, \quad t_2 = \bar{t} - \frac{(T - \bar{t})(d_2 - U)}{(U - d_1)}.$$

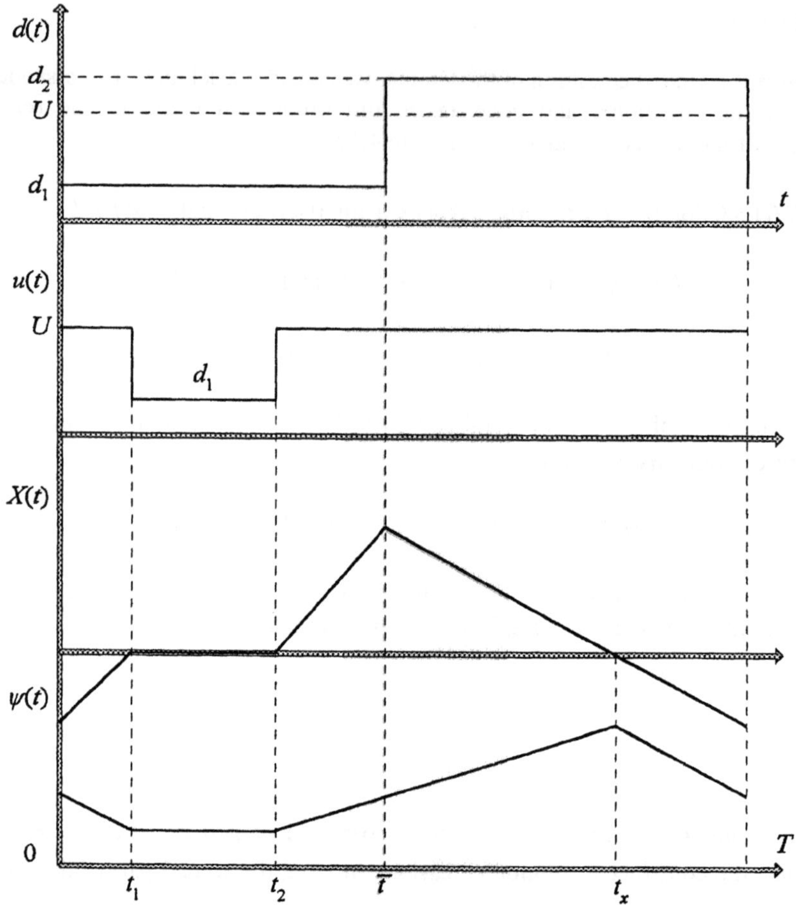

Figure 4.12. Demand $d(t)$, optimal control $u(t)$, state $X(t)$ and co-state $\psi(t)$ variables for the $1|1|u(t)\in[0,U], X(0)=X(T)\,|d(t)|\,L_u, L_x$ problem with a two-level, piece-wise constant demand.

No Backlog

The solution of the $1|1|u(t)\in[0,U], X(t)\geq 0, X(0)=X(T)|d(t)|\,L_u, L_x$ problem coincides with the solution of the $1|1|u(t)\in[0,U], X(0)=X(T)|d(t)|\,L_u, L_x, c_x^+=0$ problem. One can verify this by assigning a very large value to c_x^- in Algorithm 4.19. The rigorous proof of this fact is left for the reader.

4.3.2 NON-LINEAR INVENTORY COST

Next we consider a linear production cost function and a non-linear inventory cost which will be approximated by a quadratic function. It will be shown below that introducing the non-linearity affects only the timing of the optimal production regimes.

Backlogs

The problem to be considered is one-machine, one-product-type scheduling with backlogs and limited production rate to minimize quadratic inventory and linear production costs over period T, that is, the $1|1|u(t)\in[0,U], X(0)=X(T)\,|d(t)|\, L_u, Q_x$ problem which is stated as follows

$$J=\int_0^T\left[c_x^+X^+(t)^2+c_x^-X^-(t)^2+c_uu(t)\right]dt\to\min, \tag{4.109}$$

subject to (4.102) and (4.103).

Dual Formulation

The Hamiltonian, H, resulting from the maximum principle is

$$H(t)=-\left[c_x^+X^+(t)^2+c_x^-X^-(t)^2+c_uu(t)\right]+\psi(t)\left[u(t)-d(t)\right]. \tag{4.110}$$

The co-state variable, $\psi(t)$ in (4.110), satisfies the following co-state differential equation with its transversality condition

$$\dot{\psi}(t)=\begin{cases}2c_x^+X(t), & \text{if } X(t)>0;\\ 2c_x^-X(t), & \text{if } X(t)\le 0;\end{cases}\qquad \psi(0)=\psi(T). \tag{4.111}$$

Maximization of the Hamiltonian implies that the control effort satisfies (4.106).

Algorithm 4.20

Step 1:

If $d(t)\le U$ for $0\le t\le T$, then
set the optimal solution as
$u(t)=d(t)$ for $0\le t\le T$, STOP.
Otherwise go to the next step.

Step 2:

If $d(t)=\begin{cases}d_1, & \text{if } 0\le t\le \bar{t},\\ d_2, & \text{if } \bar{t}<t\le T,\end{cases}$ $d_1<U$, $d_2>U$ and $U\ge\frac{1}{T}\int_0^T d(t)dt$, then

calculate $t_2\in[0,\bar{t}]$ as a root of the quadratic equation

$$c_x^+(t_2-\bar{t})^2+2c_x^+(t_x-\bar{t})\big(\bar{t}(d_2-d_1)-t_2(U-d_1)\big)+$$
$$+c_x^+(U-d_2)(t_x^2-\bar{t}^2)-c_x^-(T+t_1-t_x)^2=0,$$

where
$$t_x=\frac{(d_1-d_2)\bar{t}+(U-d_1)t_2}{U-d_2} \text{ and } t_1=t_2-T+\frac{(T-\bar{t})(d_2-d_1)}{U-d_1}.$$

Set the optimal solution as
$u(t)=U$ for $0\le t<t_1$,
$u(t)=d(t)$ for $t_1\le t<t_2$,
$u(t)=U$ for $t_2\le t\le T$; STOP.

Discussion

Algorithm 4.20, similar to Algorithm 4.19 which was developed for the linear performance measure, finds the same sequence of optimal production regimes. However, the switching points between the regimes are now calculated on the basis of a quadratic equation rather than a system of linear equations. This, of course, does not influence the computational complexity of Algorithm 4.20.

Recall, that due to boundary inventory constraint $X(0)=X(T)$, no feasible solution exists unless $U\ge\frac{1}{T}\int_0^T d(t)dt$ and that the production never stops no matter what the production cost c_u is.

LEMMA 4.35. (Correctness of Algorithm 4.20) Given problem (4.109), (4.102) and (4.103), the optimal control is determined as follows

- if $d(t)\le U$ for $0\le t\le T$, then $u(t)=d(t)$ for $0\le t\le T$,
- if $d(t)=\begin{cases} d_1, & \text{if } 0\le t\le\bar{t}, \\ d_2, & \text{if } \bar{t}<t\le T, \end{cases}$ $d_1<U$, $d_2>U$ and $U\ge\frac{1}{T}\int_0^T d(t)dt$, then

$u(t)=U$ for $0\le t<t_1$, $u(t)=d(t)$ for $t_1\le t<t_2$, $u(t)=U$ for $t_2\le t\le T$, and $t_2\in[0,\bar{t}]$ is a root of the following equation

$$c_x^+(t_2-\bar{t})^2+2c_x^+(t_x-\bar{t})\big(\bar{t}(d_2-d_1)-t_2(U-d_1)\big)+c_x^+(U-d_2)(t_x^2-\bar{t}^2)-c_x^-(T+t_1-t_x)^2=0,$$

where $t_x=\dfrac{(d_1-d_2)\bar{t}+(U-d_1)t_2}{U-d_2}$ and $t_1=t_2-T+\dfrac{(T-\bar{t})(d_2-d_1)}{U-d_1}$.

PROOF. The proof is similar to that of Lemma 4.34. ■

Special Cases of Zero Costs

The $1|1|u(t)\in[0,U]$, $X(0)$=$X(T)$|$d(t)$| $L_u, Q_x, c_x^+=0$ problem with no inventory cost can be solved for arbitrary demand by setting a constant production rate equal to the average demand, $u(t)=\frac{1}{T}\int_0^T d(t)dt$, $t\in[0,T]$. This solution follows from the fact that $\psi(t)$ must be constant ($\psi(t)$ is periodic, $\psi(0)=\psi(T)$, and cannot increase because $c_x^+=0$). The value of $X(0)$ is chosen such that no backlog can occur along the period,

$$X(0)=\max_{t\in[0,T]}\left[\int_0^t d(\tau)d\tau-\frac{t}{T}\int_0^T d(\tau)d\tau\right].$$

The solution of Algorithm 4.20 does not include the production cost coefficient and thus does not change for the special case of the zero production cost problem, $1|1|u(t)\in[0,T]$,$X(0)$=$X(T)$|$d(t)$|Q_x.

No Backlog

When backlogs are disallowed, i.e. for the $1|1|u(t)\in[0,U]$, $X(t)\geq 0$,$X(0)$=$X(T)$||$d(t)$| L_u, Q_x problem, the solution algorithm significantly simplifies and the demand functions it can cope with are extended. This extension, however, requires that for the problem to have a feasible solution, the demand must be moderate

$$\frac{1}{t}\int_0^t d(\tau)d\tau\leq U \quad \text{for } 0<t\leq T.$$

If, in particular, the demand is a non-decreasing function of time, it is sufficient to verify the last condition at a single point $t=T$, that is $U\geq\frac{1}{T}\int_0^T d(t)dt$.

Dual Formulation

Now problem (4.109), (4.102) and (4.103) has the additional constraint $X(t)\geq 0$ which affects neither Hamiltonian (4.110) nor optimality conditions (4.106), but changes the co-state differential equation

$$d\psi(t)=c_x^+dt-d\mu(t),\quad \psi(T+0)=\psi(0),$$

where Lagrange multiplier $d\mu(t) \geq 0$ corresponds to the no-backlog constraint and satisfies the complementary slackness condition

$$\int_0^T X(t)d\mu(t) = 0.$$

Algorithm 4.21
Step 1:

If $d(t) \leq U$ for $0 \leq t \leq T$, then
set the optimal solution as
$u(t) = d(t)$ for $0 \leq t \leq T$, STOP.
Otherwise go to the next step.

Step 2:

If the demand is a non-decreasing function of time and $U \geq \frac{1}{T}\int_0^T d(t)dt$, then calculate time point t_1 from the following equation

$$U = \frac{1}{T - t_1}\int_{t_1}^T d(t)dt.$$

Set the optimal solution as
$u(t) = d(t)$ for $0 \leq t < t_1$,
$u(t) = U$ for $t_1 \leq t \leq T$; STOP.

Example 4.8
Consider the $1|1|u(t) \in [0, U], X(t) \geq 0, X(0)=X(T)||d(t)|\, L_u, Q_x$ problem with two-level, piece-wise constant demand, $d(t) = \begin{cases} d_1, & \text{if } 0 \leq t \leq \bar{t}, \\ d_2, & \text{if } \bar{t} < t \leq T, \end{cases}$ $d_1<U$, $d_2>U$ and $U \geq \frac{1}{T}\int_0^T d(t)dt$.

Since the conditions of Step 1 are not satisfied, Algorithm 4.21 proceeds to Step 2. Step 2 finds the root of $U = \frac{1}{T - t_1}\int_{t_1}^T d(t)dt = \frac{1}{T - t_1}\left(d_2(T - \bar{t}) + d_1(\bar{t} - t_1)\right)$ as $t_1 = \bar{t} - \frac{(T - \bar{t})(d_2 - U)}{U - d_1}$ and sets the optimal solution as

$u(t) = d_1$ for $0 \le t < t_1$,
$u(t) = U$ for $t_1 \le t \le T$.
■

Discussion

Algorithm 4.21 is of $O(1)$ complexity if the demand function can be integrated analytically. It is similar to Algorithm 4.19 developed for the linear performance measure. The only difference is that in Step 2 it finds the sequence of optimal production regimes characterized by only one unknown switching point when the fractional production regime is replaced with the full production regime.

LEMMA 4.36. (Correctness of Algorithm 4.21) Given problem (4.109), (4.102) and (4.103) with backlog disallowed, the optimal control is determined as follows

- if $d(t) \le U$ for $0 \le t \le T$, then $u(t) = d(t)$ for $0 \le t \le T$,
- if the demand is a non-decreasing function of time and $U \ge \frac{1}{T}\int_0^T d(t)dt$, then

$u(t) = d(t)$ for $0 \le t < t_1$, $u(t) = U$ for $t_1 \le t \le T$,

where time point t_1 satisfies $U = \frac{1}{T-t_1}\int_{t_1}^T d(t)dt$.

PROOF. First note, that if $d(t) \le U$ for $0 \le t \le T$, the solution

$$X(t) \equiv 0, \quad \psi(t) \equiv c_u, \quad u(t) \equiv d(t)$$

is always feasible for the system of the state and co-state equations and optimality condition (4.106).

Next, to prove the second condition of the lemma, we consider a solution which includes the fractional and full production regimes with time point t_1 when the regimes change over (see Figure 4.13)

$$X(t) = 0, \quad \psi(t) = c_u, \quad u(t) = d(t), \quad d\mu(t) = c_x^+ dt \quad \text{for } 0 < t < t_1,$$

$$X(t) > 0, \quad \psi(t) > c_u, \quad u(t) = U, \quad d\mu(t) = 0 \quad \text{for } t_1 \le t \le T,$$

$$d\mu(T) = \psi(T) - c_u.$$

Thus, the unknown time point t_1 must satisfy equation $X(t_1) = X(T)$ which, by integrating the state differential equation (4.102), transforms into the following form

$$U = \frac{1}{T - t_1} \int_{t_1}^{T} d(t)dt . \qquad (4.112)$$

Since time point t_1 found from (4.112) is such that $0 \le t_1 < T$ always holds, the constructed solution is feasible, proving the last part of the lemma.
■

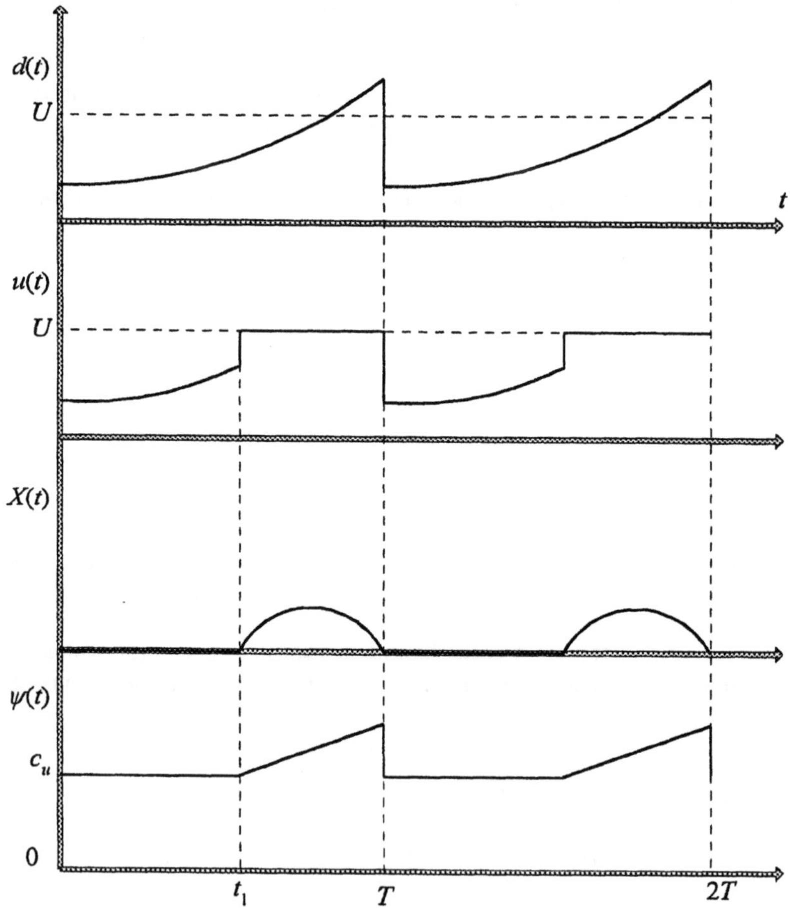

Figure 4.13. Demand $d(t)$, optimal control $u(t)$, state $X(t)$ and co-state $\psi(t)$ variables for the $1|1|u(t) \in [0, U], X(t) \ge 0, X(0)=X(T)||d(t)| L_u, Q_x$ problem with non-decreasing demand.

4.3.3 NON-LINEAR PRODUCTION COST

In this section, we consider a combination of a piece-wise linear inventory/backlog cost function and a non-linear production cost function. There are only a few solvable cases which are characterized by special conditions imposed on the demand. In what follows, we present two solvable cases appearing when the demand is either weak or two-level, piece-wise constant.

Backlogs

The problem to be considered is one-machine, one-product-type scheduling with backlogs and limited production rate to minimize the piece-wise inventory cost and quadratic production cost over period T, that is, the $1|1|u(t)\in[0,U]$,$X(0)=X(T)|d(t)|Q_u, L_x$ problem which is stated as follows

$$J=\int_0^T\left[c_x^+X^+(t)+c_x^-X^-(t)+c_u u(t)^2\right]dt\to\min, \tag{4.113}$$

subject to (4.102) and (4.103).

Dual Formulation

The Hamiltonian, H, resulting from the maximum principle is

$$H(t)=-\left[c_x^+X^+(t)+c_x^-X^-(t)+c_u u(t)^2\right]+\psi(t)\left[u(t)-d(t)\right]. \tag{4.114}$$

The co-state variable, $\psi(t)$ in (4.114), satisfies the co-state differential equation (4.105).

Maximization of the Hamiltonian implies that the optimal control function $u(t)$ satisfies (4.84).

Algorithm 4.22

Step 1:

Run Algorithm 3.16. If $u(t)\le U$, then the found solution is optimal, STOP.

Otherwise go to the next step.

Step 2:

If the demand function $d(t)$ is two-level, piece-wise constant

$$d(t)=\begin{cases}d_1, & \text{if } 0\le t\le\bar{t},\\ d_2, & \text{if } \bar{t}<t\le T,\end{cases}\qquad U\ge\frac{1}{T}\int_0^T d(t)dt,$$

$$\frac{c_x^+c_x^-}{c_x^++c_x^-}(d_2-U)(T-\bar{t})\le c_u(U-d_1)^2 \text{ and}$$

$$T \geq \frac{c_u(c_x^+ + c_x^-)}{c_x^+ c_x^-}(U - d_1) + \frac{d_2 - d_1}{U - d_1}(T - \bar{t})\text{, then}$$

calculate $t_2 = \bar{t} + \frac{c_u}{c_x^+}(U - d_1) - \frac{c_x^-(d_2 - U)(T - \bar{t})}{(c_x^- + c_x^+)(U - d_1)}$, $t_3 = \frac{c_x^+ \bar{t} + c_x^- T}{c_x^+ + c_x^-}$,

$t_1 = t_2 - \frac{2c_u}{c_x^+}(U - d_1)$, $t_4 = T - \frac{c_x^+}{c_x^-}(t_2 - \bar{t})$, $t_0 = t_4 - T + \frac{2c_u}{c_x^-}(U - d_1)$.

Set the optimal solution as

$u(t) = d_1 + \frac{c_x^-}{2c_u}(t_0 - t)$ for $0 \leq t \leq t_0$;

$u(t) = d_1$ for $t_0 \leq t \leq t_1$;

$u(t) = d_1 + \frac{c_x^+}{2c_u}(t - t_1)$ for $t_1 \leq t \leq t_2$;

$u(t) = U$ for $t_2 \leq t \leq t_4$;

$u(t) = U - \frac{c_x^-}{2c_u}(t - t_4)$ for $t_4 \leq t \leq T$; STOP.

Example 4.9

Consider the $1|1|u(t) \in [0, U]$,$X(0)$=$X(T)|d(t)|\,Q_u, L_x$ problem with parameters T=10hr, U=3unit/hr, c_u=0.1\$hr/unit², c_x^+=0.2\$/hr.unit, c_x^-=1\$/hr.unit and two-level piece-wise constant demand $d(t) = \begin{cases} 1, & \text{if } 0 \leq t \leq 6 \\ 3.5, & \text{if } 6 < t \leq 10 \end{cases}$ unit/hr.

Step 1 of Algorithm 4.22 runs Algorithm 3.16, which returns the optimal control that exceeds U before the end of the period. Therefore, the algorithm proceeds to Step 2 where it verifies conditions

$$0 \geq \frac{c_x^+ c_x^-}{c_x^+ + c_x^-}(d_2 - U)(T - \bar{t}) - c_u(U - d_1)^2 = -0.07 \text{ and}$$

$$T \geq \frac{c_u(c_x^+ + c_x^-)}{c_x^+ c_x^-}(U - d_1) + \frac{d_2 - d_1}{U - d_1}(T - \bar{t}) = 6.2\text{. Then Step 2 calculates}$$

switching points $t_2 = 6.17$ hr, $t_3 = 9.33$ hr, $t_1 = 1.97$ hr, $t_4 = 9.97$ hr and $t_0 = 0.37$ hr, and sets the optimal solution as

$u(t) = (2.85 - 5t)$ unit/hr for $0 \leq t \leq 0.37$;

$u(t) = 1$ unit/hr for $0.37 \le t \le 1.97$;
$u(t) = (t - 0.97)$ unit/hr for $1.97 \le t \le 6.17$;
$u(t) = 3$ unit/hr for $6.17 \le t \le 9.97$;
$u(t) = (52.85 - 5t)$ unit/hr for $9.97 \le t \le 10$.
■

Discussion

Algorithm 4.22 Step 1 finds the exact optimal solution for the case of a weak demand by running Algorithm 3.16 developed for the uncapacitated problem. Algorithm 4.22 Step 2 finds the closed form optimal solution which is characterized by a sequence of partial, full, and again, partial production regimes. Within the partial production regime, the optimal control effort changes continuously, from the maximum level $u(t) = U$ down to the constant level $u(t) = d_1$ and then back to the maximum level.

Due to boundary inventory constraint $X(0) = X(T)$, no feasible solution exists unless $U \ge \frac{1}{T}\int_0^T d(t)dt$. This implies that for the two-level, non-weak, piece-wise constant demand dealt with at Step 2, one of the levels must be less than the capacity, while the other must be greater than it. Since the demand is periodic, we further assume that $d_1 < U < d_2$ without loss of generality.

LEMMA 4.37. (Correctness of Algorithm 4.22) Given problem (4.113), (4.102) and (4.103), if the demand function *d(t)* is two-level, piece-wise constant

$$d(t) = \begin{cases} d_1, & \text{if } 0 \le t \le \bar{t}, \\ d_2, & \text{if } \bar{t} < t \le T, \end{cases} \quad U \ge \frac{1}{T}\int_0^T d(t)dt \ \text{ and } \ T \ge \frac{c_u(c_x^+ + c_x^-)}{c_x^+ c_x^-}(U - d_1) + \frac{d_2 - d_1}{U - d_1}(T - \bar{t}),$$

then the optimal solution is

$$u(t) = d_1 - \frac{c_x^-}{2c_u}(t - t_0) \ \text{ for } 0 \le t \le t_0; \qquad u(t) = d_1 \ \text{ for } t_0 \le t \le t_1;$$

$$u(t) = d_1 + \frac{c_x^+}{2c_u}(t - t_1) \ \text{ for } t_1 \le t \le t_2; \qquad u(t) = U \ \text{ for } t_2 \le t \le t_4;$$

$$u(t) = U - \frac{c_x^-}{2c_u}(t - t_4) \ \text{ for } t_4 \le t \le T;$$

where $t_3 = \dfrac{c_x^+ \bar{t} + c_x^- T}{c_x^+ + c_x^-}$, $t_1 = -\dfrac{c_u}{c_x^+}(U - d_1) + \dfrac{(U - d_2)t_3 + (d_2 - d_1)\bar{t}}{U - d_1}$,

$$t_2 = t_1 + \frac{2c_u}{c_x^+}(U - d_1),\ t_0 = \frac{c_u}{c_x^-}(U - d_1) + \frac{d_2 - U}{U - d_1}(T - t_3),\ t_4 = t_0 + T - \frac{2c_u}{c_x^-}(U - d_1).$$

PROOF. Let there exist time points t_0, t_1, t_2, t_3 and t_4 such that the optimal state and co-state variables evolution is

$$X(t) < 0 \text{ for } 0 \le t < t_0,\ X(t) = 0 \text{ for } t_0 \le t \le t_1,\ X(t) > 0 \text{ for } t_1 < t < t_4,$$

$$X(t_4) = 0,\ X(t) < 0 \text{ for } t_4 < t \le T; \tag{4.115}$$

$$2c_u d_1 < \psi(t) < 2c_u U \text{ for } 0 < t < t_0,\ t_1 < t < t_2 \text{ and } t_3 < t < T,$$

$$\psi(t) = 2c_u d_1 \text{ for } t_0 \le t \le t_1,\ 2c_u U < \psi(t) \text{ for } t_2 < t < t_3, \tag{4.116}$$

$$\psi(t_2) = \psi(t_3) = 2c_u U.$$

By integrating state (4.102) and co-state (4.105) differential equations with boundary conditions from (4.115) and (4.116) respectively, we obtain state-variable-related equations:

$$d_1(\bar{t} - t_1) + d_2(t_4 - \bar{t}) = U(t_4 - t_2) + \frac{1}{2}(U + d_1)(t_2 - t_1) - \frac{c_x^-}{4c_u}(t_4 - t_3)^2,$$

$$d_1 t_0 + d_2(T - t_4) = d_1(t_0 + T - t_4) + \frac{c_x^-}{4c_u}(t_0 + T - t_4)^2;$$

and co-state-variable-related equations:

$$2c_u d_1 + c_x^+(t_2 - t_1) = 2c_u U,\ 2c_u d_1 + c_x^-(t_0 + T - t_4) = 2c_u U,\ c_x^+(t_4 - t_1) = c_x^-(t_0 + T - t_4).$$

The solution of this system of five linear algebraic equations in five unknown time points proves the lemma if $t_1 \ge t_0$.

■

Note that the maximum principle does not provide enough information about the optimal solution for exactly solving the $1|1|u(t) \in [0, U], X(0)=X(T)|d(t)|Q_u, L_x$ problem with demand as an arbitrary function of time. At the same time, the numerical approximation algorithms suggested so far in Section 3.1 cannot be effectively implemented. Indeed, these algorithms use one-dimensional dichotomous search to locate roots of monotone slack functions. Since new boundary conditions $X(0)=X(T)$ and $\psi(0) = \psi(T)$ cause the corresponding slack functions to become two dimensional, which results in much less efficient approximation algorithms, they are not presented in this book.

Special Cases of Zero Costs

The problem with no production cost, $1|1|u(t)\in[0,U]$,$X(0)$=$X(T)$|$d(t)$| L_x, has been solved in Section 4.3.1.

The problem with no-inventory cost, $1|1|u(t)\in[0,U]$,$X(0)$=$X(T)$|$d(t)$| $Q_u, L_x, c_x^+=0$ is easily solved for a moderate demand such that $\frac{1}{t}\int_0^t d(\tau)d\tau \le \frac{1}{T}\int_0^T d(\tau)d\tau \le U$ for $0<t\le T$. Indeed, since the co-state variable cannot increase when $c_x^+=0$, the solution periodicity causes the co-state variable to be constant. This is only possible if no backlog occurs over the demand period. Therefore, the optimal control must be constant and the only control regime is the partial production $u(t)\equiv\frac{1}{T}\int_0^T d(t)dt$. This value obtained from the periodicity of the state variable, $X(T)=X(0)$, ensures that there is no backlog along the demand period.

No Backlog

When backlogs are disallowed, i.e. the $1|1|u(t)\in[0,U]$, $X(t)\ge 0$,$X(0)$=$X(T)$|$d(t)$| Q_u, L_x problem is solvable with the aid of Algorithm 4.22 by setting c_x^- very large. In this case, the solution includes constant partial production $u(t)=d_1$ for $0\le t\le t_1$, linearly increasing partial production $u(t)=d_1+\frac{c_x^+}{2c_u}(t-t_1)$ for $t_1\le t\le t_2$ and full production $u(t)=U$ for $t_2\le t\le T$, where the switching points are as follows: $t_1=-\frac{c_u}{c_x^+}(U-d_1)+\frac{(U-d_2)T+(d_2-d_1)\bar{t}}{U-d_1}$, $t_2=t_1+\frac{2c_u}{c_x^+}(U-d_1)$. The formal proof for optimality of this solution is left as an exercise for the reader.

4.3.4 NON-LINEAR INVENTORY AND PRODUCTION COSTS

Both costs entering the objective are now non-linear. This further complicates the optimal behavior of the system. However, an exact optimal solution is still possible for a case of polynomial demands. An effective approximation algorithm is also available for an arbitrary demand function.

Backlogs

The problem to be considered is one-machine, one-product-type scheduling with backlogs and limited production capacity to minimize the quadratic inventory and production costs over period T, that is, the $1|1|u(t)\in[0,U], X(0)=X(T)|d(t)|Q_u, Q_x$ problem which is stated as follows

$$J=\int_0^T\left[c_x^+X^+(t)^2+c_x^-X^-(t)^2+c_uu(t)^2\right]dt\to\min, \tag{4.117}$$

subject to (4.102) and (4.103).

Dual Formulation

The Hamiltonian, H, resulting from the maximum principle is

$$H(t)=-\left[c_x^+X^+(t)^2+c_x^-X^-(t)^2+c_uu(t)^2\right]+\psi(t)\left[u(t)-d(t)\right]. \tag{4.118}$$

The co-state variable $\psi(t)$ satisfies the co-state differential equation (4.111). Maximization of the Hamiltonian as a function of $u(t)$ results, as in the previous section, in

$$u(t)=\begin{cases}U, \text{ if } \psi(t)>2c_uU,\\ 0, \text{ if } \psi(t)<0,\\ \dfrac{\psi(t)}{2c_u}, \text{if } 0\le\psi(t)\le 2c_uU,\end{cases} \tag{4.119}$$

Algorithm 4.23

Step 1:

Run Algorithm 3.18. If $u(t)\le U$ for $0\le t\le T$, then the optimal solution found by Algorithm 3.18 is valid, STOP.
Otherwise go to the next step.

Step 2:

(i) Set accuracy ξ required for approximating the objective of the optimal solution. Calculate parameters $g=\sqrt{\dfrac{\max\{c_x^+,c_x^-\}}{c_u}}$,

$$A=\max\left\{1,\frac{\max_t d(t)}{U}\right\},\ \varepsilon=\left(2c_ugTU^2\left(1+g^2T^2\right)\left(gTch(gT)+Ash(gT)\right)\right)^{-1}\xi,$$

$\varepsilon_1=2UTc_ug^2\varepsilon$ and $\varepsilon_2=UA\varepsilon$.

(ii) Set inventory levels $X^{\max} = UT$, $X^{\min} = -UT$.

(iii) Calculate $X_0 = 0.5(X^{\max} + X^{\min})$, set co-state levels $\psi^{\max} = 2c_u U(1 + g^2 T^2)$, $\psi^{\min} = 0$ and $\psi_0 = \psi^{\max}$.

(iv) Integrate the following system of ordinary differential equations

$$\dot{\psi}(t) = \begin{cases} 2c_x^+ X(t), & \text{if } X(t) > 0; \\ 2c_x^- X(t), & \text{if } X(t) \le 0; \end{cases} \qquad \psi(0) = \psi_0,$$

$$\dot{X}(t) = \begin{cases} \dfrac{\psi(t)}{2c_u} - d(t), & \text{if } 0 \le \psi(t) \le 2c_u U, \\ U - d(t), & \text{if } 2c_u U \le \psi(t), \end{cases} \qquad X(0) = X_0 \tag{4.120}$$

over interval [0,*T*].
If $\psi(T) < \psi(0)$, then $X^{\max} = X_0$ and go to (viii). Otherwise, assign $\psi_0 = \psi^{\min}$ and integrate (4.120).
If $\psi(T) > \psi(0)$, then $X^{\min} = X_0$ and go to (viii).
Otherwise go to (v).

(v) Assign $\psi_0 = 0.5(\psi^{\min} + \psi^{\max})$ and integrate (4.120).
If $\psi(T) > \psi(0)$, then $\psi^{\max} = \psi_0$ and go to (vi);
If $\psi(T) < \psi(0)$, then $\psi^{\min} = \psi_0$ and go to (vi);

(vi) If $\psi^{\max} - \psi^{\min} > \varepsilon_1$, then go to (v); Otherwise go to (vii).

(vii) If $X(T) > X(0)$, then $X^{\max} = X_0$ and go to (viii);
If $X(T) < X(0)$, then $X^{\min} = X_0$ and go to (viii);
(viii) If $X^{\max} - X^{\min} > \varepsilon_2$ then go to (iii). Otherwise,
set the optimal solution as

$$u(t) = \begin{cases} \dfrac{\psi(t)}{2c_u}, & \text{if } 0 \le \psi(t) \le 2c_u U, \\ U, & \text{if } 2c_u U \le \psi(t). \end{cases} \qquad \text{STOP.}$$

Discussion

Step 1 of Algorithm 4.23 gives a closed form optimal control by running Algorithm 3.18 developed for the corresponding uncapacitated problem. Step 2 of Algorithm 4.23 presents a dichotomous-search-based procedure for approximating the optimal solution. This is due to the fact that localization of the partial and full production regimes is reduced to finding a root of specially constructed slack functions. In what follows, we a) define slack functions, b) prove their monotonicity and c) find boundaries for the functions' arguments within which the roots fall.

Definition 4.6. Given system (4.120), F_{11} is a function which maps $\psi(0) \to \psi(T) - \psi(0)$.

Definition 4.7. Given system (4.120), F_{12} is a function which maps $X(0) \to X(T) - X(0)$.

LEMMA 4.38. Slack functions F_{11} and F_{12} are non-decreasing functions of their arguments.

PROOF. Since the proof for both slack functions is very similar, we consider here function $F_{11} = \psi(T) - \psi(0) = \int_0^T \dot{\psi}(t)dt$. By substituting co-state equation (4.111) into the last expression, we find

$$F_{11} = 2\int_0^T c_x(X(t))X(t)dt, \tag{4.121}$$

where $c_x(X) = \begin{cases} c_x^+, & \text{if } X \geq 0, \\ c_x^-, & \text{if } X < 0. \end{cases}$

Let $(X_1(t), \psi_1(t))$ and $(X_2(t), \psi_2(t))$ be two solutions of system (4.120) satisfying the initial conditions $X_1(0) = X_2(0)$ and $\psi_1(0) > \psi_2(0)$. Then we prove that $X_1(t) \geq X_2(t)$ for $0 \leq t \leq T$. It follows from this that F_{11} determined by (4.121) is a non-decreasing function of its argument.

This is accomplished by considering a small interval of time $[0, \hat{t}]$ at which $\psi_1(t) > \psi_2(t)$. According to (4.119), $u(t)$ is a non-decreasing function of $\psi(t)$, i.e. $u_1(t) \geq u_2(t)$, $t \in [0, \hat{t}]$. From equation (4.102), it immediately follows that $\dot{X}_1(t) \geq \dot{X}_2(t)$ at the same time interval. Because $X_1(0) = X_2(0)$ and $\dot{X}_1(t) \geq \dot{X}_2(t)$, then $X_1(t) \geq X_2(t)$. Consequently, from the co-state equation (4.111), we conclude that

$\dot{\psi}_1(t) \geq \dot{\psi}_2(t)$. Because $\psi_1(0) > \psi_2(0)$ and $\dot{\psi}_1(t) \geq \dot{\psi}_2(t)$, then $\psi_1(\hat{t}) - \psi_2(\hat{t}) \geq \psi_1(0) - \psi_2(0)$. Thus, it has been proved that if $\psi_1(0) > \psi_2(0)$, then $X_1(\hat{t}) \geq X_2(\hat{t})$ and $\psi_1(\hat{t}) - \psi_2(\hat{t}) > 0$. Therefore, the proof can be continued in the same way on subsequent time intervals, each of which is no shorter than $[0, \hat{t}]$ until the entire period is covered. Thus, $X_1(t) \geq X_2(t)$, for $0 \leq t \leq T$ which means that F_{11} is a non-decreasing function.
■

The following lemma sets bounds for the state and co-state variables to be utilized in the dichotomous search performed in Step 2 of Algorithm 4.23.

LEMMA 4.39. Given problem (4.117), (4.102) and (4.103), the state variable is bounded as $X(t) \in [-UT, UT]$ and the co-state variable $\psi(t)$ is bounded as $\psi(t) \in [0, \hat{\psi}]$, $t \in [0, T]$, where $\hat{\psi} = 2c_u U(1 + g^2 T^2)$ and $g = \sqrt{\dfrac{\max\{c_x^+, c_x^-\}}{c_u}}$.

PROOF. The bounds of the state variable are proved by contradiction. Assume that there exists an optimal solution $(X(t), u(t))$ with a point of time $\bar{t}$ where $X(\bar{t}) = UT + \delta$, $\delta > 0$. Then, with respect to (4.102), (4.103) and recalling the feasibility condition $U \geq \dfrac{1}{T}\displaystyle\int_0^T d(t)dt$, we conclude that $X(t) \geq \delta$ for $t \in [0, T]$. Consider now another solution $(X(t) - \delta, u(t))$. This solution is feasible since it satisfies constraints (4.102) and (4.103). Moreover, this solution obviously provides a better objective value (4.117) that contradicts the assumption that $(X(t), u(t))$ is optimal, that is $X(t) \leq UT$ for $t \in [0, T]$. Similarly, one can prove the lower bound for the state variable.

The lower bound for the co-state variable is also proved by contradiction. Contrary to the lemma, let $\psi(t) < 0$ at an interval of time $t \in [t_1, t_2]$. If, at a sub-interval of this interval $X(t) \leq 0$, then according to the co-state equation (4.111) $\psi(t)$ does not increase along the interval. Therefore, the control effort, which is zero at the interval (see (4.119)) retains this value since $\psi(t)$ is negative. Under zero production rate, the buffer level $X(t)$ will not become positive, forcing $\psi(t)$ to decrease along the infinite planning horizon. This contradicts the periodicity of the optimal solution.

Likewise, assuming that $X(t)>0$ at an interval where $\psi(t) < 0$, we conclude that $X(t)$ does not increase before this interval. This again contradicts the periodicity of the optimal solution.

The upper bound of the co-state variable is found by considering two marginal cases. The first case is due to the positive inventory all over the demand period $\psi(t)=\psi(0)+2c_x^+\int_0^t X(\tau)d\tau$. Having $\psi(0)=2c_uU$ and $X(t)\le UT$, we find $\psi(t)\le 2c_uU+2c_x^+UT^2$. The second case is due to the negative inventory all over the demand period $\psi(t)=\psi(T)+2c_x^-\int_T^t X(\tau)d\tau$. Having $\psi(T)=2c_uU$ and $X(t)\ge -UT$, we find $\psi(t)\le 2c_uU+2c_x^-UT^2$. By choosing the worst of the two estimates, we prove the lemma.

■

THEOREM 4.12. Problem (4.117), (4.102) and (4.103) is solvable with accuracy ξ in $O\left(\frac{R}{\xi}\left(\log\frac{R}{\xi}\right)^2\right)$ time,

where $R=2c_ugT^2U^2\left(1+g^2T^2\right)\left(gTch(gT)+Ash(gT)\right)$, $g=\sqrt{\frac{\max\left\{c_x^+,c_x^-\right\}}{c_u}}$ and

$$A=\max\left\{1,\frac{\max\limits_t d(t)}{U}\right\}.$$

PROOF. The first step of Algorithm 4.23 runs Algorithm 3.18 which exactly solves the problem for the case of polynomial demand of order P in $O(P)$ time. The second step approximates the solution by searching for the roots of monotone slack functions F_{11} and F_{12} with accuracy ε_1 and ε_2 respectively. Since obtaining the slack function values involves integration of system (4.120) with accuracy ε, the total complexity of Algorithm 4.23 Step 2 is $O\left(\frac{T}{\varepsilon}\log\frac{\hat{\psi}}{\varepsilon_1}\log\frac{2UT}{\varepsilon_2}\right)$.

To estimate the relationship between computational accuracy of the objective ξ, accuracy of integration ε and accuracy of the dichotomous search ε_1 and ε_2, we vary the objective function (4.117)

$$\xi\le 2T\max\{c_x^+\max_t X^+(t),c_x^-\max_t X^-(t)\}\max_t\left|\Delta X(t)\right|+2Tc_u\max_t u(t)\max_t\left|\Delta u(t)\right|. \quad (4.122)$$

The value of $\max_t u(t)$ is evidently U. The maximum inventory and backlog have been estimated in Lemma 4.39 as UT. The value of $\max_t |\Delta u(t)|$ is estimated from (4.119) as $\frac{1}{2c_u}\max_t |\Delta\psi(t)|$. To estimate the error of the co-state variable, we solve the following differential equation

$$\Delta\ddot{\psi}(t) = g^2 \Delta\psi(t),\quad \Delta\psi(0) = \varepsilon_1 \ ,\quad \Delta\dot{\psi}(0) = 2c_u g^2 \varepsilon_2\,, \tag{4.123}$$

where ε_1 and ε_2 are determined as

$$\varepsilon_1 = \max_t \dot{\psi}(t)\varepsilon = 2\max\{c_x^+ \max_t X^+(t), c_x^- \max_t X^-(t)\}\varepsilon = 2UTc_u g^2 \varepsilon\,. \tag{4.124}$$

$$\varepsilon_2 = \max_t |\dot{X}(t)|\varepsilon = \max\{U, \max_t d(t)\}\varepsilon\,. \tag{4.125}$$

Thus, solving (4.123), we find $\Delta\psi(t) = \varepsilon_1 ch(gt) + 2c_u g\varepsilon_2 sh(gt)$, and $\max_t |\Delta\psi(t)|$ is achieved at t=T. The upper estimates for $\max_t |\Delta u(t)|$ and $\max_t |\Delta X(t)|$ are $\max_t |\Delta u(t)| = \frac{1}{2c_u}\max_t |\Delta\psi(t)|$ and $\max_t |\Delta X(t)| = \frac{T}{2c_u}\max_t |\Delta\psi(t)|$ respectively. By substituting the found estimates as well as expressions (4.124) and (4.125) into (4.122), we obtain maximum error ξ as a function of ε

$$\xi = 2c_u gTU^2\left(1+g^2T^2\right)\left(gTch(gT) + Ash(gT)\right)\varepsilon\,, \tag{4.126}$$

where $A = \max\left\{1, \frac{\max_t d(t)}{U}\right\}$.

Finally, to prove the theorem, we substitute (4.126) into the complexity expression and introduce parameter R to simplify the presentation.

■

Special Cases of Zero Costs

The problem with no production cost, i.e. the $1|1|u(t)\in[0,U]$,$X(0)$=$X(T)|d(t)|Q_x$ problem has been solved in Section 4.3.2. The problem with no-inventory cost, $1|1|u(t)\in[0,U]$, $X(0)$=$X(T)|d(t)|Q_u, Q_x, c_x^+=0$ is equivalent to the $1|1|u(t)\in[0,U]$, $X(0)$=$X(T)|d(t)|Q_u, L_x, c_x^+=0$ problem considered in Section 4.3.3.

No Backlog

The $1|1|u(t)\in[0,U], X(t)\geq 0, X(0)=X(T)|d(t),T|Q_u, Q_x$ problem is polynomially solvable if either the demand is two-level, piece-wise constant $d(t)=\begin{cases} d_1, \text{ if } 0\leq t\leq \bar{t}, \\ d_2, \text{ if } \bar{t}<t\leq T, \end{cases}$ $d_1<U<d_2$, or it is such that the production rate does not reach the maximum level.

Dual Formulation

Now problem (4.117), (4.102) and (4.103) has the additional constraint $X(t)\geq 0$ which affects neither Hamiltonian (4.118) nor optimality condition (4.119), but changes the co-state differential equation

$$d\psi(t)=c_x^+X(t)dt-d\mu(t),\quad \psi(T+0)=0, \tag{4.127}$$

where Lagrange multiplier $d\mu(t)\geq 0$ corresponds to the no-backlog constraint and satisfies the complementary slackness condition (4.16).

Algorithm 4.24

Step 1:

Run Algorithm 3.19. If $u(t)\leq U$ for $0\leq t\leq T$, then the optimal solution found by Algorithm 3.19 is valid, STOP.
Otherwise go to the next step.

Step 2:

If the demand function $d(t)$ is two-level, piece-wise constant

$$d(t)=\begin{cases} d_1, \text{ if } 0\leq t\leq \bar{t}, \\ d_2, \text{ if } \bar{t}<t\leq T, \end{cases} \quad \text{and} \quad U\geq d_2+\frac{Bsh(yT)}{2c_u}\left(1-cth(yT)ch(yT)\right),$$

where $B=2c_u(d_2-d_1)sh(y\bar{t})$, $A=B\big(cth(y\bar{t})-cth(yT)\big)$, $y=\sqrt{\dfrac{c_x^+}{c_u}}$,

then set the optimal solution as

$$u(t)=d_1+\frac{A}{2c_u}ch(yt) \qquad \text{for } 0\leq t\leq \bar{t};$$

$$u(t)=d_2+\frac{Bsh(yt)}{2c_u}\left(1-cth(yT)ch(yt)\right) \qquad \text{for } \bar{t}\leq t\leq T;$$

STOP.
Otherwise go to the next step.

Step 3:

(i) If the demand function $d(t)$ is two-level, piece-wise constant

$$d(t)=\begin{cases} d_1, \text{ if } 0\leq t\leq \bar{t}, \\ d_2, \text{ if } \bar{t}<t\leq T, \end{cases} \quad \text{and} \quad U<d_2+\frac{Bsh(yT)}{2c_u}\left(1-cth(yT)ch(yT)\right),$$

then set accuracy ξ required for approximating the objective of the optimal solution, calculate accuracy of dichotomous search, $\varepsilon = \left(4c_u T^2 U^2 y^2 \left(1 + y^2 T^2\right) ch(yT)\right)^{-1} \xi$ and set time points $t_1 = 0$, $t_2 = \bar{t}$.

(ii) Calculate $t_y = 0.5\left(t_1 + t_2\right)$, and

$$f = UT - d_1\bar{t} - d_2(T - \bar{t}) + \frac{U - d_1}{y}\left(th(yt_y) - yt_y\right).$$

If $f > 0$, then $t_1 = t_y$ and go to (iii);

if $f < 0$, then $t_2 = t_y$ and go to (iii).

(iii) If $t_1 - t_2 > \varepsilon$, then go to (ii). Otherwise, check if $t_y \le \bar{t}$, then set the optimal solution as

$$u(t) = d_1 + \frac{U - d_1}{ch(yt_y)} ch(yt) \quad \text{for } 0 \le t < t_y,$$

$u(t) = U$ for $t_y \le t \le T$, STOP.

Otherwise STOP.

Discussion

According to Algorithm 4.24, problem (4.117), (4.102) and (4.103) is not always solvable. Step 1 runs Algorithm 3.19 developed for approximating the optimal solution to the corresponding uncapacitated problem. A solvable case is found when demand is two-level, piece-wise constant and only a partial production regime is involved. Step 2 determines the exact optimal solution in $O(1)$ time for this case. Step 3 deals with the case when the two-level, piece-wise constant demand induces a switch from partial to full production at time point $t_y \le \bar{t}$. This case is approximately solved with accuracy ξ in $O\left(\log \frac{R}{\xi}\right)$ time. The same complexity is needed to solve the problem if $t_y > \bar{t}$. The development of the solution for this case is left as an exercise for the reader.

LEMMA 4.40. Given problem (4.117), (4.102) and (4.103), the co-state variable $\psi(t)$ is non-negative, $\psi(t) \ge 0$ for $t \in [0, T]$.

PROOF. The proof of this lemma is similar to the proof of the lower bound for the co-state variable in Lemma 4.39 and therefore is omitted.

■

LEMMA 4.41. (Correctness of Algorithm 4.24 Steps 2 and 3) Given problem (4.117), (4.102) and (4.103), and parameter $y = \sqrt{\dfrac{c_x^+}{c_u}}$,

- if the demand function *d(t)* is two-level, piece-wise constant $d(t) = \begin{cases} d_1, & \text{if } 0 \le t \le \bar{t}, \\ d_2, & \text{if } \bar{t} < t \le T, \end{cases}$

and $U \ge d_2 + \dfrac{Bsh(yT)}{2c_u}(1 - cth(yT)ch(yT))$, then the optimal solution is

$u(t) = d_1 + \dfrac{A}{2c_u} ch(yt)$ for $0 \le t \le \bar{t}$; $u(t) = d_2 + \dfrac{Bsh(yt)}{2c_u}(1 - cth(yT)ch(yt))$ for $\bar{t} \le t \le T$,

where $B = 2c_u(d_2 - d_1)sh(y\bar{t})$, $A = B(cth(y\bar{t}) - cth(yT))$;

- otherwise, the optimal solution is

$u(t) = d_1 + \dfrac{U - d_1}{ch(yt_y)} ch(yt)$ for $0 \le t < t_y$; $u(t) = U$ for $t_y \le t \le T$,

where point $t_y \le \bar{t}$ satisfies the following algebraic equation

$$UT - d_1\bar{t} - d_2(T - \bar{t}) + \frac{U - d_1}{y}\left(th(yt_y) - yt_y\right) = 0.$$

PROOF. First, we prove by contradiction that the partial production regime $u(t) = d(t)$, $X(t) = 0$ can never occur at an interval of time. If this is indeed the case, then this regime must be maintained until the end of the demand period. Since $d_2 > U$, the backlog would be unavoidable. This contradicts the no-backlog constraint. Moreover, *X(t)* can become zero only at points $t = 0$ and $t = T$ because, due to the two-level piece-wise constant demand, *X(t)* always increases before $t = \bar{t}$ and always decreases after $t = \bar{t}$.

Thus, the inventory is positive within the demand period and the co-state differential equation (4.127) simplifies to

$$\dot{\psi}(t) = 2c_x^+ X(t). \tag{4.128}$$

By differentiation (4.128) and taking into account (4.102), (4.119) and Lemma 4.40, we obtain

$$\ddot{\psi}(t) = \begin{cases} 2c_x^+(U - d(t)), & \text{if } 2c_u U \le \psi(t) \\ \dfrac{c_x^+}{c_u}\psi(t) - 2c_x^+ d(t), & \text{if } 0 \le \psi(t) \le 2c_u U \end{cases}, \qquad 0 \le t \le T. \tag{4.129}$$

Thus, the system of the state and co-state differential equations has been presented as a second-order ordinary differential equation (4.129). The boundary conditions for (4.129) are the periodicity conditions

$$\dot{\psi}(0)=\dot{\psi}(T)=0. \tag{4.130}$$

$$\psi(0)=\psi(T+0)=0. \tag{4.131}$$

Due to the jump of the demand function $d(t)$ at point $\bar{t}$, additional boundary constraints for the co-state equation (4.129) are

$$\psi(\bar{t}-0)=\psi(\bar{t}+0), \tag{4.132}$$

$$\dot{\psi}(\bar{t}-0)=\dot{\psi}(\bar{t}+0). \tag{4.133}$$

Consider the first case stated in the lemma, when the optimal control is characterized by the partial production regime over the entire demand period. In such a case, introducing parameter $y=\sqrt{\dfrac{c_x^+}{c_u}}$ causes equation (4.129) to take the following form

$$\ddot{\psi}(t)=y^2\psi(t)-2c_x^+d(t), \qquad 0\le t\le T. \tag{4.134}$$

$$d\mu(t)=0,\ \ 0<t<T,\ d\mu(T)=\psi(T)-\psi(0).$$

By analytically solving equation (4.134) with boundary conditions (4.130), (4.132) and (4.133), we obtain

$$\psi(t)=2c_u d_1+Ach(yt) \text{ for } 0\le t\le \bar{t};$$

$$\psi(t)=2c_u d_2+Bsh(yt)(1-cth(yT)ch(yt)) \text{ for } \bar{t}\le t\le T,$$

where $B=2c_u(d_2-d_1)sh(y\bar{t})$, $A=B\big(cth(y\bar{t})-cth(yT)\big)$. This solution is feasible if $\psi(T)\le 2c_uU$. This proves the first condition of the lemma.

Consider the second case stated in the lemma, when the optimal control is characterized by the partial and full production regimes and switching point $t_y\le\bar{t}$ between these two regimes (see Figure 4.14). In this case, equation (4.129) takes the following form

$$\ddot{\psi}(t)=2c_x^+U-d(t), \qquad 0\le t<t_y, \tag{4.135}$$

$$\ddot{\psi}(t) = y^2\psi(t) - d(t), \quad t_y < t \le T, \tag{4.136}$$

$$d\mu(t) = 0, \;\; 0 < t < T, \; d\mu(T) = \psi(T) - \psi(0).$$

The solution of equations (4.135) and (4.136) with boundary conditions (4.130), (4.132) and (4.133) leads to the algebraic equation in unknown t_y stated in the second condition of the lemma. This solution is feasible if $t_y \le \bar{t}$.

■

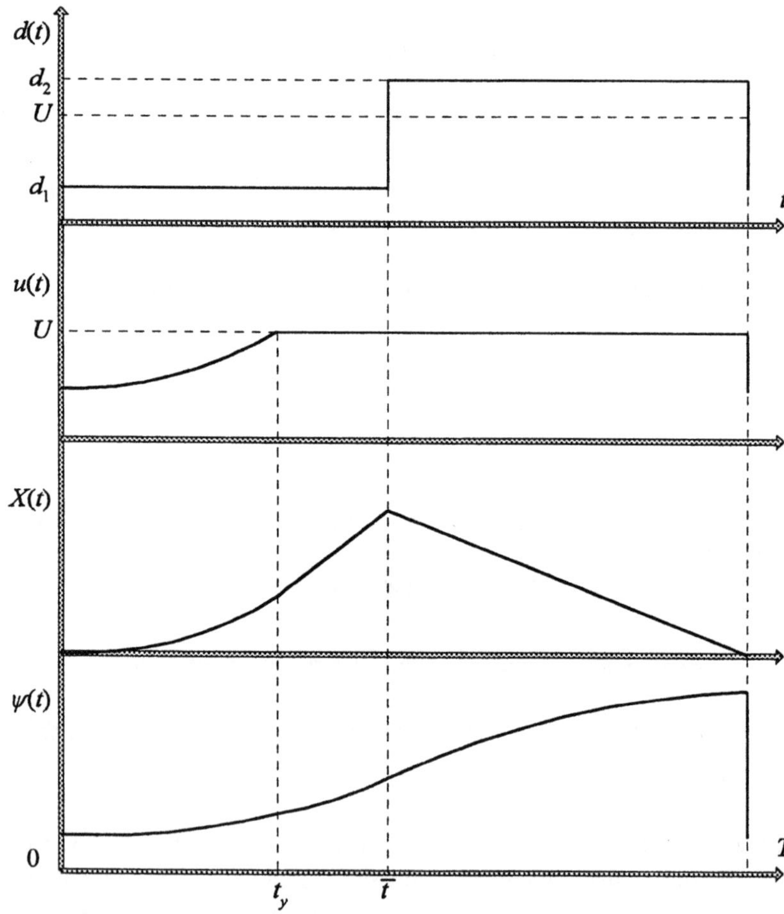

Figure 4.14. Demand $d(t)$, optimal control $u(t)$, state $X(t)$ and co-state $\psi(t)$ variables for the $1|1|u(t) \in [0,U], X(t) \ge 0, X(0)=X(T)|d(t),T|Q_u, Q_x$ problem with two-level piece-wise constant demand.

THEOREM 4.13. If the demand function $d(t)$ is two-level, piece-wise constant $d(t)=\begin{cases} d_1, \text{ if } 0\le t\le \bar{t}, \\ d_2, \text{ if } \bar{t}<t\le T, \end{cases}$ then problem (4.117), (4.102) and (4.103) is solvable with accuracy ξ in $O\left(\log\frac{R}{\xi}\right)$ time, where $R=4c_uT^3U^2y^2\left(1+y^2T^2\right)ch(yT)$.

PROOF. Step 2 of Algorithm 4.24 finds the solution in $O(1)$ time if the optimal control does not exceed the capacity. If, however, this is not the case, Step 3 approximates the optimal solution by the dichotomous search with accuracy ε for the root of non-increasing function $f(t_y)=UT-d_1\bar{t}-d_2(T-\bar{t})+\frac{U-d_1}{y}\left(th(yt_y)-yt_y\right)$.

This requires $O\left(\log\frac{T}{\varepsilon}\right)$ operations.

To estimate the relationship between computational accuracy of the objective ξ and accuracy of the dichotomous search ε, we vary the objective function (4.117)

$$\xi\le 2Tc_x^+\max_t X(t)\max_t\left|\Delta X(t)\right|+2c_uUT\max_t\left|\Delta u(t)\right|. \qquad (4.137)$$

The maximum inventory is

$$\max_t X(t)=T\max_t u(t)=TU\ .$$

The value of $\max_t|\Delta u(t)|$ is estimated from (4.119) as $\frac{1}{2c_u}\max_t|\Delta\psi(t)|$. To estimate the error of the co-state variable, we solve the following differential equation

$$\Delta\ddot{\psi}(t)=y^2\Delta\psi(t)\,,\quad \Delta\psi(t_y)=2\hat{\varepsilon}\ ,\quad \Delta\dot{\psi}(0)=0\,, \qquad (4.138)$$

where $\hat{\varepsilon}$ is determined by

$$\hat{\varepsilon}=\max_t\left|\dot{\psi}(t)\right|\varepsilon=2c_x^+\max_t X(t)\varepsilon=2UTc_uy^2\varepsilon\,. \qquad (4.139)$$

Since $\max_t\Delta\psi(t)$ found as a solution of (4.138) is less than $2ch(yT)\hat{\varepsilon}$ (see proof of Theorem 3.3), the upper estimates for $\max_t|\Delta u(t)|$ and $\max_t|\Delta X(t)|$ are

$\max_t|\Delta u(t)|=\frac{1}{2c_u}\max_t|\Delta\psi(t)|=\frac{\hat{\varepsilon}}{c_u}ch(yT)$ and $\max_t|\Delta X(t)|=\frac{T}{2c_u}\max_t|\Delta\psi(t)|=\frac{T\hat{\varepsilon}}{c_u}ch(yT)$

respectively. By substituting the found estimates into (4.137), we obtain maximum error ξ as a function of $\hat{\varepsilon}$

$$\xi = 2TUch(yT)\left(1+y^2T^2\right)\hat{\varepsilon}. \tag{4.140}$$

To obtain the final estimates of the complexity and accuracy for Algorithm 4.24 Step 3, we express $\hat{\varepsilon}$ from (4.140) and ε from (4.139), substitute them into the complexity of Step 3 and introduce parameter R to simplify the presentation.

■

4.3.5 MAINTAINING STABLE INVENTORY

Backlogs

The problem to be considered is one-machine, one-product-type production control, with limited rate to maintain stable inventory, under unknown, arbitrary demand functions, that is the $1|1|u(t)\in[0,U]\,|d(t)|\,\overline{X}$ problem which is stated as follows

$$\sup_{0\le t<\infty} |X(t)| < \infty, \tag{4.141}$$

$$\dot{X}(t) = u(t) - d(t), \quad X(0) = 0, \tag{4.142}$$

$$0 \le u(t) \le U. \tag{4.143}$$

The feedback control rules suggested in this section are based on those presented in Section 3.3.5 for the uncapacitated problem.

Feedback Rule 4.1

Proportional, limited capacity, feedback control:

$$\text{set } u(X(t)) = \begin{cases} U, & \text{if } X(t) < \overline{X} - \dfrac{U}{G}, \\ G(\overline{X} - X(t)), & \text{if } \overline{X} - \dfrac{U}{G} \le X(t) \le \overline{X}, \\ 0, & \text{if } X(t) > \overline{X}. \end{cases} \text{ at every } t.$$

Feedback Rule 4.2

Bang/bang (limited capacity, impulse) feedback control:

$$\text{set } u(X(t)) = \begin{cases} U, & \text{if } X(t) \le 0, \\ 0, & \text{if } X(t) \ge \overline{X}, \\ \text{no change}, & \text{if } 0 < X(t) < \overline{X}. \end{cases} \qquad \text{at every } t.$$

Discussion

Feedback Rules 4.1 and 4.2 are of $O(1)$ complexity at every time t and are natural extensions of corresponding Rules 3.1 and 3.3, developed for the uncapacitated problems. Feedback Rule 3.2 is not considered here because its straightforward extensions for the case of limited capacity result in either special cases of Rules 4.1 and 4.2, or chattering with uncountable switches between the marginal values of control, $u(t) \in \{0, U\}$.

Although Rules 4.1 and 4.2 are simple to use, inventory stability is not guaranteed when dealing with arbitrary demands. In other words, given a specific feedback rule, the inventory can grow unlimitedly. Therefore, given a limited production capacity, it is even more critical, in comparison to the uncapacitated problem, to determine the class of demand functions for which inventory stability is guaranteed. The following two lemmas identify the demand-related conditions which are sufficient for bounding the inventory when the two respective feedback rules are applied.

LEMMA 4.42. (Stability of Feedback Rule 4.1) Given problem (4.141)-(4.143), if the proportional, limited capacity, feedback control is applied with $\overline{X} \ge 0$ and demand is such that $\lim_{t\to\infty} \frac{1}{t}\int_0^t d(\tau)d\tau < U$, then $X(t) \le \overline{X}$ for all t and $\inf_{0 \le t < \infty} X(t) > -\infty$.

PROOF. First note that since $X(0) \le \overline{X}$, the inventory level can never exceed $\overline{X}$. Indeed, by definition of the proportional feedback control, once $X(t)$ reaches $\overline{X}$, production stops and demand again drains inventory below $\overline{X}$, i.e.

$$X(t) \le \overline{X}. \tag{4.144}$$

Since the second condition of the proportional feedback control is applied only when the inventory is bounded, $\overline{X} - \frac{U}{G} \le X(t) \le \overline{X}$, to estimate the lower bound of $X(t)$, we substitute the first condition of the rule, $u(t)=U$, into the state equation (4.142)

$$\dot{X}(t) = U - d(t), \quad X(0) = 0. \tag{4.145}$$

The solution of this equation is

$$X(t) = Ut - \int_0^t d(\tau)d\tau \,. \tag{4.146}$$

From (4.146), it follows that the inventory $X(t)$ is bounded from below if

$$\lim_{t\to\infty} \frac{1}{t} \int_0^t d(\tau)d\tau < U \,. \tag{4.147}$$

■

LEMMA 4.43. (Stability of Feedback Rule 4.2) Given problem (4.141)-(4.143), if bang/bang feedback control is applied with $\overline{X} \geq 0$ and demand is such that $\lim_{t\to\infty} \frac{1}{t} \int_0^t d(\tau)d\tau < U$, then $X(t) \leq \overline{X}$ for all t and $\inf_{0\leq t<\infty} X(t) > -\infty$.

PROOF. First note that since $X(0) \leq \overline{X}$, the inventory level can never exceed $\overline{X}$. Indeed, by definition of the bang/bang control (the second condition), once $X(t)$ reaches $\overline{X}$, production stops and demand again drains inventory below $\overline{X}$, i.e. $X(t) \leq \overline{X}$.

Since the third condition of the bang/bang feedback control presents the case when the inventory is bounded, to estimate the lower bound of $X(t)$, we substitute the first condition of the rule, $u(t)=U$, into the state equation (4.142)

$$\dot{X}(t) = U - d(t) \,, \quad X(0) = 0 \,.$$

The solution of this equation is

$$X(t) = Ut - \int_0^t d(\tau)d\tau \,. \tag{4.148}$$

From (4.148) it follows that the inventory $X(t)$ is bounded from below, as stated in the lemma.

■

Example 4.10

Consider the $1|1|u(t) \in [0,U]\,|d(t)|\,\overline{X}$ problem with a constant weak demand $d(t) = d < U$. Feedback Rules 4.1 and 4.2 are applied with unit parameter G

and $\overline{X} > \frac{d}{G}$. Since the demand is weak, both feedback rules provide stable inventories bounded as $0 \le X(t) \le \overline{X}$.

Furthermore, if Rule 4.1 is applied, the inventory level converges as though the problem is uncapacitated (see Example 3.9), while Rule 4.2 forces the inventory level to evolve cyclically. Each cycle of length T_{per} includes a production interval of length τ and no-production interval $T_{per} - \tau$. This implies (see Figure 4.15) that

$$\tau = \frac{\overline{X}}{U-d}, \qquad dT_{per} = U\tau,$$

and thus $T_{per} = \frac{U\overline{X}}{d(U-d)}$.

■

No Backlog

For the case when backlogs are disallowed, i.e. for the $1|1|u(t) \in [0,U], X(t) \ge 0\,|d(t)|\,\overline{X}$ problem, both Feedback Rules 4.1 and 4.2 are evidently stable and provide no backlog if demand is weak, $d(t) \le U$.

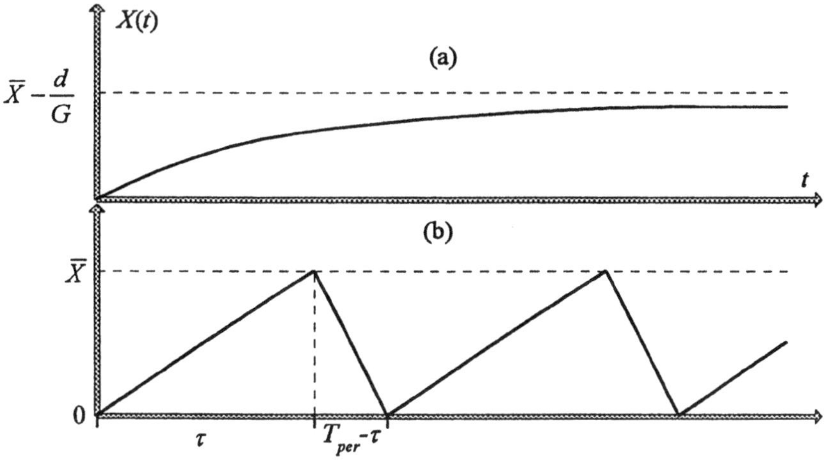

Figure 4.15. Evolution of inventory $X(t)$ under a constant, weak demand and Rules 4.1 (a) and 4.2 (b) for the $1|1|u(t) \in [0,U]\,|d(t)|\,\overline{X}$ problem.

4.4 VARIABLE PLANNING HORIZON

This section addresses the scheduling problems characterized by an amount of product to be completed by the end of a planning horizon. Unlike the previous sections, since in this section the length of the planning horizon is unknown *a priori*, it will be minimized along with the inventory and production costs.

4.4.1 LINEAR PERFORMANCE MEASURE

Consider one-machine, one-product-type scheduling with limited production rate to minimize the total linear cost of inventory, production and the planning horizon, that is, the $1|1|u(t)\in[0,U], X(T)=X_T \,\|\, L_u, L_x, T$ problem which is stated as follows

$$J=\int_0^T\left[c_x^+X(t)+c_uu(t)\right]dt+c_TT\to\min,$$

subject to

$$\dot{X}(t)=u(t),\qquad X(0)=0,\qquad X(T)=X_T, \tag{4.149}$$

$$0\le u(t)\le U. \tag{4.150}$$

Note that the above statement is characterized by no backlogs along the overall planning horizon because the planning horizon ends up as soon as the necessary amount of the product, $X(T)=X_T$, is completed. Next, the objective can be simplified by substituting the state equation (4.149) as follows

$$J=\int_0^T\left[c_x^+X(t)+c_uu(t)\right]dt+c_TT=c_uX_T+\int_0^T c_x^+X(t)dt+c_TT.$$

Since the total production cost c_uX_T is a constant which does not influence the optimization, the objective of the problem takes the form

$$J=\int_0^T c_x^+X(t)dt+c_TT\to\min. \tag{4.151}$$

The fact that the total production cost is independent of the control function results from the linearity of the production cost and the fixed terminal inventory.

Dual Formulation

The Hamiltonian, H, resulting from the maximum principle is

$$H(t) = -c_x^+ X(t) + \psi(t)u(t) = c_T. \tag{4.152}$$

The co-state variable, $\psi(t)$ in (4.152), satisfies the following co-state differential equation with no transversality condition

$$\dot{\psi}(t) = c_x^+. \tag{4.153}$$

Maximization of the Hamiltonian implies that

$$u(t) = \begin{cases} U, & \text{if } \psi(t) > 0, \\ 0, & \text{if } \psi(t) < 0, \\ b, \ b \in [0,U], & \text{if } \psi(t) = 0. \end{cases} \tag{4.154}$$

Algorithm 4.25

Step 1:

Set $T = \dfrac{X_T}{U}$.

Set the optimal solution as

$u(t) = U$ for $0 \le t \le T$,

STOP.

Discussion

From Algorithm 4.25, one can observe that the new terminal constraint $X(T) = X_T$ dictates the required amount of the product to be completed as fast as possible, regardless of the inventory and production costs.

The complexity of Algorithm 4.25 is $O(1)$ and its correctness is verified by solving the dual formulation in the following lemma.

LEMMA 4.44. (Correctness of Algorithm 4.25) Given problem (4.149)-(4.151), then $T = \dfrac{X_T}{U}$ and the optimal control is $u(t) = U$ for $0 \le t \le T$.

PROOF. Consider a solution to the problem in the form

$$u(t) = U, \quad X(t) = Ut \quad \text{and} \quad \psi(t) = \psi(0) + c_x^+ t \quad \text{for } 0 \le t \le T. \tag{4.155}$$

This solution satisfies the state (4.149) and co-state (4.153) equations as well as the optimality conditions (4.154) if $\psi(0) \geq 0$. To prove that $\psi(0) \geq 0$, we substitute solution (4.155) into (4.152)

$$H(t) = -c_x^+ Ut + \left(\psi(0) + c_x^+ t\right)U = c_T,$$

from where $\psi(0) = \dfrac{c_T}{U} \geq 0$.

■

Special Cases of Zero Costs

Since the solution that Algorithm 4.25 finds is independent of any cost coefficient, the problems: $1|1|u(t) \in [0,U], X(T) = X_T \| L_x, T$, $1|1|u(t) \in [0,U], X(T) = X_T \| L_u, L_x, T, c_x^+ = 0$ and $1|1|u(t) \in [0,U], X(T) = X_T \| L_u, L_x, T, c_T = 0$ are solvable by this algorithm with no modification.

4.4.2 NON-LINEAR INVENTORY COST

The problem to be considered is one-machine, one-product-type scheduling with limited capacity to minimize the quadratic inventory cost as well as the linear production and planning horizon costs, that is, the $1|1|u(t) \in [0,U], X(T) = X_T \| L_u, Q_x, T$ problem which is stated as follows

$$J = \int_0^T \left[c_x^+ X(t)^2 + c_u u(t)\right] dt + c_T T \rightarrow \min,$$

subject to (4.149) and (4.150).

As in the previous section, since the total production cost is constant, it does not influence the optimization. Therefore, the objective simplifies to

$$J = \int_0^T c_x^+ X(t)^2 \, dt + c_T T \rightarrow \min. \tag{4.156}$$

Dual Formulation

The Hamiltonian, *H*, resulting from the maximum principle is

$$H(t) = -c_x^+ X(t)^2 + \psi(t)u(t) = c_T. \tag{4.157}$$

The co-state variable, $\psi(t)$ in (4.157), satisfies the following co-state differential equation

$$\dot{\psi}(t) = 2c_x^+ X(t). \tag{4.158}$$

Maximization of the Hamiltonian implies that control $u(t)$ satisfies (4.154).

Algorithm

The algorithm is identical to Algorithm 4.25.

Discussion

Algorithm 4.25 is correct for both linear and quadratic inventory costs. The latter is proven in the following lemma.

LEMMA 4.45. (Correctness of Algorithm 4.25 for the case of the quadratic inventory cost) Given problem (4.156), (4.149) and (4.150), then $T = \frac{X_T}{U}$ and the optimal control is $u(t) = U$ for $0 \le t \le T$.

PROOF. Consider a solution in the form

$$u(t) = U, \quad X(t) = Ut \quad \text{and} \quad \psi(t) = \psi(0) + c_x^+ Ut^2 \quad \text{for } 0 \le t \le T. \tag{4.159}$$

This solution satisfies the state (4.149) and co-state (4.158) equations as well as the optimality conditions (4.154) if $\psi(0) \ge 0$. To prove that $\psi(0) \ge 0$, we substitute solution (4.159) into (4.157)

$$H(t) = -c_x^+ U^2 t^2 + \left(\psi(0) + c_x^+ Ut^2\right)U = c_T,$$

from where $\psi(0) = \frac{c_T}{U} \ge 0$.

■

Special Cases of Zero Costs

As with the case of the linear inventory cost problem, since production, inventory or planning horizon cost coefficients do not influence the optimal solution, no modification is required to Algorithm 4.25.

4.4.3 NON-LINEAR PRODUCTION COST

The next problem to be considered is one-machine, one-product-type scheduling with limited production rate to minimize the quadratic production cost as well as the linear inventory and planning horizon costs, that is, the
$1|1|u(t) \in [0, U], X(T) = X_T \,||\, Q_u, L_x, T$ problem which is stated as follows

$$J = \int_0^T \left[c_x^+ X(t) + c_u u(t)^2\right] dt + c_T T \rightarrow \min, \tag{4.160}$$

subject to (4.149) and (4.150).

Dual Formulation
The Hamiltonian, H, resulting from the maximum principle is

$$H(t) = -\left[c_x^+ X(t) + c_u u(t)^2\right] + \psi(t) u(t) = c_T. \tag{4.161}$$

The co-state variable, $\psi(t)$ in (4.161), satisfies the co-state differential equation (4.153).

Maximization of the Hamiltonian implies that the control function $u(t)$ satisfies

$$u(t) = \begin{cases} U, & \text{if } \psi(t) > 2c_u U, \\ 0, & \text{if } \psi(t) < 0, \\ \dfrac{\psi(t)}{2c_u}, & \text{if } 0 \le \psi(t) \le 2c_u U. \end{cases} \tag{4.162}$$

Algorithm 4.26
Step 1:

Calculate $T = \dfrac{2}{c_x^+}\left(\sqrt{c_u c_T + c_x^+ c_u X_T} - \sqrt{c_u c_T}\right)$.

If $2c_u U > 2\sqrt{c_u c_T} + c_x^+ T$, then

set the optimal solution as

$u(t) = \dfrac{1}{2c_u}\left(2\sqrt{c_u c_T} + c_x^+ t\right)$ for $0 \le t \le T$, STOP.

Otherwise go to the next step.

Step 2:

If $U \ge \sqrt{\dfrac{c_T}{c_u}}$, then calculate $t_x = \dfrac{2\left(c_u U - \sqrt{c_u c_T}\right)}{c_x^+}$ and

$T = t_x + \dfrac{1}{U}\left(X_T - \sqrt{\dfrac{c_T}{c_u}} t_x - \dfrac{c_x^+}{4c_u} t_x^2\right)$.

Set the optimal solution as

$$u(t) = \frac{1}{2c_u}\left(2\sqrt{c_u c_T} + c_x^+ t\right) \text{ for } 0 \le t \le t_x \text{ and}$$

$$u(t) = U \text{ for } t_x \le t \le T \text{ . STOP.}$$

Otherwise go to the next step.

Step 3:

Calculate $T = \frac{X_T}{U}$ and set the optimal solution as

$$u(t) = U \text{ for } 0 \le t \le T \text{; STOP.}$$

Example 4.11

Consider the $1|1|u(t) \in [0,U], X(T) = X_T \| Q_u, L_x, T$ problem with parameters U=4unit/hr, c_u=0.1\$hr/unit^2, c_x^+=0.2\$/hr.unit, c_T=1\$/hr and X_T = 13unit.

Step 1 of Algorithm 4.26 calculates $T = \frac{2}{c_x^+}\left(\sqrt{c_u c_T + c_x^+ c_u X_T} - \sqrt{c_u c_T}\right)$=2.84hr and proceeds to Step 2 because $2c_u U > 2\sqrt{c_u c_T} + c_x^+ T$ is not satisfied.

Step 2 verifies that $U \ge \sqrt{\frac{c_T}{c_u}}$, calculates $t_x = \frac{2\left(c_u U - \sqrt{c_u c_T}\right)}{c_x^+}$=0.84hr and $T = t_x + \frac{1}{U}\left(X_T - \sqrt{\frac{c_T}{c_u}} t_x - \frac{c_x^+}{4c_u} t_x^2\right)$=3.34hr. It then sets the optimal solution as

$u(t) = (3.16 + t)$ unit/hr for $0 \le t \le 0.84$ and
$u(t) = 4$ unit/hr for $0.84 \le t \le 3.34$.
■

Discussion

The optimal solution to this problem is characterized by a sequence of partial and full production regimes with switching time point t_x. Since one of the regimes may vanish, Algorithm 4.26 introduces three analytical cases of the optimal solution. These cases involve one of the following: partial production only (Step 1); full production following partial production (Step 2); full production only (Step 3). These cases are proved in the following lemma.

LEMMA 4.46. (Correctness of Algorithm 4.26) Given problem (4.160), (4.149) and (4.150), the optimal control is determined as follows

- if $2c_uU > 2\sqrt{c_uc_T} + c_x^+T$, then $u(t) = \frac{1}{2c_u}\left(2\sqrt{c_uc_T} + c_x^+t\right)$ for $0 \le t \le T$,

where $T = \frac{2}{c_x^+}\left(\sqrt{c_uc_T + c_x^+c_uX_T} - \sqrt{c_uc_T}\right)$;

- if $2\sqrt{c_uc_T} \le 2c_uU \le 2\sqrt{c_uc_T} + c_x^+T$, then

$u(t) = \frac{1}{2c_u}\left(2\sqrt{c_uc_T} + c_x^+t\right)$ for $0 \le t \le t_x$ and $u(t) = U$ for $t_x \le t \le T$,

where $t_x = \frac{2\left(c_uU - \sqrt{c_uc_T}\right)}{c_x^+}$ and $T = t_x + \frac{1}{U}\left(X_T - \sqrt{\frac{c_T}{c_u}}t_x - \frac{c_x^+}{4c_u}t_x^2\right)$;

- if $2c_uU < 2\sqrt{c_uc_T}$, then $u(t) = U$ for $0 \le t \le T$, where $T = \frac{X_T}{U}$.

PROOF. First we note that the production always starts from the very beginning of the planning horizon. The proof of this fact is identical to that of Lemma 3.38. Therefore, we consider a solution which starts with partial production followed by full production

$$\psi(t) = \psi(0) + c_x^+t,\quad u(t) = \frac{\psi(t)}{2c_u},\quad X(t) = \int_0^t u(\tau)d\tau \quad \text{for } 0 \le t \le t_x.$$

$$\psi(t) = \psi(0) + c_x^+t,\quad u(t) = U,\quad X(t) = \int_0^t u(\tau)d\tau \quad \text{for } t_x \le t \le T.$$

By taking into account the terminal inventory condition (4.149), switching condition $\psi(t_x) = 2c_uU$ emanating from (4.152) and Hamiltonian equation (4.151), we respectively find

$$\frac{\psi(0)}{2c_u}t_x + \frac{c_x^+}{4c_u}t_x^2 + (T - t_x)U = X_T, \tag{4.163}$$

$$\psi(0) + c_x^+t_x = 2c_uU, \tag{4.164}$$

$$\psi(0) = 2\sqrt{c_uc_T}. \tag{4.165}$$

The solution of system (4.163)-(4.165) in three unknowns $\psi(0)$, t_x and T proves the second condition of the lemma if $\psi(0) \le 2c_u U \le \psi(T)$. The proof for the remaining two conditions of this lemma is similar.

■

Special Cases of Zero Costs

The $1|1|u(t) \in [0,U], X(T) = X_T \| L_x, T,$ problem has been solved in Section 4.4.1. The case of zero inventory cost and quadratic production cost, i.e. the $1|1|u(t) \in [0,U], X(T) = X_T \| Q_u, L_x, T, c_x^+ = 0$ problem deserves special consideration.

Algorithm 4.27

Step 1:

If $U > \sqrt{\frac{c_T}{c_u}}$, then set $T = X_T \sqrt{\frac{c_u}{c_T}}$.

Set the optimal solution as

$$u(t) = \frac{X_T}{T} \text{ for } 0 \le t \le T, \text{ STOP.}$$

Otherwise, $T = \frac{X_T}{U}$.

Set the optimal solution as

$$u(t) = U \text{ for } 0 \le t \le T, \text{ STOP.}$$

Discussion

Algorithm 4.27 presents a simple analytical solution which is determined by the relation between the cost of production at the maximal rate per time unit $c_u U^2$ and the cost of prolonging the planning horizon by one time unit c_T. As a result, the value of the optimal production control is constant in time as proven in the following lemma.

LEMMA 4.47. (Correctness of Algorithm 4.27) Given problem (4.160), (4.149) and (4.150) with $c_x^+ = 0$, if $U > \sqrt{\frac{c_T}{c_u}}$, then the optimal solution is $u(t) = \frac{X_T}{T}$, $T = X_T \sqrt{\frac{c_u}{c_T}}$; otherwise, $u(t) = U$, $T = \frac{X_T}{U}$ for $0 \le t \le T$.

PROOF. Since $c_x^+ = 0$, the co-state variable $\psi(t)$ is constant in time, i.e. $\psi(t) = \psi(0) = 2\sqrt{c_u c_T}$ (see Lemma 4.46). Consequently, the optimal control $u(t)$ is also constant (see optimality conditions (4.162)). Thus, partial production $u(t) = \frac{X_T}{T}$ is optimal if $\psi(t) < 2c_u U$, and full production $u(t) = U$ is otherwise optimal.

■

4.4.4 NON-LINEAR INVENTORY AND PRODUCTION COSTS

In this section, we show that although inventory and production costs are now considered non-linear, the problem is still analytically solvable in a way similar to that of the previous section.

The problem to be considered is one-machine, one-product-type scheduling with limited production rate to minimize the quadratic costs of inventory and production and the linear cost of the planning horizon, that is, the $1|1|u(t) \in [0, U], X(T) = X_T \| Q_u, Q_x, T$ problem which is stated as follows

$$J = \int_0^T \left[c_x^+ X(t)^2 + c_u u(t)^2\right] dt + c_T T \to \min, \tag{4.166}$$

subject to (4.149) and (4.150).

Dual Formulation

The Hamiltonian, *H*, resulting from the maximum principle is

$$H(t) = -\left[c_x^+ X(t)^2 + c_u u(t)^2\right] + \psi(t)u(t) = c_T. \tag{4.167}$$

The co-state variable $\psi(t)$ satisfies the co-state differential equation (4.158). Maximization of the Hamiltonian as a function of $u(t)$ results in (4.162)

Algorithm 4.28

Step 1:

If $U \geq \sqrt{\frac{c_T}{c_u}}$ and $X_T \leq \frac{1}{y}\sqrt{U^2 - \frac{c_T}{c_u}}$, then calculate

$T = \frac{1}{y}\ln\left(\sqrt{\frac{c_x^+}{c_T}}X_T + \sqrt{\frac{c_x^+}{c_T}X_T^2 + 1}\right)$ and

set the optimal solution as

$u(t) = yX_T \frac{ch(yt)}{sh(yT)}$ for $0 \leq t \leq T$, STOP.

Otherwise go to the next step.

Step 2:

If $U \geq \sqrt{\frac{c_T}{c_u}}$ and $X_T > \frac{1}{y}\sqrt{U^2 - \frac{c_T}{c_u}}$, then calculate

$$t_x = \frac{1}{y}\ln\left(\sqrt{\frac{c_u}{c_T}}U + \sqrt{\frac{c_u}{c_T}U^2 - 1}\right) \text{ and } T = t_x + \frac{X_T}{U} - \frac{1}{y}\sqrt{1 - \frac{c_T}{c_u U^2}}.$$

Set the optimal solution as

$$u(t) = U\frac{ch(yt)}{ch(yt_x)} \quad \text{for } 0 \leq t \leq t_x,$$

$$u(t) = U \quad \text{for } t_x \leq t \leq T \text{ , STOP.}$$

Otherwise go to the next step.

Step 3:

Calculate $T = \frac{X_T}{U}$ and set the optimal solution as

$$u(t) = U \quad \text{for } 0 \leq t \leq T\text{; STOP.}$$

Discussion

Algorithm 4.28 is similar to Algorithm 4.26. It also presents closed form analytical solutions for those cases when the optimal solution includes partial production followed by full production, as well as those cases when one of these two regimes vanishes.

LEMMA 4.48. (Correctness of Algorithm 4.28) Given problem (4.166), (4.149) and (4.150),

- if $U \geq \sqrt{\frac{c_T}{c_u}}$ and $X_T \leq \frac{1}{y}\sqrt{U^2 - \frac{c_T}{c_u}}$, then the optimal solution is

$$u(t) = yX_T\frac{ch(yt)}{sh(yT)} \quad \text{for } 0 \leq t \leq T\text{, where } T = \frac{1}{y}\ln\left(\sqrt{\frac{c_x^+}{c_T}}X_T + \sqrt{\frac{c_x^+}{c_T}X_T^2 + 1}\right);$$

- if $U \geq \sqrt{\frac{c_T}{c_u}}$ and $X_T > \frac{1}{y}\sqrt{U^2 - \frac{c_T}{c_u}}$, then the optimal solution is

$$u(t) = U\frac{ch(yt)}{ch(yt_x)} \quad \text{for } 0 \leq t \leq t_x \text{ and } u(t) = U \quad \text{for } t_x \leq t \leq T \text{ , where}$$

$$t_x = \frac{1}{y}\ln\left(\sqrt{\frac{c_u}{c_T}}U + \sqrt{\frac{c_u}{c_T}U^2 - 1}\right) \text{ and } T = t_x + \frac{X_T}{U} - \frac{1}{y}\sqrt{1 - \frac{c_T}{c_u U^2}};$$

- if $U < \sqrt{\frac{c_T}{c_u}}$, then the optimal solution is $u(t) = U$ for $0 \le t \le T$, where $T = \frac{X_T}{U}$.

PROOF. Note that the production always starts from the very beginning of the planning horizon. The proof of this fact is identical to that of Lemma 3.38. Therefore, we consider a solution which starts with partial production and switches to full production at a point $t = t_x$ (see Figure 4.16)

$$\psi(t) \le 2c_u U, \quad u(t) = \frac{\psi(t)}{2c_u}, \quad X(t) = \int_0^t u(\tau)d\tau \quad \text{for } 0 \le t \le t_x;$$

$$\psi(t) \ge 2c_u U, \quad u(t) = U, \quad X(t) = \int_0^t u(\tau)d\tau \quad \text{for } t_x \le t \le T.$$

By differentiating the co-state differential equation (4.158) at interval $[0, t_x]$, we find

$$\ddot{\psi}(t) - \frac{c_x^+}{c_u}\psi(t) = 0, \quad \dot{\psi}(0) = 0, \quad \psi(t_x) = 2c_u U. \tag{4.168}$$

The solution to the second order differential equation (4.168) is (see Section 2.3.1)

$$\psi(t) = 2c_u U \frac{ch(yt)}{ch(yt_x)},$$

where $y = \sqrt{\frac{c_x^+}{c_u}}$. By taking into account terminal inventory condition (4.149) and Hamiltonian equation (4.167), we respectively find

$$\frac{U}{y}\frac{sh(yt_x)}{ch(yt_x)} + (T - t_x)U = X_T, \tag{4.169}$$

$$ch(yt_x) = \sqrt{\frac{c_u}{c_T}}U. \tag{4.170}$$

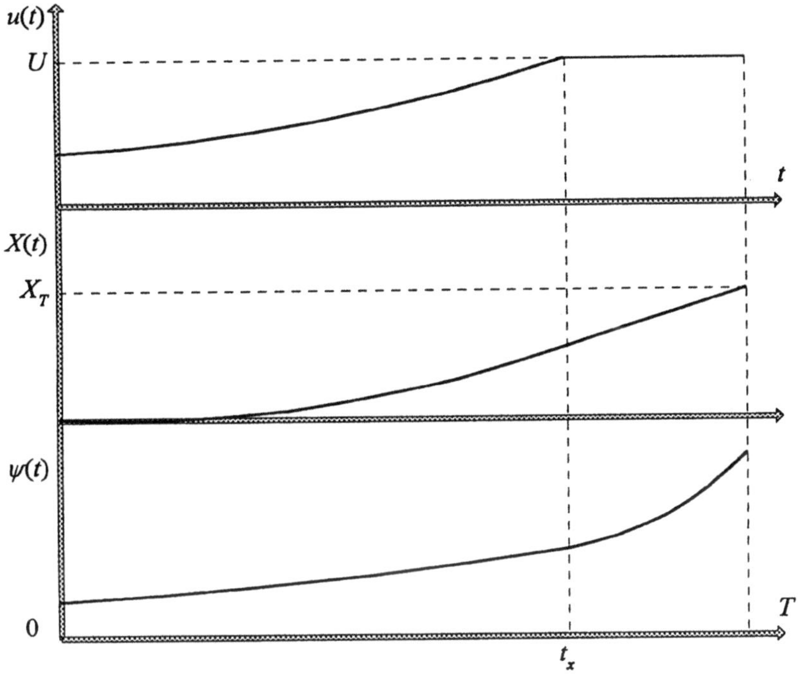

Figure 4.16. Demand $d(t)$, optimal control $u(t)$, state $X(t)$ and co-state $\psi(t)$ variables for the $1|1|u(t)\in[0,U], X(T)=X_T \| Q_u, Q_x, T$ problem.

The solution of system (4.169)-(4.170) in two unknowns, t_x and T, proves the second condition of the lemma if $X(t_x) < X(T)$. The proof for the remaining two conditions of this lemma is similar.

■

Special Cases of Zero Costs

The $1|1|u(t)\in[0,U], X(T)=X_T \| Q_x, T$ problem has been solved in Section 4.4.2, while the $1|1|u(t)\in[0,U], X(T)=X_T \| Q_u, Q_x, T, c_x^+=0$ problem is solved by setting $u(t)=\min\left\{U,\sqrt{\frac{c_T}{c_u}}\right\}$, $t\in[0,T]$. This solution follows from the fact that $\psi(t)$ must be constant since $c_x^+=0$. The optimal value of T is $T = X_T \Big/ \min\left\{U,\sqrt{\frac{c_T}{c_u}}\right\}$.

Finally, by letting $c_T \to 0$ in Algorithm 4.28, one can observe that T tends to infinity and optimal control $u(t)$ tends to zero at each t, that is the $1|1|u(t) \in [0,U], X(T) = X_T \| Q_u, Q_x, T, c_T = 0$ problem has no optimal solution.

PART 3

MULTIPLE PRODUCT-TYPE SCHEDULING

5

ONE-MACHINE PROBLEMS

Scheduling multiple product types with unlimited production capacity is straightforwardly decomposed into a number of problems, each of which deals with a single product. The number of such problems is equal to the number of products to be scheduled. Such one-machine, one-product-type problems have been discussed in the previous part of the book. In contrast, scheduling multiple product types, with limited capacity, cannot be decomposed. This is because different products compete for the common capacity in a different, changing-in-time manner. Because the multi-product, capacitated scheduling problem is much harder than the problems considered in the previous chapters, the objective of this chapter is to study the optimal behavior of production systems under only linear performance measures.

5.1 DEMAND OVER FINITE PLANNING HORIZON

This section addresses multiple-product-type capacitated scheduling problems characterized by a demand function specified over a finite planning horizon.

Backlogs

The problem to be considered is one-machine, N-product-type scheduling with backlogs and bounded production rate to minimize the total linear cost of inventory and production, that is, the $1|N|\,u(t)\in[0,U]\,|d(t),T|\,L_u, L_x$ problem which is stated as follows

$$J=\int_0^T\sum_n\left[c_{xn}^{+}X_n^{+}(t)+c_{xn}^{-}X_n^{-}(t)+c_{un}u_n(t)\right]dt\rightarrow\min, \tag{5.1}$$

$$X_n^+(t) = \max\{X_n(t),0\}\,,\; X_n^-(t) = \max\{-X_n(t),0\}$$

subject to

$$\dot{X}_n(t) = u_n(t) - d_n(t)\,, \qquad X_n(0) = 0\,, \tag{5.2}$$

$$0 \le u_n(t) \le U_n\,, \tag{5.3}$$

$$\sum_n \frac{u_n(t)}{U_n} \le 1\,. \tag{5.4}$$

Note that constraint (5.4) ensures that the total loading of the machine does not exceed hundred percent capacity.

Dual Formulation

The Hamiltonian, H, resulting from the maximum principle is

$$H = -\sum_n \left[c_{xn}^+ X_n^+(t) + c_{xn}^- X_n^-(t) + c_{un} u_n(t)\right] + \sum_n \psi_n(t)\left(u_n(t) - d_n(t)\right). \tag{5.5}$$

The co-state variables, $\psi_n(t)$ in (5.5), satisfy the following co-state differential equation with corresponding transversality condition

$$\dot{\psi}_n(t) = \begin{cases} c_{xn}^+, \text{ if } X_n(t) > 0; \\ -c_{xn}^-, \text{ if } X_n(t) < 0; \\ a,\; a \in [-c_{xn}^-, c_{xn}^+], \text{ if } X_n(t) = 0; \end{cases} \qquad \psi_n(T) = 0\,. \tag{5.6}$$

Maximization of the Hamiltonian at every point of time with respect to admissible controls $u_n(t)$ implies that

$$u_n(t) = \begin{cases} U_n, \text{ if } \psi_n(t) > c_{un} \text{ and } U_n\left(\psi_n(t) - c_{un}\right) > U_{n'}\left(\psi_{n'}(t) - c_{un'}\right), \forall n' \neq n \\ 0, \text{ if } \psi_n(t) < c_{un}, \\ b,\; b \in [0, U_n], \text{ if } \psi_n(t) \ge c_{un} \text{ and } U_n\left(\psi_n(t) - c_{un}\right) \ge U_{n'}\left(\psi_{n'}(t) - c_{un'}\right), \forall n' \neq n. \end{cases} \tag{5.7}$$

The first condition in equation (5.7) defines the production of product n at the maximum rate (full production regime). The second condition defines a no-production regime of product n. Finally, the last condition in (5.7) presents the fractional regime which allows production at an intermediate rate to track the demand. According to conditions (5.7), optimal production regimes are completely determined by the value of $U_n\left(\psi_n(t) - c_{un}\right)$, which is further referred to as a *dual indicator*.

Algorithm 5.1

Step 1:

Calculate $t_n = T - \frac{c_{un}}{c_{xn}^-}$ for n=1,...,N. If $t_n \le 0$, then do not produce product n, exclude it from further consideration and set N=N-1. Re-number the remaining products, n=1,...,N.

If $\sum_n \frac{d_n(t)}{U_n} \le 1$ for $0 \le t \le T$, then set the optimal solution as

$u_n(t) = d_n(t)$ for $0 \le t \le t_n$,

$u_n(t) = 0$ for $t_n < t \le T$; STOP.

Otherwise go to the next step.

Step 2:

If $d_n(t) \le U_n$ for $0 \le t \le T$ and costs are agreeable in terms of Definition 5.1 below, then go to (i).

(i) Sort products in the non-decreasing order of $c_{xn}^- U_n$ and re-number them in this order. Assign i=1.

(ii) If $\sum_{n=i}^{n=N} \frac{d_n(t)}{U_n} > 1$ and $\sum_{n=i+1}^{n=N} \frac{d_n(t)}{U_n} \le 1$ for $0 \le t \le T$, then go to (iii).

Otherwise assign i=i+1 and go to the beginning of (ii).

(iii) Set the optimal solution as

$u_n(t) = 0$ for $0 \le t \le T$, n=1,...,i-1,

$$u_i(t) = U_i \left(1 - \sum_{n'=i+1}^{N} \frac{d_{n'}(t)}{U_{n'}} \right) \text{ for } 0 \le t \le t_i,$$

$u_n(t) = d_n(t)$ for $0 \le t \le t_n$, n=i+1,...,N,

$u_n(t) = 0$ for $t_n < t \le T$, n=i,...,N; STOP.

Definition 5.1. The production and backlog costs are *agreeable* with respect to products n and n', $n \neq n'$ if $c_{xn}^{-}U_n \leq c_{xn'}^{-}U_{n'}$ implies $\frac{c_{un}}{c_{xn}^{-}} \geq \frac{c_{un'}}{c_{xn'}^{-}}$.

Definition 5.2. S-fractional-production regime is the fractional production characterized by a set of products S for which the dual indicators $U_n(\psi_n(t) - c_{un}) > 0$ are equal to each other and are greater than the dual indicators of the other products.

Definition 5.3. Z-fractional-production regime is the fractional production characterized by a set of products Z for which the dual indicators $U_n(\psi_n(t) - c_{un})$ are equal to zero and are greater than the dual indicators of the other products.

Example 5.1
Consider the $1|4|u(t) \in [0, U]\,|d(t), T|\, L_u, L_x$ problem, with the length of the planning horizon, T, set at 10hr, the demand depicted in Figure 5.1 and the production parameters shown in Table 5.1.

At Step 1, Algorithm 5.1 calculates the switching points $t_n = T - \frac{c_{un}}{c_{xn}^{-}}$: $t_1 = -1$, $t_2 = 8$, $t_3 = 5$, $t_4 = 9$ (hr.). Since $t_1 < 0$, we set $u_1(t) = 0$ for $0 \leq t \leq 10$ and exclude this product type from further consideration. Next, from Figure 5.1 and Table 5.1, one can observe that $d_n(t) \leq U_n$ for n=1,...,4, and the total demand exceeds the capacity of the machine, $\sum_{n=2}^{4} \frac{d_n(t)}{U_n} > 1$ for $0 \leq t \leq 10$.
Since the costs specified in Table 5.1 are agreeable in terms of Definition 5.1, the algorithm proceeds to Step 2, which in this case results in an optimal solution. To find the solution, Step 2(i) sorts the products in the non-decreasing order of $c_{xn}^{-}U_n$. The resultant sequence of the product types is: 3-2-4. At Step 2(ii), product i=2, for which $\frac{d_2(t)}{U_2} + \frac{d_4(t)}{U_4} > 1$ and $\frac{d_4(t)}{U_4} \leq 1$, $0 \leq t \leq 10$, is identified. Finally, Step 2(iii) sets the optimal solution as

$$u_1(t) = 0,\ u_3(t) = 0 \text{ for } 0 \leq t \leq 10,$$

$$u_2(t) = U_2\left(1 - \frac{d_4(t)}{U_4}\right) = 2 - \frac{2d_4(t)}{3} \text{ for } 0 \leq t \leq 8, \text{ and } u_2(t) = 0$$

$$\text{for } 8 < t \leq 10,$$

$$u_4(t) = d_4(t) \text{ for } 0 \le t \le 9 \text{ and } u_4(t) = 0 \text{ for } 9 < t \le 10.$$

This solution is depicted in Figure 5.1 along with the corresponding behavior of dual indicators $U_n(\psi_n(t) - c_{un})$ and inventory levels $X_n(t)$, n=1,...,4.

■

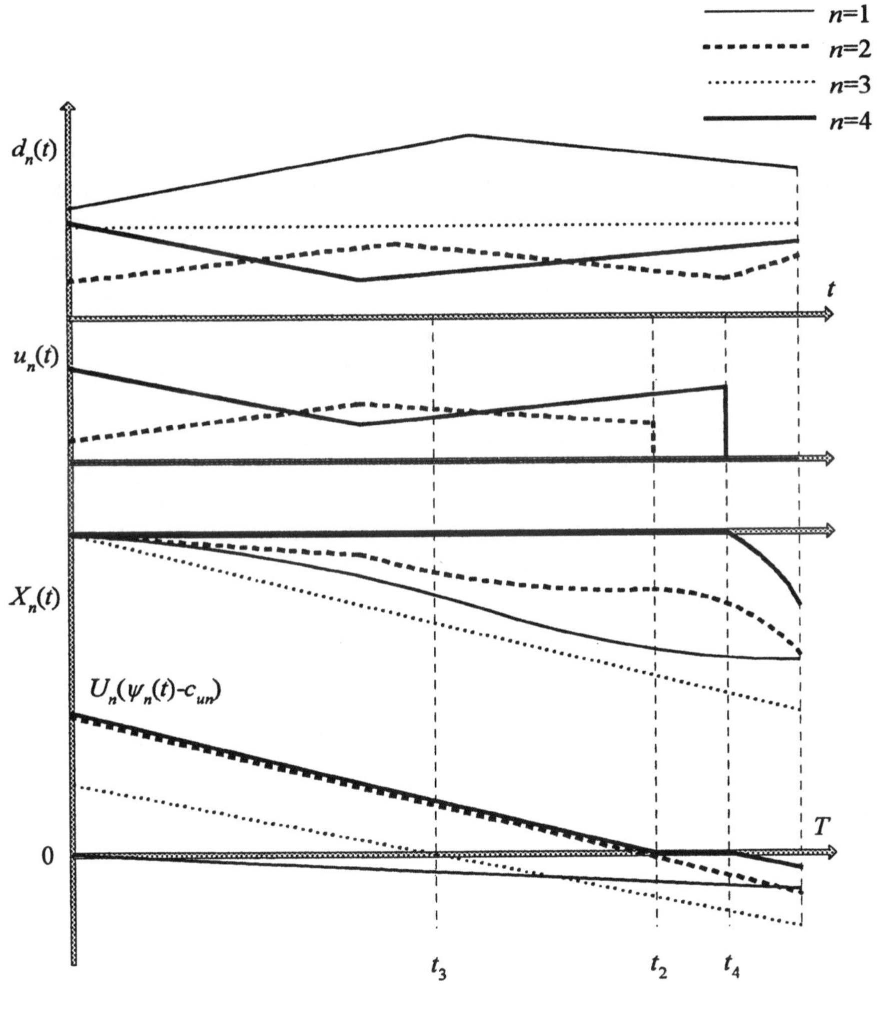

Figure 5.1. Demand, optimal control, state variables and dual indicators for the $1|N|u(t) \in [0,U]|d(t),T|L_u, L_x$ problem of Example 5.1.

Table 5.1. Maximum production rates and unit costs of the system

n	c^{+}_{xn}	c^{-}_{xn}	c_{un}	U_n
1	0.01	0.1	1.1	4.0
2	0.02	0.3	0.6	2.0
3	0.02	0.2	1.0	2.5
4	0.01	0.6	0.6	3.0

Discussion

This section presents two solvable cases. Step 1 of Algorithm 5.1 solves the problem in $O(N)$ time for the case of the weak demand. Step 2 of Algorithm 5.1 deals with the case of the pressing demand under the condition that each individual demand $d_n(t)$ does not exceed the corresponding capacity U_n. This problem is solvable in $O(N\log N)$ time if costs are agreeable with respect to Definition 5.1. For example, if production is identical for different products in terms of production costs $c_{un} = c_{un'}$ and maximum production rates $U_n = U_{n'}$ $\forall n, n'$, the costs are always agreeable. Note that if the costs are agreeable, then the production is organized with respect to the Weighted Least Production Rate rule. In contrast to combinatorial scheduling rules, where only one product with the least production rate is produced at a time, this rule assigns a number of products to be produced concurrently without preemption. These products are selected in accordance with the least production rates within the available capacity. The remaining products are not produced at all. Since the production rate is inversely proportional to the production time, this rule agrees with the Weighted Longest Processing Time rule (WLPT), well-known in combinatorial scheduling literature.

LEMMA 5.1. (Correctness of Algorithm 5.1 Step 1) Given problem (5.1)-(5.4) and $\sum_n \frac{d_n(t)}{U_n} \le 1$ for $0 \le t \le T$, the optimal solution is $u_n(t) = d_n(t)$ for $0 \le t \le t_n$,

$u_n(t) = 0$ for $t_n < t \le T$, if $t_n = T - \frac{c_{un}}{c^{-}_{xn}} \ge 0$; otherwise, $u_n(t) = 0$ for $0 \le t \le T$.

PROOF. Consider, first, the case when the demand is weak and construct a feasible solution with respect to both primal and dual problems. Since the primal problem is linear, such a solution must be optimal. The solution

$$\psi_n(t) = c_{un},\ u_n(t) = d_n(t),\ X_n(t) = 0 \text{ for } 0 \le t \le t_n,$$

$$\psi_n(t) = c_{xn}^-(T-t)\,, \quad u_n(t) = 0\,, \quad X_n(t) = -\int_0^t d_n(\tau)d\tau \quad \text{for } t_n < t \le T$$

is evidently feasible with respect to constraints (5.2)-(5.4) and (5.6)-(5.7) if the demand is weak, $\sum_n \frac{d_n(t)}{U_n} \le 1$ for $0 \le t \le T$ and switching points $t_n = T - \frac{c_{un}}{c_{xn}^-}$ are non-negative as illustrated in Figure 5.1.

■

To prove the correctness of Step 2 Algorithm 5.1, we first study the fractional production regimes in the following two lemmas.

LEMMA 5.2. Given *S*-fractional-production regime at an interval of time, the following holds at this interval:

$u_n(t) = d_n(t)\,,\quad X_n(t) = 0$ for all $n \in S$, $n \ne n^*$;

$u_{n^*}(t) = U_{n^*}\left(1 - \sum_{n \in S;\, n \ne n^*} \frac{d_n(t)}{U_n}\right)$, where n^* is such that $U_{n^*}c_{xn^*}^+ = \min_{n \in S} U_n c_{xn}^+$, if

$X_{n^*}(t) > 0$ and n^* is such that $U_{n^*}c_{xn^*}^- = \min_{n \in S} U_n c_{xn}^-$ if $X_{n^*}(t) < 0$;

$u_n(t) = 0$ for all $n \notin S$.

PROOF. We study the *S*-fractional-production regime, which according to Definition 5.2, satisfies

$$U_n\big(\psi_n(t) - c_{un}\big) = U_{n'}\big(\psi_{n'}(t) - c_{un'}\big) > 0\,, \quad \text{for all } n, n' \in S\,. \tag{5.8}$$

$$U_n\big(\psi_n(t) - c_{un}\big) > U_{n'}\big(\psi_{n'}(t) - c_{un'}\big)\,, \quad \text{for all } n \in S, n' \notin S\,. \tag{5.9}$$

By differentiating condition (5.8), we obtain

$$U_n \dot{\psi}_n(t) = U_{n'} \dot{\psi}_{n'}(t)\,. \tag{5.10}$$

Since U_n, c_{xn}^+ and c_{xn}^- are arbitrarily given constants and $\dot{\psi}_n$ is described by (5.6), equation (5.10) can be met only if $X_{n^*}(t) \neq 0$, $n^* \in S$ and $X_n(t) = 0$ for all $n \in S$, $n \neq n^*$. In turn, for this condition to hold, it is necessary that

$$U_{n^*} c_{xn^*}^+ = \min_{n \in S} U_n c_{xn}^+ \text{ if } X_{n^*}(t) > 0; \; U_{n^*} c_{xn^*}^- = \min_{n \in S} U_n c_{xn}^- \text{ if } X_{n^*}(t) < 0. \quad (5.11)$$

By differentiating condition $X_n(t) = 0$ for all $n \in S$, $n \neq n^*$, and substituting there the state equation (5.2), we obtain

$$u_n(t) = d_n(t), \text{ for all } n \in S, \; n \neq n^*. \quad (5.12)$$

By taking into account conditions (5.7) and (5.9), we find that the optimal control for the remaining products at the considered time interval is

$$u_n(t) = 0 \text{ for } n \notin S \text{ and } u_{n^*}(t) = U_{n^*} \left(1 - \sum_{\substack{n \in S, \\ n \neq n^*}} \frac{d_n(t)}{U_n} \right). \quad (5.13)$$

■

LEMMA 5.3. Given Z-fractional-production regime at an interval of time, the following holds at this interval: $u_n(t) = d_n(t)$, $X_n(t) = 0$ for all $n \in Z$ and $u_n(t) = 0$ for all $n \notin Z$.

PROOF. We study the Z-fractional-production regime, which, according to Definition 5.3, satisfies

$$\psi_n(t) - c_{un} = 0 \quad \text{for all } n \in Z \quad (5.14)$$

and

$$\psi_n(t) - c_{un} < 0 \quad \text{for all } n \notin Z.$$

According to (5.7), the last condition implies $u_n(t) = 0$ for all $n \notin Z$. By differentiating condition (5.14), we obtain

$$\dot{\psi}_n(t) = 0. \quad (5.15)$$

With respect to the co-state equation (5.6), equation (5.10) can be met only if $X_n(t) = 0$ for all $n \in Z$. By differentiating the last condition and substituting there the state equation (5.2), we obtain

$$u_n(t) = d_n(t), \text{ for all } n \in Z. \quad (5.16)$$

■

The next lemma proves the optimality of the concurrent WLPR rule with no preemption.

LEMMA 5.4. (Correctness of Algorithm 5.1 Step 2) Given problem (5.1)-(5.4), $d_n(t) \le U_n$, product i such that $\sum_{n=i}^{n=N} \frac{d_n(t)}{U_n} > 1$ and $\sum_{n=i+1}^{n=N} \frac{d_n(t)}{U_n} \le 1$ for $0 \le t \le T$, agreeable costs, the products numbered in the non-decreasing order of $c_n^- U_n$ and $t_n = T - \frac{c_{un}}{c_{xn}^-} > 0$, then the optimal solution is

$u_n(t) = 0$ for $0 \le t \le T$, $n=1,\dots,i-1$; $u_i(t) = U_i\left(1 - \sum_{n=i+1}^{N} \frac{d_n(t)}{U_n}\right)$ for $0 \le t \le t_i$;

$u_n(t) = d_n(t)$ for $0 \le t \le t_n$, $n=i+1,\dots,N$; $u_n(t) = 0$ for $t_n < t \le T$, $n=i,\dots,N$.

PROOF. Given product i such that $\sum_{n=i}^{n=N} \frac{d_n(t)}{U_n} > 1$ and $\sum_{n=i+1}^{n=N} \frac{d_n(t)}{U_n} \le 1$ for $0 \le t \le T$, consider the solution including the S-fractional-production regime with $S = \{i,\dots,N\}$ and N-i Z-fractional-production regimes with $Z = \{i+n,\dots,N\}$ at the n-th regime. Denote by t_n the time points where the regimes change over, $n = 1,\dots,N-i$. Furthermore, products $n = 1,\dots,i-1$ are never produced and no products are produced at $t \in [t_N, T]$ (no-production regime)

$$u_n(t) = 0,\ X_n(t) < 0,\ \psi_n(t) = c_{xn}^-(T-t) \text{ for } 0 < t \le T,\quad n=1,\dots,i-1,$$

$$u_n(t) = d_n(t),\ X_n(t) = 0,\ U_n\big(\psi_n(t) - c_{un}\big) = U_{n'}\big(\psi_{n'}(t) - c_{un'}\big) \text{ for } 0 \le t \le t_i,\ n,n' \in S,$$

$$u_i(t) = U_i\left(1 - \sum_{n=i+1}^{N} \frac{d_n(t)}{U_n}\right),\quad X_i(t) < 0 \quad \text{for } 0 \le t \le t_i, \tag{5.17}$$

$$u_n(t) = d_n(t),\ X_n(t) = 0,\ \psi_n(t) - c_{un} = 0 \text{ for } t_i \le t \le t_n,\ n=i,\dots,N,$$

$$u_n(t) = 0,\ X_n(t) < 0,\ \psi_n(t) = c_{xn}^-(T-t) \text{ for } t_n < t \le T,\ n=i,\dots,N.$$

Switching points t_n in (5.17) are found from the co-state boundary conditions $\psi_n(t_n) = c_{un} = c_{xn}^-(T - t_n)$. Solution (5.17) satisfies optimality condition (5.7) and Lemmas 5.2-5.3 if $c_{xn}^- U_n \le c_{xn+1}^- U_{n+1}$ and points t_n are feasible, i.e.

$$t_1 \geq 0\,,\;\; t_n \leq t_{n+1} \;\text{ for } n=1,\ldots,N-1. \tag{5.18}$$

By taking into account that

$$t_n = T - \frac{c_{un}}{c_{xn}^-}\,, \tag{5.19}$$

condition (5.18) immediately results in $t_n \geq 0$ and $\frac{c_{un}}{c_{xn}^-} \geq \frac{c_{un+1}}{c_{xn+1}^-}$. We, thus, conclude that for solution (5.17) to be optimal, the costs must be agreeable, as stated in the lemma.
■

Special Cases of Zero Costs

Since cost coefficients c_{xn}^+ are not involved in Algorithm 5.1 developed for special cases of the $1|N|\,u(t) \in [0,U]\,|d(t),T|\,L_u, L_x$ problem, the algorithm also applies to the $1|N|\,u(t) \in [0,U]\,|d(t),T|\,L_u, L_x, c_x^+ = 0$ problem.

The algorithm for the $1|N|\,u(t) \in [0,U]\,|d(t),T|\,L_x$ problem is straightforwardly obtained by substituting $c_u = 0$ into the corresponding conditions and formulas of Algorithm 5.1. As a result, the solution simplifies to

$$u_n(t) = 0 \;\text{ for } 0 \leq t \leq T\,,\; n=1,\ldots,i\text{-}1; \quad u_i(t) = U_i\left(1 - \sum_{n'=i+1}^{N} \frac{d_{n'}(t)}{U_{n'}}\right) \;\text{ for } 0 \leq t \leq T\,;$$

$$u_n(t) = d_n(t) \;\text{ for } 0 \leq t \leq T\,,\;\; n=i+1,\ldots,N.$$

No Backlog

The $1|N|\,u(t) \in [0,U], X(t) \geq 0\,|d(t),T|\,L_u, L_x$ problem has a feasible solution only if the average demand does not exceed both the individual and total capacity

$$\frac{1}{t}\int_0^t d_n(\tau)d\tau \leq U_n\,,\;\; \frac{1}{t}\int_0^t \sum_n \frac{d_n(\tau)}{U_n} d\tau \leq 1 \;\text{ for } 0 < t \leq T\,. \tag{5.20}$$

Dual Formulation

Now problem (5.1)-(5.4) has the additional constraints $X_n(t) \geq 0$ which affects neither Hamiltonian (5.5) nor optimality conditions (5.7), but changes the co-state differential equation

$$d\psi_n(t) = c_{xn}^+ dt - d\mu_n(t),\;\; \psi_n(T+0) = 0\,, \tag{5.21}$$

where Lagrange multiplier $d\mu_n(t) \geq 0$ corresponds to the no-backlog constraint and satisfies the complementary slackness condition

$$\int_0^T X_n(t)d\mu_n(t) = 0. \tag{5.22}$$

Algorithm 5.2
Step 1:

If $\sum_n \frac{d_n(t)}{U_n} \leq 1$ for $0 \leq t \leq T$, then set the optimal solution as

$u_n(t) = d_n(t)$ for $0 \leq t \leq T$.

Otherwise go to the next step.

Step 2:

If $d_n(t) \leq U_n$ for $0 \leq t \leq T$, $\sum_n \frac{d_n(t)}{U_n} \leq 1$ for $0 \leq t \leq t_x$ and

$\sum_{n \neq n'} \frac{d_n(t)}{U_n} < 1 < \sum_n \frac{d_n(t)}{U_n}$ for $t_x < t \leq T$, $n' = 1,\ldots,N$, then find

product index $i = \arg\min_n c_{xn}^{+} U_n$ and time point t_1 as a root of equation

$$\int_{t_1}^T U_i \left(1 - \sum_{n \neq i} \frac{d_n(t)}{U_n}\right) dt = \int_{t_1}^T d_i(t)dt.$$

Set the optimal solution as

$u_n(t) = d_n(t)$ for $0 \leq t \leq T$, $n \neq i$,

$u_i(t) = d_i(t)$ for $0 \leq t \leq t_1$,

$$u_i(t) = U_i \left(1 - \sum_{n' \neq i} \frac{d_{n'}(t)}{U_{n'}}\right) \text{ for } t_1 \leq t \leq T.$$

Discussion

Step 1 of Algorithm 5.2 determines a closed form solution for those problem characterized by a demand function not exceeding the individual and total capacity. Algorithm 5.2 Step 2 solves the problem in $O(N)$ time for a class of demand functions such that the switching point t_1 can be analytically obtained from equation $\int_{t_1}^{T} U_i\left(1-\sum_{n\neq i}\frac{d_n(t)}{U_n}\right)dt = \int_{t_1}^{T} d_i(t)dt$. In addition, the machine capacity must be sufficient to produce all products.

LEMMA 5.5. (Correctness of Algorithm 5.2) Given problem (5.1)-(5.4) with backlogs disallowed,

- If $\sum_n \frac{d_n(t)}{U_n} \le 1$ for $0 \le t \le T$, then the optimal solution is $u_n(t) = d_n(t)$ for $0 \le t \le T$, n=1,...,N.

- If $d_n(t) \le U_n$ for $0 \le t \le T$, $\sum_n \frac{d_n(t)}{U_n} \le 1$ for $0 \le t \le t_x$ and $\sum_{n\neq n'} \frac{d_n(t)}{U_n} < 1 < \sum_n \frac{d_n(t)}{U_n}$ for $t_x < t \le T$, n'=1,...,N, then the optimal solution is

$u_n(t) = d_n$ for $0 \le t \le T$ for $n \neq i$,

$u_i(t) = d_i(t)$ for $0 \le t \le t_1$, $u_i(t) = U_i\left(1-\sum_{n'\neq i}\frac{d_{n'}(t)}{U_{n'}}\right)$ for $t_1 \le t \le T$,

where $i = \arg\min_n c^+_{xn} U_n$ and t_1 is a root of equation

$$\int_{t_1}^{T} U_i\left(1-\sum_{n\neq i}\frac{d_n(t)}{U_n}\right)dt = \int_{t_1}^{T} d_i(t)dt .$$

PROOF. It is easy to observe that if $\sum_n \frac{d_n(t)}{U_n} \le 1$ then $d_n(t) \le U_n$, $0 \le t \le T$ for all n, and the solution

$$u_n(t) = d_n(t),\ X_n(t) = 0,\ \psi_n(t) = c_{un} \text{ for } 0 \le t \le T,$$

$$d\mu_n(t) = 0 \text{ for } 0 \le t < T,\ d\mu_n(T) = c_{un} \text{ and } \psi_n(T+0) = 0$$

always meets the state (5.2)-(5.4) and co-state (5.6), (5.7), (5.21)-(5.22) equations. Therefore, it is optimal.

To prove the second part of the lemma, we consider the solution which includes the Z-fractional regime followed, from point t_1, by the S-fractional regime with $Z = S = \{1,\ldots,N\}$

$$u_n(t) = d_n,\; X_n(t) = 0\;,\; \psi_n(t) = c_{un} \text{ for } 0 \le t \le t_1,\; n=1,\ldots,N,$$

$$u_n(t) = d_n(t),\; X_n(t) = 0 \text{ for } n \ne i,\qquad u_i(t) = U_i\left(1 - \sum_{n'=i+1}^{N}\frac{d_{n'}(t)}{U_{n'}}\right),\text{ for } t_1 \le t \le T,$$

$$U_n\left(\psi_n(t) - c_{un}\right) = U_{n'}\left(\psi_{n'}(t) - c_{un'}\right) \text{ for } t_1 \le t \le T,\, n=1,\ldots,N,\; n'=1,\ldots,N,$$

$$d\mu_n(t) = 0 \text{ for } 0 \le t < T,\; d\mu_n(T) = c_{un} + c_{xi}^{+}(T - t_1) \text{ and } \psi_n(T+0) = 0\,.$$

This solution is feasible if $X_i(T) = 0$ and index i is chosen such that $i = \arg\min_n c_{xn}^{+} U_n$ (see Lemma 5.2). By integrating the state variable from equation (5.2) over interval $[t_1, T]$ and substituting it into $X_i(T) = 0$, we obtain the equation for unknown t_1, as stated in the lemma.

■

5.2 DUE DATE

This section addresses multi-product scheduling problems characterized by limited capacity and demand concentrated at a due date.

Backlogs
The problem to be considered is one-machine, N-product-type scheduling with limited production rate and backlogs to minimize the total linear cost of inventory and production, that is, the $1|M|\,u(t) \in [0,U]\,|D,T|\,L_u, L_x$ problem which is stated as follows

$$J = \int_0^T \sum_n \left[c_{xn}^{+} Z_n^{+}(t) + c_{xn}^{-} Z_n^{-}(t) + c_{un} u_n(t)\right] dt \to \min\,, \tag{5.23}$$

$$Z_n^{+}(t) = \max\{X_n(t) - D_n(t), 0\}\,,\; Z_n^{-}(t) = \max\{D_n(t) - X_n(t), 0\}$$

subject to

$$\dot{X}_n(t) = u_n(t), \quad X_n(0) = 0, \tag{5.24}$$

$$0 \le u_n(t) \le U_n, \tag{5.25}$$

$$\sum_n \frac{u_n(t)}{U_n} \le 1. \tag{5.26}$$

where

$$D_n(t) = \begin{cases} D_n, & \text{if } t \ge t_{dn}; \\ 0, & \text{otherwise}; \end{cases} \tag{5.27}$$

and D_n is the amount of product type n to be produced by due date t_{dn} .

Dual Formulation

The Hamiltonian, H, resulting from the maximum principle is

$$H = -\sum_n \left[c_{xn}^+ Z_n^+(t) + c_{xn}^- Z_n^-(t) + c_{un} u_n(t)\right] + \sum_n \psi_n(t) u_n(t). \tag{5.28}$$

The co-state variable, $\psi(t)$ in (5.28), satisfies the following co-state differential equation with its transversality condition

$$\dot{\psi}_n(t) = \begin{cases} c_{xn}^+, & \text{if } X_n(t) > D_n(t); \\ -c_{xn}^-, & \text{if } X_n(t) < D_n(t); \\ a,\ a \in [-c_{xn}^-, c_{xn}^+], & \text{if } X_n(t) = D_n(t); \end{cases} \qquad \psi_n(T) = 0. \tag{5.29}$$

Maximization of the Hamiltonian at every point of time implies that

$$u_n(t) = \begin{cases} U_n, & \text{if } \psi_n(t) > c_{un} \text{ and } U_n\left(\psi_n(t) - c_{un}\right) > U_{n'}\left(\psi_{n'}(t) - c_{un'}\right),\ \forall n' \ne n \\ 0, & \text{if } \psi_n(t) < c_{un}, \\ b,\ b \in [0, U_n], & \text{if } \psi_n(t) \ge c_{un} \text{ and } U_n\left(\psi_n(t) - c_{un}\right) \ge U_{n'}\left(\psi_{n'}(t) - c_{un'}\right),\ \forall n' \ne n. \end{cases} \tag{5.30}$$

Algorithm 5.3

Step 1:

(i) Assign n=1.

(ii) Run Algorithm 4.12 and assign $t_{n1} = t_1$ and $t_{n2} = t_2$.

(iii) If $n<N$, then assign $n=n+1$ and go to (ii). Otherwise go to (iv).
(iv) Check by sorting whether the intervals (t_{n1}, t_{n2}) are disjoint, $\bigcap_n (t_{n1}, t_{n2}) = \varnothing$. If this is true, then set the optimal solution as
$u_n(t) = 0$ for $0 \le t < t_{n1}$ and $t_{n2} < t \le T$,
$u_n(t) = U_n$ for $t_{n1} \le t \le t_{n2}$; STOP.

Algorithm 5.4

Step 1:

If the due date is common for all products, $t_{dn} = t_d$, costs are agreeable in terms of Definition 5.4 below and $\sum_n \frac{D_n}{U_n} \le t_d \le T - \sum_n \frac{D_n}{U_n}$, then sort and re-number product types $n=1,\ldots,N$ in the non-decreasing order of $c_n^+ U_n$. Set $k=1$ and go to the next step. Otherwise, STOP.

Step 2:

Calculate time points

$$t_k = t_d - \frac{\sum_{n=k}^{N} c_{xn}^- D_n - \sum_{n=1}^{k-1} c_{xn}^+ D_n}{\left(c_{xk}^+ + c_{xk}^-\right) U_k}, \quad t_n = t_k - \sum_{j=n}^{k-1} \frac{D_j}{U_j}, \text{ for } n=1,\ldots,k\text{-}1$$

and $t_n = t_k + \sum_{j=k}^{n-1} \frac{D_j}{U_j}$ for $n=k+1,\ldots,N+1$.

Step 3:

If $t_k < t_d \le t_{k+1}$, then set the optimal solution as
$u_n(t) = U_n$ for $t_n \le t < t_{n+1}$,
$u_n(t) = 0$ for $0 \le t < t_n$ and $t_{n+1} \le t \le T$, $n=1,\ldots,N$; STOP.
Otherwise, set $k=k+1$ and go to Step 2.

Example 5.2

Consider the $1|4|\,u(t) \in [0, U]\,|D,T|\, L_u, L_x$ problem with the length of the planning horizon, T, set at 10hr, the due date $t_{dn} = t_d = 5$ hr and the production parameters shown in Table 5.2.

At Step 1, Algorithm 5.4 verifies that $\sum_n \frac{D_n}{U_n} = 4 \le t_d \le 6 = T - \sum_n \frac{D_n}{U_n}$, and that the costs are agreeable. Then the algorithm sorts the products as 4-2-1-3, and identifies at Steps 2 and 3 the key-product as n=3. The switching time points are: $t_1 = 4.22$, $t_2 = 2.72$, $t_3 = 4.72$, $t_4 = 1.52$, $t_5 = 5.52$ (hr). The optimal solution is set as

$u_1(t) = 4$ unit/hr for $4.22 \le t < 4.72$, $u_1(t) = 0$ otherwise;
$u_2(t) = 2$ unit/hr for $2.72 \le t < 4.22$, $u_2(t) = 0$ otherwise;
$u_3(t) = 2.5$ unit/hr for $4.72 \le t < 5.52$, $u_3(t) = 0$ otherwise;
$u_4(t) = 3$ unit/hr for $1.52 \le t < 2.72$, $u_4(t) = 0$ otherwise.
■

Definition 5.4. The inventory and backlog costs are agreeable with respect to products n and n', $n \neq n'$ if $c_n^- U_n > c_{n'}^- U_{n'}$ implies $c_n^+ U_n \le c_{n'}^+ U_{n'}$.

Definition 5.5. The key product type is a product type k, $1 \le k \le N$, for which $u_k(t) = U_k$ during an interval of time $\tau = [\hat{t}_1, \hat{t}_2]$, such that $\hat{t}_1 < t_d$, $\hat{t}_2 \ge t_d$.

Discussion

Algorithm 5.3 solves the problem in $O(N\log N)$ time by decomposing it into N one-product sub-problems, each of which is solved by Algorithm 4.12. The decomposition is valid if the time intervals at which the production rates are positive, do not overlap. This is verified by sorting the intervals. If the problem cannot be decomposed, Algorithm 5.4 solves it in $O(N^2)$ time for the case involving a common due date and agreeable costs in terms of Definition 5.4.

LEMMA 5.6. (Correctness of Algorithm 5.3) Given problem (5.23)-(5.27) and time points t_{n1} and t_{n2} found by Algorithm 4.12 for every product n separately, if $\bigcap_n (t_{n1}, t_{n2}) = \varnothing$, then the optimal control is

$$u_n(t) = 0 \text{ for } 0 \le t < t_{n1} \text{ and } t_{n2} < t \le T; \quad u_n(t) = U_n \text{ for } t_{n1} \le t \le t_{n2}.$$

PROOF. Condition $\bigcap_n (t_{n1}, t_{n2}) = \varnothing$ ensures that the production intervals found separately (see Lemma 4.26) for different products do not overlap. Therefore, the total capacity is never exceeded (5.26) and the determined control is optimal.
■

The following lemma proves that there always exists an optimal solution without any intermediate control values. The production is thus characterized by binary controls $u_n(t) \in \{0, U_n\}$.

Table 5.2. Demand, maximum production rates and unit costs of the system

n	D_n	c_{xn}^{+}	c_{xn}^{-}	c_{un}	U_n
1	2	0.01	0.1	1.1	4.0
2	3	0.02	0.3	0.6	2.0
3	2	0.02	0.1	1.0	2.5
4	3.6	0.01	0.6	0.6	3.0

LEMMA 5.7. *(Integrality Property)* Given problem (5.23)-(5.27), there always exists an optimal solution with binary control values, i.e. $u_n(t) \in \{0, U_n\}$.

PROOF. Since the first and second conditions in (5.30) explicitly and uniquely determine the binary optimal control values, we need to resolve the ambiguity of only fractional regimes defined by the third condition in (5.30). This condition can be detailed in two possible cases.

1. The first case is characterized by a few dual indicators $U_n\left(\psi_n(t) - c_{un}\right)$ which are positive, equal to each other, and greater than the others at an interval of time. Consider a couple of such dual indicators at an interval of time

$$U_n\left(\psi_n(t) - c_{un}\right) = U_{n'}\left(\psi_{n'}(t) - c_{un'}\right). \tag{5.31}$$

Then, by differentiating equation (5.31) at this time interval and taking into account the co-state equation (5.29), we obtain

$$U_n \dot{\psi}_n(t) = U_{n'} \dot{\psi}_{n'}(t). \tag{5.32}$$

Since U_n, c_{xn}^{+} and c_{xn}^{-} are arbitrarily given constants and $\dot{\psi}_n$ is described by (5.29), equation (5.32) can be met only if one of the following two conditions holds

$$X_n(t) = D_n(t), \text{ or } X_{n'}(t) = D_{n'}(t). \tag{5.33}$$

However, due to the step-wise demand (5.27) and production process (5.24), condition (5.33) is met only if $u_n(t) = 0$. Thus, no intermediate optimal control value can be found in this case.

2. The second case is characterized by one or more dual indicators equal to zero, when the others are negative

$$\psi_n(t) = c_{un}.$$

By differentiating this condition, one obtains that it can hold only if $X_n(t) = D_n(t)$ and, hence, $u_n(t) = 0$.

■

Lemma 5.7 reduces preemptive, continuous-rate production scheduling with a common due date to a binary-rate scheduling, or more precisely, a sequencing problem. The following two lemmas treat sequencing of the products before and after the due date for arbitrary and special ratios between parameters of the problem which induce the existence of non-preemptive optimal solutions. To simplify the presentation, we use notions of a key product type and of agreeable production costs which have been defined in Definitions 5.5 and 5.4 respectively.

LEMMA 5.8. (*Sequencing Before The Key Product Type*) Given the optimal solution with key product type k and $t_{dn} = t_d$, n=1,...,N, if product type $n2$ is scheduled immediately after $n1$ and both are scheduled before the key product, then

$$c_{xn1}^{+} U_{n1} \le c_{xn2}^{+} U_{n2}. \tag{5.34}$$

PROOF. The proof for the sequencing is by contradiction. Let $c_{xn1}^{+} U_{n1} > c_{xn2}^{+} U_{n2}$, and product types $n1$, $n2$ be produced in this order at time intervals $[t_1, t_{12}]$ and $[t_{12}, t_2]$ respectively. Thus,

$$X_{n2}(t) \ge D_n(t) \text{ for } t_1 \le t \le t_{12},\ X_{n2}(t) > D_n(t) \text{ for } t_{12} < t \le t_d,$$

$$X_{n1}(t) > D_n(t) \text{ for } t_1 < t \le t_d, \tag{5.35}$$

$$\dot\psi_{n1}(t) = c_{xn1}^{+} \text{ for } t_1 < t \le t_d \text{ and } \dot\psi_{n2}(t) = c_{xn2}^{+} \text{ for } t_{12} < t \le t_d. \tag{5.36}$$

Next, from optimality conditions (5.30), we have

$$U_{n1}\left(\psi_{n1}(t) - c_{un1}\right) \ge U_{n2}\left(\psi_{n2}(t) - c_{un2}\right),\ t \in [t_1, t_{12}]$$

$$U_{n1}\left(\psi_{n1}(t) - c_{un1}\right) < U_{n2}\left(\psi_{n2}(t) - c_{un2}\right),\ t \in [t_{12}, t_2].$$

By taking into account that $\dot\psi_{n1}(t)$ and $\dot\psi_{n2}(t)$ are constants at $t \in [t_{12}, t_2]$, we conclude that the last two inequalities hold, if

$$\dot\psi_{n1}(t) > \dot\psi_{n2}(t) \frac{U_{n2}}{U_{n1}},\quad t \in (t_{12}, t_2].$$

Thus, by substituting (5.36) into the last inequality, we find $c^{+}_{xn1} > c^{+}_{xn2}\dfrac{U_{n2}}{U_{n1}}$ that contradicts assumption $c^{+}_{xn1}U_{n1} > c^{+}_{xn2}U_{n2}$.

■

LEMMA 5.9. (*No Preemption Before The Key Product Type*) Given the optimal solution and $t_{dn} = t_d$, n=1,...,N, no product is preempted before the due date.

PROOF. The proof for non-preemption before the due date immediately follows from Lemma 5.8. Indeed, once production of a product starts, it can be halted when its dual indicator is passed by a dual indicator of another product. Since the indicator is a linear function of time and the demand is zero before the due date, the by-passed indicator will never regain its priority.

■

LEMMA 5.10. (*Sequencing After The Key Product Type*) Given the optimal solution with key product type k and $t_{dn} = t_d$, n=1,...,N, if product type $n2$ is scheduled immediately after $n1$ and both are scheduled after the key product, then

$$c^{-}_{xn1}U_{n1} \geq c^{-}_{xn2}U_{n2}. \tag{5.37}$$

PROOF. The proof is similar to that for Lemma 5.8. ■

LEMMA 5.11. (*No Preemption After The Key Product Type*) Given the optimal solution and $t_{dn} = t_d$, n=1,...,N, no product is preempted after the due date.

PROOF. The proof is similar to that for Lemma 5.9. ■

Note that sequencing (5.34) proves the Weighted Least Production Rate rule (WLPR) to be optimal for ordering product types before the common due date. In comparison to Section 5.1, where the WLPR rule was applied concurrently to a number of products with the least production rate, now the rule is applied consecutively so that only one product is produced at a time.

It is important to note that if all demands $D_n = 1$, the WLPR rule is straightforwardly interpreted as the well-known Weighted Longest Processing Time (WLPT) rule found to be optimal for a number of classical, non-preemptive, due-date scheduling problems. On the other hand, sequencing (5.37) proves the Weighted Highest Production Rate rule (WHPR) to be optimal for ordering product types after the common due date. This is also interpreted as the well-known Weighted Shortest Processing Time (WSPT) rule. Moreover, both rules applied together result in V-shaped schedules which are optimal for problems with the objective of minimizing total earliness and tardiness.

Having proved Lemmas 5.8-5.11, we now conclude that if the costs are agreeable with respect to Definition 5.4, the sequencing of the product types

$$c_{x1}^{+}U_1 \le c_{x2}^{+}U_2 \le \ldots \le c_{xN}^{+}U_N \tag{5.38}$$

is optimal no matter which product is chosen to be the key product type. Moreover, preemptions cannot reduce the objective value because sequence (5.38) ensures that $c_{x1}^{-}U_1 \ge c_{x2}^{-}U_2 \ge \ldots \ge c_{xN}^{-}U_N$.

LEMMA 5.12. (*Non-Preemptive Timing*) Given problem (5.23)-(5.27) with $t_{dn} = t_d$, n=1,...,N, agreeable costs in terms of Definition 5.4 and $\sum_n \frac{D_n}{U_n} \le t_d \le T - \sum_n \frac{D_n}{U_n}$, if k is the key product type, then the switching times are determined as follows

$$t_k = t_d - \frac{\sum_{n=k}^{N} c_{xn}^{-} D_n - \sum_{n=1}^{k-1} c_{xn}^{+} D_n}{\left(c_{xk}^{+} + c_{xk}^{-}\right) U_k}, \qquad t_n = t_k - \sum_{j=n}^{k-1} \frac{D_j}{U_j}, \qquad \text{for } n=1,\ldots,k-1 \text{ and}$$

$$t_n = t_k + \sum_{j=k}^{n-1} \frac{D_j}{U_j} \quad \text{for } n=k+1,\ldots,N+1.$$

PROOF. Given key-product k, consider time points

- t_n, n=2,...,N, where production of product n-1 switches off and product n switches on;
- t_1, where production of the first product switches on;
- t_{N+1}, where production of the last product switches off.

Then, the following solution satisfies the state and co-state equations (see Figure 5.2 for the case of N=4 and k=3)

$$\psi_n(t) = \psi_{n-1}(t_n) + c_{xn}^{+}(t - t_n) \quad \text{for } t_n \le t \le t_{n+1},\ n=2,\ldots,k,$$

$$\psi_1(t) = c_{u1} + c_{x1}^{+}(t - t_1) \quad \text{for } t_1 \le t \le t_2,$$

$$\psi_n(t) = \psi_{n-1}(t_n) - c_{xn}^{-}(t - t_n) \quad \text{for } n=k+1,\ldots,N,\quad t_n \le t \le t_{n+1}, \tag{5.39}$$

$$\psi_k(t) = \psi_k(t_d) - c_{xk}^{-}(t - t_d) \quad \text{for } t_d \le t \le t_{k+1},$$

$$\psi_N(t_{N+1}) = c_{uN},$$

$$X_n(t) = U_n(t - t_n) \text{ for } t_n \le t \le t_{n+1}, \quad n=1,\ldots,N. \tag{5.40}$$

By considering system (5.39) and optimality condition (5.30) at point $t = t_d$, we find

$$\sum_{n=1}^{k-1} c_{xn}^{+} U_n\left(t_{n+1} - t_n\right) + c_{xk}^{+} U_k\left(t_d - t_k\right) = \sum_{n=k+1}^{N} c_{xn}^{-} U_n\left(t_{n+1} - t_n\right) + c_{xk}^{-} U_k\left(t_{k+1} - t_d\right). \tag{5.41}$$

Next, by substituting $t_{n+1} = t_n + \frac{D_n}{U_n}$ from (5.40) into (5.41), we obtain unknowns t_n n=1,...,N+1, as stated in the lemma.

■

THEOREM 5.1. Problem (5.23)-(5.27) is solvable in $O(N^2)$ time if $t_{dn} = t_d$, n=1,...,N, costs are agreeable in terms of Definition 5.4 and $\sum_n \frac{D_n}{U_n} \le t_d \le T - \sum_n \frac{D_n}{U_n}$.

PROOF. Algorithm 5.4 constructs a solution which is based on Lemmas 5.8-5.12. Thus, the maximum principle is satisfied and the solution is globally optimal.

The complexity of the algorithm is estimated straightforwardly. Sorting at the first step of Algorithm 5.4 requires $O(N \log N)$ time. Given a key-product, Step 2 calculates timing of the solution in $O(N)$ time. Step 3 of the algorithm (which needs $O(1)$ operations) ensures that the timing at Step 2 will be repeated at most N times trying each product as the key-product.

■

Special Cases of Zero Costs

Both the $1|N|\,u(t) \in [0,U]\,|D,T|\,L_x$ and $1|N|\,u(t) \in [0,U]\,|D,T|\,L_u, L_x, c_x^{+} = 0$ problems are straightforwardly solved by Algorithms 5.3 and 5.4. Furthermore, Algorithm 5.4 simplifies to require $O(N \log N)$ time instead of $O(N^2)$ time. Indeed, by Definition 5.4, no-inventory cost means that all costs are agreeable and, therefore, Algorithm 5.4 can be applied to construct the optimal solution. At Step 2, timing is now trivial because no cost for inventory implies that production should be started as early as possible, i.e. at the beginning of the planning horizon, $t_1 = 0$, $t_n = \sum_{j=1}^{n-1} \frac{D_j}{U_j}$ for n=2,...,N+1.

No Backlog

The case when backlogs are disallowed, i.e. the $1|N|u(t)\in[0,U], X(t)\geq D(t)|D,T|L_u,L_x$ problem, is solvable in $O(N\log N)$. Indeed, the costs are always agreeable in this case and a feasible solution always exists if $t_d \geq \sum_n \frac{D_n}{U_n}$. Therefore, the solution algorithm can be obtained by an easy modification to Algorithm 5.4. In contrast to the case of zero inventory cost, timing at Step 2 is calculated backward from the due date because no product can be late: $t_{N+1}=t_d$, $t_n = t_{N+1} - \sum_{j=n}^{N}\frac{D_j}{U_j}$ for n=1,...,N.

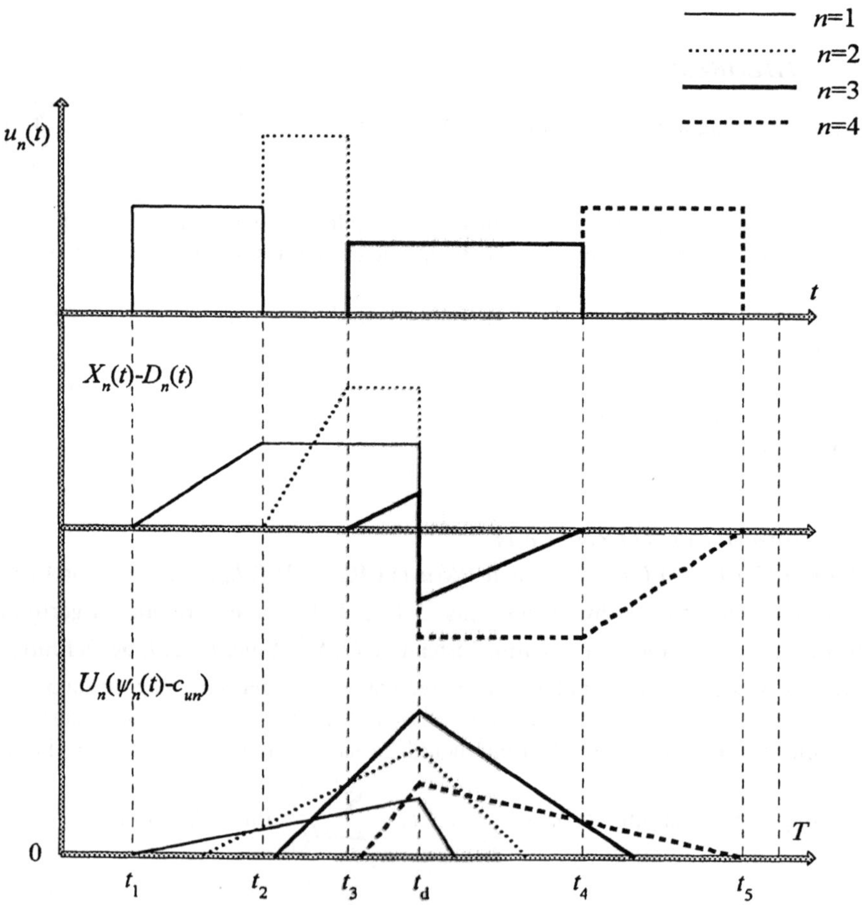

Figure 5.2. Optimal control, state variables and dual indicators for the $1|N|u(t)\in[0,U]|D,T|L_u,L_x$ problem.

5.3 DEMAND OVER INFINITE PLANNING HORIZON

This section addresses the scheduling problems characterized by periodic demand as a function of time specified over an infinite planning horizon. The periodic form of the demand causes the optimal solution to be also periodic unless the initial inventory is fixed. Since the period of the optimal solution is assumed to be equal to the demand period, $X_n(t) = X_n(t+T)$ for every product n=1,...,N, where T is the demand period. Clearly, to provide the periodicity of the solution, it is sufficient to consider the problem over the finite planning horizon, T, with boundary condition $X_n(0) = X_n(T)$.

5.3.1 LINEAR PERFORMANCE MEASURE

Backlogs

The problem to be considered is one-machine, N-product-type scheduling with backlogs and limited production rate to minimize the linear cost of inventory and production over period T, that is, the $1|M|\, u(t) \in [0, U], X(0) = X(T)\, |d(t)|\, L_u, L_x$ problem which is stated as follows

$$J = \int_0^T \sum_n \left[c_{xn}^+ X_n^+(t) + c_{xn}^- X_n^-(t) + c_{un} u_n(t)\right] dt \to \min, \tag{5.42}$$

$$X_n^+(t) = \max\{X_n(t), 0\},\ X_n^-(t) = \max\{-X_n(t), 0\}$$

subject to

$$\dot{X}_n(t) = u_n(t) - d_n(t), \quad X_n(0) = X_n(T) \tag{5.43}$$

$$0 \le u_n(t) \le U_n, \tag{5.44}$$

$$\sum_n \frac{u_n(t)}{U_n} \le 1. \tag{5.45}$$

Dual Formulation

The Hamiltonian, H, resulting from the maximum principle is

$$H = -\sum_n \left[c_{xn}^+ X_n^+(t) + c_{xn}^- X_n^-(t) + c_{un} u_n(t)\right] + \sum_n \psi_n(t)\left(u_n(t) - d_n(t)\right). \tag{5.46}$$

The co-state variables, $\psi_n(t)$ in (5.46), satisfy the following co-state differential equation with corresponding transversality condition

$$\dot{\psi}_n(t) = \begin{cases} c_{xn}^+, \text{ if } X_n(t) > 0; \\ -c_{xn}^-, \text{ if } X_n(t) < 0; \\ a, \; a \in [-c_{xn}^-, c_{xn}^+], \text{ if } X_n(t) = 0; \end{cases} \qquad \psi_n(0) = \psi_n(T). \tag{5.47}$$

Maximization of the Hamiltonian implies that

$$u_n(t) = \begin{cases} U_n, \text{ if } \psi_n(t) > c_{un} \text{ and } U_n\left(\psi_n(t) - c_{un}\right) > U_{n'}\left(\psi_{n'}(t) - c_{un'}\right), \; \forall n' \neq n \\ 0, \text{ if } \psi_n(t) < c_{un}, \\ b, \; b \in [0, U_n], \text{ if } \psi_n(t) \geq c_{un} \text{ and } U_n\left(\psi_n(t) - c_{un}\right) \geq U_{n'}\left(\psi_{n'}(t) - c_{un'}\right), \; \forall n' \neq n. \end{cases} \tag{5.48}$$

Algorithm 5.5

Step 1:

If $\sum_n \frac{d_n(t)}{U_n} \leq 1$ for $0 \leq t \leq T$, then set the optimal solution as $u_n(t) = d_n(t)$ for $0 \leq t \leq T$, STOP.

Otherwise go to the next step.

Step 2:

Sort products in the non-decreasing order of $c_{xn}^- U_n$ and re-number them accordingly. If demand of product N is two-level, piece-wise constant

$$d_N(t) = \begin{cases} d_{N1}, \text{ if } 0 \leq t \leq \bar{t}, \\ d_{N2}, \text{ if } \bar{t} < t \leq T, \end{cases} \qquad d_{N1} < U_N, \quad d_{N2} > U_N$$

and demand of the other products is constant

$d_n(t) = d_n$ for $0 \leq t \leq T$, $n < N$,

$\sum_n \frac{d_n(t)}{U_n} \leq 1$ for $0 \leq t < \bar{t}$ and $\sum_{n<N} \frac{d_n(t)}{U_n} \leq 1 \leq \sum_n \frac{d_n(t)}{U_n}$ for $\bar{t} \leq t \leq T$, and costs are agreeable in terms of Definition 5.4, then solve the following system of $2N+1$ linear equations in $2N+1$ unknown time points t_n^1, t_n^2 and t_x

$$U_n\left(1-\sum_{n'=n+1}^{N-1}\frac{d_{n'}}{U_{n'}}-\frac{d_{N1}}{U_N}\right)\left(t_{n+1}^1-t_n^1\right)=d_n\left(t_n^2-t_n^1\right),\ \text{for } n=1,\ldots,N\text{-}1;$$

$$U_N\left(t_N^2-t_{N-1}^2\right)+U_N\sum_{n=2}^{N-1}\left(1-\sum_{n'=n}^{N-1}\frac{d_{n'}}{U_{n'}}\right)\left(t_{n-1}^2-t_n^2\right)+$$

$$+U_N\left(1-\sum_{n'=1}^{N-1}\frac{d_{n'}}{U_{n'}}\right)\left(t_N^2-t_1^2\right)=d_{N1}\left(\bar{t}-t_N^1\right)+d_{N2}\left(t_N^2-\bar{t}\right);$$

(5.49)

$$c_{xn+1}^+U_{n+1}\left(t_{n+1}^2-t_{n+1}^1\right)=c_{xn}^+U_n\left(t_n^2-t_{n+1}^1\right)+c_{xN}^-U_N\left(t_n^2-t_{n+1}^2\right),\ n=1,\ldots,N\text{-}1;$$

$$\sum_{n=1}^{N-1}c_{xn}^+U_n\left(t_{n+1}^1-t_n^1\right)+c_{xN}^+U_N\left(t_x-t_N^1\right)=c_{xN}^-U_N\left(t_N^2-t_x\right);$$

$$U_N\left(t_x-t_N^1\right)=d_{N1}\left(\bar{t}-t_N^1\right)+d_{N2}\left(t_x-\bar{t}\right).$$

If $t_1^1\ge t_N^2-T$, then set the optimal solution as

$$u_n(t)=d_n(t)\ \text{ for } t_1^1\le t\le t_n^1 \text{ and } t_n^2\le t\le T+t_1^1,\qquad n=1,\ldots,N,$$

$$u_n(t)=U_n\left(1-\sum_{n'=n+1}^{N}\frac{d_{n'}(t)}{U_{n'}}\right)\ \text{ for } t_n^1\le t\le t_{n+1}^1,\qquad n=1,\ldots,N\text{-}1,$$

$$u_N(t)=U_N\ \text{ for } t_N^1\le t\le t_{N-1}^2,$$

$$u_N(t)=U_N\left(1-\sum_{n'=n}^{N}\frac{d_{n'}(t)}{U_{n'}}\right)\ \text{ for } t_n^2\le t\le t_{n+1}^2,\qquad n=1,\ldots,N\text{-}1,$$

$$u_n(t)=0\ \text{ for } t_{n+1}^1\le t\le t_n^2,\qquad n=1,\ldots,N\text{-}1,$$

STOP.

Otherwise go to the next step.

Step 3:

If demand of product k, $k=\arg\min_n c_{xn}^+U_n=\arg\min_n c_{xn}^-U_n$, is two-level, piece-wise constant

$$d_k(t) = \begin{cases} d_{k1}, \text{ if } 0 \le t \le \bar{t}, \\ d_{k2}, \text{ if } \bar{t} < t \le T, \end{cases}$$

demand for the other products is constant

$$d_n(t) = d_n \text{ for } 0 \le t \le T,\ n \ne k,\ \sum_n \frac{d_n(t)}{U_n} \le 1 \text{ for } 0 \le t < \bar{t},$$

$$\sum_n \frac{d_n(t)}{U_n} \ge 1 \text{ for } \bar{t} \le t \le T \text{ and } \frac{(T-\bar{t})(d_{k2}-d_{k1})}{U_k\left(1-\sum_{n\ne k}\frac{d_n}{U_n}\right)-d_{k1}} \le T,$$

then calculate

$$t_1 = \frac{t_x\left(U_k\left(1-\sum_{n\ne k}\frac{d_n}{U_n}\right)-d_{k2}\right)+\bar{t}(d_{k2}-d_{k1})}{U_k\left(1-\sum_{n\ne k}\frac{d_n}{U_n}\right)-d_{k1}},$$

$$t_2 = \frac{t_x\left(U_k\left(1-\sum_{n\ne k}\frac{d_n}{U_n}\right)-d_{k2}\right)+T(d_{k2}-d_{k1})}{U_k\left(1-\sum_{n\ne k}\frac{d_n}{U_n}\right)-d_{k1}}, \text{ where } t_x = \frac{\bar{t}c_{xk}^{+}+Tc_{xk}^{-}}{c_{xk}^{+}+c_{xk}^{-}}.$$

Set the optimal solution as

$$u_n(t) = d_n \text{ for } t_1 \le t \le t_1 + T,\quad n \ne k,$$

$$u_k(t) = U_k\left(1-\sum_{n\ne k}\frac{d_n}{U_n}\right) \text{ for } t_1 \le t < t_2,$$

$u_k(t) = d_k(t)$ for $t_2 \le t < t_1 + T$; STOP.

Example 5.3
Consider the $1|N|u(t)\in[0,U], X(0)=X(T)\,|d(t)|\,L_u, L_x$ problem with: the length of the planning horizon, T, set at 10hrs; the demand depicted in Figure 5.3; and the production parameters shown in Table 5.3.

Since the capacity constraint of Step 1 is not met after point $\bar{t}$ =5hr and the costs are not agreeable in terms of Definition 5.4, Algorithm 5.5 proceeds to Step 3. At Step 3, it finds k=4 and verifies the capacity constraints. Then, the algorithm calculates time points $t_x=9.5$, $t_1=3.5$, $t_2=10.17$ (hr.) and sets the optimal solution as

$u_1(t)=1$ unit/hr, $u_2(t)=0.7$ unit/hr, $u_3(t)=0.25$ unit/hr for $3.5\le t\le 13.5$,
$u_4(t)=0.9$ unit/hr for $3.5\le t<10.17$,
$u_4(t)=0.6$ unit/hr for $10.17\le t<13.5$.

This solution is depicted in Figure 5.3 along with the corresponding behavior of dual indicators $U_n(\psi_n(t)-c_{un})$ and inventory levels $X_n(t)$, n=1,...,4. From Figure 5.3, one can observe that all dual indicators coincide along the period and that the inventories are zero except for the product, k=4, at interval $3.5\le t<10.17$.
■

Discussion

This section presents three solvable cases. Step 1 of Algorithm 5.5 solves the problem for the case when the total demand does not exceed the total production capacity. Then, the optimal solution requiring $O(N)$ time is simply to exactly trace the demand over the period. Step 2 of Algorithm 5.5 deals with the case when the costs are agreeable in terms of Definition 5.4 and demand is constant for all products except for product $k=\arg\max_n c_{xn}^{+}U_n=\arg\min_n c_{xn}^{-}U_n$. The demand for this product is two-level piece-wise constant. In such a case, the concurrent WLPR rule discussed in Section 5.1 becomes optimal for non-preemptive scheduling before key product k, while the concurrent WHPR rule is optimal for resuming production after the key product. As a result, the key product is not preempted, while every other product is preempted only once. Since timing for this problem is reduced to a system of linear algebraic equations, the problem is solvable in $O(N^2)$ time.

Finally, the last step of the algorithm solves the problem when the key product is $k=\arg\min_n c_{xn}^{+}U_n=\arg\min_n c_{xn}^{-}U_n$. This problem is solvable in $O(N)$ time because sorting of products is no longer required and timing is determined by only two switching points.

Table 5.3. Maximum production rates and unit costs of the system

n	c^{+}_{xn}	c^{-}_{xn}	c_{un}	U_n
1	0.01	0.1	1.1	4.0
2	0.02	0.3	0.6	2.0
3	0.02	0.2	1.0	2.5
4	0.01	0.1	0.6	3.0

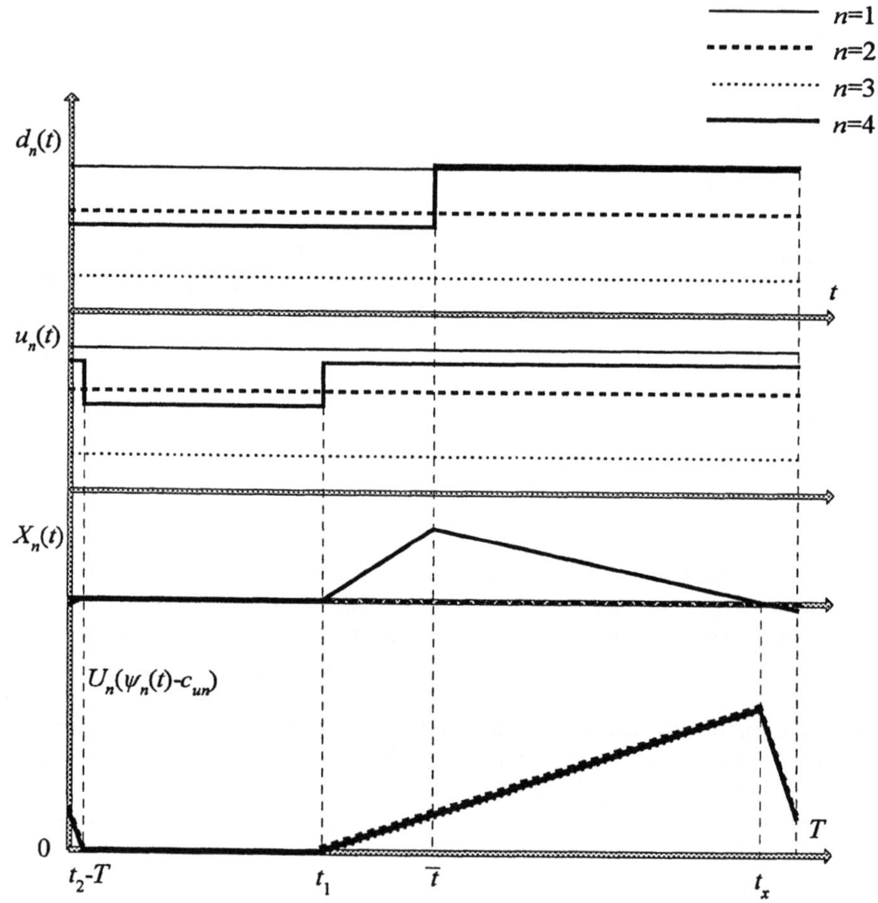

Figure 5.3. Demand, optimal control, state variables and dual indicators for the $1|M|\, u(t)\in[0,U]\,, X(0)=X(T)\,|d(t)|\, L_u, L_x$ problem of Example 5.3.

LEMMA 5.13. (Correctness of Algorithm 5.5 Step 1) Given problem (5.42)-(5.45), if $\sum_n \frac{d_n(t)}{U_n} \le 1$ for $0 \le t \le T$, then the optimal solution is $u_n(t) = d_n(t)$ for $0 \le t \le T$, n=1,...,N.

PROOF. Since the demand does not exceed the capacity $\sum_n \frac{d_n(t)}{U_n} \le 1$ for $0 \le t \le T$, then the following solution

$$\psi_n(t) = c_{un}, \quad u_n(t) = d_n(t), \quad X_n(t) = 0 \quad \text{for } 0 \le t \le T,$$

is always feasible.

■

To prove the correctness of Step 2 Algorithm 5.5, we make use of the *S*-fractional-production and *Z*-fractional-production regimes (see Definitions 5.2 and 5.3).

LEMMA 5.14. (Correctness of Algorithm 5.5 Step 2) Let costs be agreeable in terms of Definition 5.4 and the products be numbered in the non-decreasing order of $c_n^+ U_n$. If $d_N(t) = \begin{cases} d_{N1}, \text{ if } 0 \le t \le \bar{t}, \\ d_{N2}, \text{ if } \bar{t} < t \le T, \end{cases}$ $d_{N1} < U_N$, $d_{N2} > U_N$; $d_n(t) = d_n$ for $0 \le t \le T$, $n < N$, $\sum_n \frac{d_n(t)}{U_n} \le 1$ for $0 \le t < \bar{t}$ and $\sum_{n<N} \frac{d_n(t)}{U_n} \le 1 \le \sum_n \frac{d_n(t)}{U_n}$ for $\bar{t} \le t \le T$,

then the optimal solution is

$u_n(t) = d_n(t)$ for $t_1^1 \le t \le t_n^1$ and $t_n^2 \le t \le T + t_1^1$, n=1,...,N,

$$u_n(t) = U_n\left(1 - \sum_{n'=n+1}^{N} \frac{d_{n'}(t)}{U_{n'}}\right) \quad \text{for } t_n^1 \le t \le t_{n+1}^1, \; n=1,\dots,N-1,$$

$u_N(t) = U_N$ for $t_N^1 \le t \le t_{N-1}^2$,

$$u_N(t) = U_N\left(1 - \sum_{n'=n}^{N} \frac{d_{n'}(t)}{U_{n'}}\right) \quad \text{for } t_n^2 \le t \le t_{n+1}^2, \; n=1,\dots,N-1$$

$u_n(t)=0$ for $t^1_{n+1}\le t\le t^2_n$, n=1,...,N-1,

where 2N+1 time points t^1_n, t^2_n and t_x are calculated from system (5.49) and $t^1_1\ge t^2_N-T$.

PROOF. Consider the V-shaped periodic solution which includes:

- Z-fractional-production regime with $Z=\{1,\dots,N\}$;
- N-1 S-fractional-production regimes with $S=\{n,\dots,N\}$ at the n-th regime and time points t^1_n where the regimes change over, $n=1,\dots,N-1$;
- full production regime of product N;
- N-1 S-fractional-production regimes with $S=\{N-n,\dots,N\}$ at the n-th regime and time points t^2_{N-n} where the regimes change over, $n=1,\dots,N-1$;
- Z-fractional-production regime with $Z=\{1,\dots,N\}$.

According to this solution, product n, $n=1,\dots,N-1$, is produced at interval $[0,t^1_{n+1}]$; then its production is halted (preempted) at interval $[t^1_{n+1},t^2_n]$; and it is resumed at interval $[t^2_n,T]$.

Consequently, the control, state and co-state variables behave as follows

- at the Z-fractional-production regime with $Z=\{1,\dots,N\}$ for $0\le t\le t^1_1$,

$$u_{n'}(t)=d_{n'}(t),\ X_{n'}(t)=0,\ \psi_{n'}(t)=c_{un'}\ \text{ for all } n'.$$

- at the N-1 S-fractional-production regimes with $S=\{n,\dots,N\}$ at the n-th regime for $t^1_n\le t\le t^1_{n+1}$, $n=1,\dots,N-1$,

$$u_n(t)=U_n\left(1-\sum_{n'=n+1}^{N}\frac{d_{n'}(t)}{U_{n'}}\right),\ X_n(t)>0,$$

$$u_{n'}(t)=d_{n'}(t),\ X_{n'}(t)=0\ \text{ for } n'=n+1,\dots,N,$$
$$u_{n'}(t)=0,\ X_{n'}(t)>0\ \text{ for } n'=1,\dots,n-1,$$

$$U_n\left(\psi_n(t)-c_{un}\right)=U_{n'}\left(\psi_{n'}(t)-c_{un'}\right)\ \text{ for } n'=n+1,\dots,N.$$

- at the full production regime of product N for $t^1_N\le t\le t^2_{N-1}$,

$$u_N(t)=U_N,\ u_{n'}(t)=0\ \text{ for } n'=1,\dots,N-1,$$

$$X_N(t)>0 \text{ for } t_N^1<t<t_x,\ X_N(t)<0 \text{ for } t_x<t<t_{N-1}^2,$$

$$X_{n'}(t)>0,\ U_N\left(\psi_N(t)-c_{uN}\right)>U_{n'}\left(\psi_{n'}(t)-c_{un'}\right) \text{ for } n'=1,\dots,N-1.$$

- at the N-1 S-fractional-production regimes with $S=\{N-n,\dots,N\}$ at the n-th regime and time points t_{N-n}^2 where the regimes change over, $n=1,\dots,N-1$ for $t_n^2 \le t \le t_{n+1}^2$,

$$u_N(t)=U_N\left(1-\sum_{n'=n}^{N}\frac{d_{n'}(t)}{U_{n'}}\right),\ X_N(t)<0,$$

$$u_{n'}(t)=d_{n'}(t),\ X_{n'}(t)=0 \text{ for } n'=n,\dots,N-1,$$

$$u_{n'}(t)=0,\ X_{n'}(t)>0 \text{ for } n'=1,\dots,n-1,$$

$$U_n\left(\psi_n(t)-c_{un}\right)=U_{n'}\left(\psi_{n'}(t)-c_{un'}\right) \text{ for } n'=n+1,\dots,N.$$

- at the Z-fractional-production regime with $Z=\{1,\dots,N\}$ for $t_N^2 \le t \le T$.

$$u_{n'}(t)=d_{n'}(t),\ X_{n'}(t)=0,\ \psi_{n'}(t)=c_{un'} \text{ for all } n'.$$

Note that this solution includes a number of S-fractional production regimes which occur twice: when the inventory is positive and when it is negative. According to Lemma 5.2, such a solution can be feasible only if the products are ordered in the non-decreasing value of $c_{xn}^{+}U_n$ provided the inventory is positive, and in the non-increasing value of $c_{xn}^{-}U_n$, otherwise. Thus, the costs must be agreeable in terms of Definition 5.4.

Switching points t_n^1, t_n^2 and t_x must satisfy the following boundary conditions of the above solution

$$X_n(t_n^2)=0,\ n=1,\dots,N,$$

$$U_n\left(\psi_n(t_n^2)-c_{un}\right)=U_N\left(\psi_N(t_n^2)-c_{uN}\right),\ n=1,\dots,N-1, \tag{5.50}$$

$$U_N\left(\psi_N(t_1^1)-c_{uN}\right)=U_N\left(\psi_N(t_N^2)-c_{uN}\right),$$

$$X_N(t_x)=0.$$

By taking into account the state (5.43) and co-state (5.47) equations, system (5.50) transforms into $2N+1$ linear algebraic equations (5.49), the solution of which is feasible if $t_1^1 \geq t_N^2 - T$.

■

LEMMA 5.15. (Correctness of Algorithm 5.5 Step 2)

If $k = \arg\min_n c_{xn}^+ U_n = \arg\min_n c_{xn}^- U_n$, $d_k(t) = \begin{cases} d_{k1}, \text{ if } 0 \leq t \leq \bar{t}, \\ d_{k2}, \text{ if } \bar{t} < t \leq T, \end{cases}$ $d_n(t) = d_n$ for $0 \leq t \leq T$, $n \neq k$, $\sum_n \frac{d_n(t)}{U_n} \leq 1$ for $0 \leq t < \bar{t}$, $\sum_n \frac{d_n(t)}{U_n} \geq 1$ for $\bar{t} \leq t \leq T$, and

$\bar{t}(d_{k2} - d_{k1}) \geq Td_{k2} - TU_k\left(1 - \sum_{n \neq k} \frac{d_n}{U_n}\right)$, then the optimal solution is

$u_n(t) = d_n$ for $0 \leq t \leq T$, $n \neq k$,

$u_k(t) = U_k\left(1 - \sum_{n \neq k} \frac{d_n}{U_n}\right)$ for $t_1 \leq t < t_2$; $u_k(t) = d_k(t)$ for $0 \leq t < t_1$ and $t_2 \leq t < T$,

where

$$t_1 = \frac{t_x\left(U_k\left(1 - \sum_{n \neq k} \frac{d_n}{U_n}\right) - d_{k2}\right) + \bar{t}(d_{k2} - d_{k1})}{U_k\left(1 - \sum_{n \neq k} \frac{d_n}{U_n}\right) - d_{k1}}, \quad t_2 = \frac{t_x\left(U_k\left(1 - \sum_{n \neq k} \frac{d_n}{U_n}\right) - d_{k2}\right) + T(d_{k2} - d_{k1})}{U_k\left(1 - \sum_{n \neq k} \frac{d_n}{U_n}\right) - d_{k1}}$$

and $t_x = \dfrac{\bar{t}c_{xk}^+ + Tc_{xk}^-}{c_{xk}^+ + c_{xk}^-}$.

PROOF. Consider a periodic solution starting from the *Z*-fractional-production regime, $Z = \{1, \ldots, N\}$, proceeding to the *S*-fractional-production regime from point t_1 to point t_2, $S = \{1, \ldots, N\}$ and terminating with the *Z*-fractional-production regime, $Z = \{1, \ldots, N\}$

$u_n(t) = d_n$, $X_n(t) = 0$, $\psi_n(t) = c_{un}$ for all n, $0 \leq t < t_1$ and $t_2 \leq t < T$;

$$u_n(t) = d_n,\ X_n(t) = 0,\ U_n\big(\psi_n(t) - c_{un}\big) = U_k\big(\psi_k(t) - c_{uk}\big) \text{ for all } n \neq k,\ t_1 \leq t \leq t_2;\ (5.51)$$

$$u_k(t) = U_k\left(1 - \sum_{n \neq k} \frac{d_n}{U_n}\right),\ X_k(t) > 0 \text{ for } t_1 < t < t_x,\ X_k(t) < 0 \text{ for } t_x < t < t_2.$$

Switching points t_1, t_2 and t_x must satisfy the following boundary conditions of solution (5.51)

$$X_k(t_x) = 0,\ X_k(t_2) = 0 \text{ and } U_k\big(\psi_k(t_1) - c_{uk}\big) = U_k\big(\psi_k(t_2) - c_{uk}\big). \qquad (5.52)$$

By taking into account solution (5.51), state (5.43) and co-state (5.47) equations, (5.52) transforms into three algebraic equations in three unknowns t_1, t_2 and t_x. The solution of these equations prove the lemma if $t_1 \geq t_2 - T$.

■

Special Cases of Zero Costs

Solution Algorithm 5.5 does not include production cost coefficients c_{un} and, therefore, does not change for the special cases of zero production cost, i.e. for the $1|N|u(t) \in [0, U], X(0){=}X(T)|d(t)|\, L_x$ problem.

The no inventory cost $1|N|u(t) \in [0, U], X(0){=}X(T)|d(t)|\, L_u, L_x, c_x^+ = 0$ problem is solved by Steps 1 and 3 of Algorithm 5.5 by substituting $c_x^+ = 0$ into the equations which determine switching points t_1 and t_2. Since Step 2 of the algorithm is now identical to Step 3, it is no longer needed. Furthermore, since the costs are now agreeable in terms of Definition 5.4, the condition for key product k is simplified as $k = \arg\min_n c_{xn}^- U_n$. Thus, the $1|N|u(t) \in [0, U], X(0){=}X(T)|d(t)|\, L_u, L_x, c_x^+ = 0$ problem is solvable in $O(N)$ time.

Additionally, from the proof of Lemma 5.15, it follows that $X(t)$ is never negative.

No Backlog

The solution of the $1|N|u(t) \in [0, U], X(t) \geq 0, X(0){=}X(T)|d(t)|\, L_u, L_x$ problem, coincides with the solution of the $1|N|u(t) \in [0, U], X(0){=}X(T)|d(t)|\, L_u, L_x, c_x^+ = 0$ problem. One can verify this by assigning a very large value to c_x^- in Algorithm 5.15. The rigorous proof of this fact is left for the reader.

5.3.2 MAINTAINING STABLE INVENTORY

Backlogs

The problem to be considered is one-machine, N-product-type production control, with limited rate, to maintain stable inventory of each product under unknown, arbitrary demand functions, that is the $1|N|\, u(t) \in [0,U]\, |d(t)|\, \overline{X}$ problem which is stated as follows

$$\sup_{0 \le t < \infty} |X_n(t)| < \infty, \tag{5.53}$$

$$\dot{X}_n(t) = u_n(t) - d_n(t), \tag{5.54}$$

$$0 \le u_n(t) \le U_n, \tag{5.55}$$

$$\sum_n \frac{u_n(t)}{U_n} \le 1. \tag{5.56}$$

Feedback control for multi-product systems is a control effort given as a function of all inventories rather than of a single buffer level as was the case in one-product systems. Generally, feedback control rules may involve several components with adjustable positive parameters G and desired inventory levels $\overline{X}_n$. The rules suggested in this section are based on those presented in Sections 3.3.5 and 4.3.5 for the single-product problems.

Feedback Rule 5.1

Proportional, multi-product, feedback control:

$$\text{set } u_n(X_1(t),\ldots,X_N(t)) = \begin{cases} U_n \dfrac{G_n(\overline{X}_n - X_n(t))}{\sum_{n'} G_{n'}(\overline{X}_{n'} - X_{n'}(t))}, & \text{if } X_n(t) < \overline{X}_n, \\ 0, \ \text{if } X_n(t) \ge \overline{X}_n. \end{cases}$$

at every t.

Feedback Rule 5.2

Fill-in-the-Lowest-Buffer-Level feedback control:

(i) select product n such that $X_n(t) = \min_{1 \le n' \le N} X_{n'}(t)$;

(ii) if $X_n(t) \le \overline{X}_n$, then set $u_n(t) = U_n$, $u_{n'}(t) = 0$ for all $n' \ne n$; otherwise, set $u_n(t) = 0$ and go to (i).

Feedback Rule 5.3
Fill-A-Fraction feedback control:

(i) select a product n such that $X_n(t) \le \frac{G}{N}\sum_{n'} X_{n'}(t)$;

(ii) if $X_n(t) \le \overline{X}_n$, then set $u_n(t) = U_n$, $u_{n'}(t) = 0$ for all $n' \ne n$; otherwise, set $u_n(t) = 0$ and go to (i).

Feedback Rule 5.4
Cost-Improving-Fill-A-Fraction feedback control:

(i) form a set of products S such that $n \in S$ if and only if $X_n(t) \le \frac{G}{N}\sum_{n'} X_{n'}(t)$;

(ii) if there exists $n' \in S$ such that $X_{n'}(t) < 0$, then select product $n = \arg\max_{n' \in S, X_{n'}(t)<0} c_{xn'}^{-} U_{n'}$; otherwise select product $n = \arg\min_{n' \in S} c_{xn'}^{+} U_{n'}$;

(iii) if $X_n(t) \le \overline{X}_n$, then set $u_n(t) = U_n$, $u_{n'}(t) = 0$ for all $n' \ne n$; otherwise, set $u_n(t) = 0$ and go to (i).

Discussion

Feedback Rule 5.1 is based on the proportional policy and Feedback Rules 5.2-5.4 are based on the bang/bang policy. The rules are of $O(N)$ complexity at every time t and are extensions of corresponding Rules 4.1 and 4.2, developed for the single-product capacitated problems. Specifically, Feedback Rule 5.2 switches the machine from product n which has reached the desired level $\overline{X}_n$ to the product whose inventory level is minimal. This policy corresponds to the well-known Clear-the-Largest-Buffer-Level (CLB) policy proven to maintain stable inventory if demand for the products does not exceed the total capacity. Feedback Rule 5.3 generalizes Feedback Rule 5.2 by switching the machine to a product whose inventory is below some average level. This rule corresponds to the Clear-A-Fraction (CAF) policy and is feasible if adjustable parameter G is bounded as $0 \le G \le 1$. Finally, Rule 5.4 elaborates Rule 5.3 for the case when a secondary cost criterion is important. Indeed, while Feedback Rule 5.3 offers a number of products that the machine can switch on, Feedback Rule 5.4 singles out which product to switch on, so that the total cost of inventory and backlog is reduced. The choice of such a product is based on the optimal control properties, formalized in the previous sections as the WLPR rule for positive inventories and as the WHPR rule for negative inventories.

Bounded evolution of inventories under Feedback Rules 5.1 and 5.2 is shown in Figure 5.4 for cases when the demand is weak most of the time. However, for arbitrary demand functions, stability of inventories is not guaranteed. The following lemmas identify demand-related conditions sufficient for the inventory to be bounded for Feedback Rules 5.1-5.4.

LEMMA 5.16. (Stability of Feedback Rule 5.1) Given problem (5.53)-(5.56), if the proportional, multi-product, feedback control is applied with $\overline{X}_n \geq 0$ and $\lim_{t\to\infty}\frac{1}{t}\int_0^t \sum_n \frac{d_n(\tau)}{U_n} d\tau < 1$, then $X_n(t) \leq \overline{X}_n$ for all t and $\inf_{0\leq t<\infty} X_n(t) > -\infty$.

PROOF. First note, that if $X_n(0) \leq \overline{X}_n$, the inventory level can never exceed $\overline{X}_n$. Indeed, by definition of the proportional feedback control, once $X_n(t)$ reaches $\overline{X}_n$, production stops and demand drains inventory again below $\overline{X}_n$, i.e.

$$X_n(t) \leq \overline{X}_n. \tag{5.57}$$

Thus, to prove that the inventory is bounded from below, we can substitute the first condition of the rule into the state equation (5.54)

$$\dot{X}_n(t) = U_n \frac{G_n(\overline{X}_n - X_n(t))}{\sum_{n'} G_{n'}(\overline{X}_{n'} - X_{n'}(t))} - d_n(t), \quad X_n(0) = 0. \tag{5.58}$$

Next, by summing up equations (5.58) for n=1,...,N, we find

$$\sum_n \frac{\dot{X}_n(t)}{U_n} = 1 - \sum_n \frac{d_n(t)}{U_n}. \tag{5.59}$$

From (5.59), it follows that inventories $X_n(t)$ are bounded from below if

$$\lim_{t\to\infty}\frac{1}{t}\int_0^t \sum_n \frac{d_n(\tau)}{U_n} d\tau < 1. \tag{5.60}$$

■

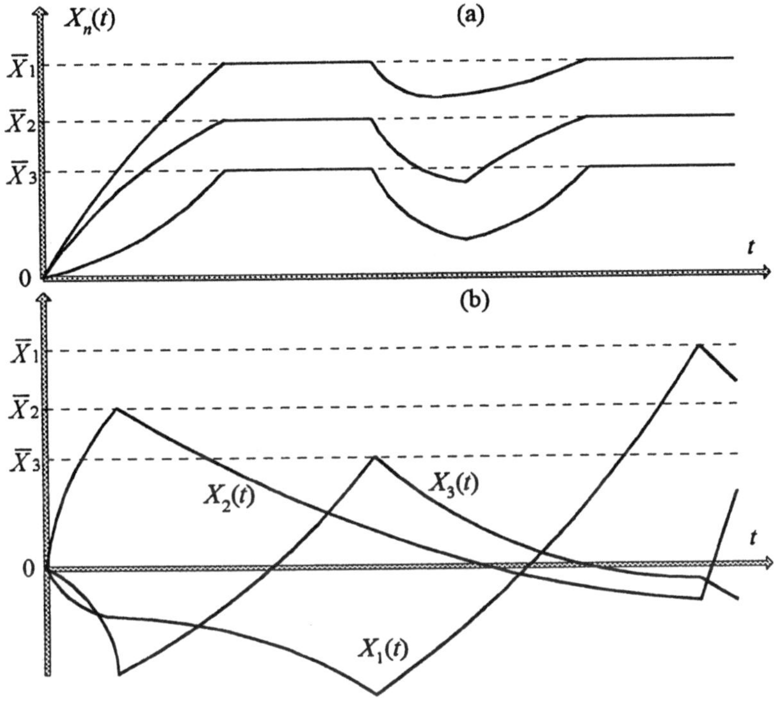

Figure 5.4. Evolution of inventory $X_n(t)$ under Rules 5.1 (a) and 5.2 (b) for the $1|N|\,u(t) \in [0, U]\,|d(t)|\,\overline{X}$ problem.

LEMMA 5.17. (Stability of Feedback Rules 5.2-5.4) Given problem (5.53)-(5.56), if Feedback Rules 5.2-5.4 are applied with $\overline{X}_n \geq 0$ and

$\lim\limits_{t\to\infty} \frac{1}{t}\int\limits_0^t \sum\limits_n \frac{d_n(\tau)}{U_n} d\tau < 1$, then $X_n(t) \leq \overline{X}_n$ for all t and $\inf\limits_{0 \leq t < \infty} X_n(t) > -\infty$.

PROOF. First note, that Feedback Rules 5.2 and 5.4 are special cases of Rule 5.3. Therefore, it is sufficient to prove stability of inventories under Rule 5.3. According to this rule, if $X_n(0) \leq \overline{X}_n$, the inventory level can never exceed $\overline{X}_n$.

To prove that the inventory is bounded from below, we consider the worst case, when the demand is so strong that machine capacity is fully utilized. Then, by summing up equations (5.54) for n=1,...,N, and substituting capacity constraint (5.56) as equality, we find

$$\sum_n \frac{\dot{X}_n(t)}{U_n} = 1 - \sum_n \frac{d_n(t)}{U_n}. \tag{5.61}$$

From (5.61), it follows that inventories $X_n(t)$ are bounded from below if

$$\lim_{t\to\infty} \frac{1}{t} \int_0^t \sum_n \frac{d_n(\tau)}{U_n} d\tau < 1.$$

■

No Backlog

For the case when backlogs are disallowed, i.e. for the $1|N|\, u(t) \in [0, U], X(t) \geq 0\, |d(t)|\, \overline{X}$ problem, if $X_n(0) = \overline{X}_n$ and demand is weak, i.e. $\sum_n \frac{d_n(t)}{U_n} < 1$ at every point of time, Feedback Rules 5.1-5.4 strictly maintain the desired inventory level $X_n(t) \equiv \overline{X}_n$. Thus, these rules are evidently stable and provide no backlog for such a case.

5.4 VARIABLE PLANNING HORIZON

This section addresses the scheduling problems characterized by an amount of product to be completed by the end of a planning horizon. The length of the planning horizon is unknown *a priori* and is to be minimized along with the inventory and production costs.

We consider one-machine, *N*-product-type scheduling with limited production rate, to minimize the total linear cost of inventory, production and the planning horizon, that is, the $1|N|\, u(t) \in [0, U], X(T) = X_T \,\|\, L_u, L_x, T$ problem which is stated as follows

$$J = \int_0^T \sum_n \left[c_{xn}^{+} X_n(t) + c_{un} u_n(t)\right] dt + c_T T \to \min,$$

subject to

$$\dot{X}_n(t) = u_n(t), \quad X_n(0) = 0, \quad X_n(T) = X_{Tn} \tag{5.62}$$

$$0 \leq u_n(t) \leq U_n, \tag{5.63}$$

$$\sum_n \frac{u_n(t)}{U_n} \leq 1. \tag{5.64}$$

Note that the above statement is characterized by non-negative inventory along the overall planning horizon because the planning horizon ends as soon as the required amount of the products, $X_n(T) = X_{Tn}$, is completed.

The objective of the problem can be simplified by substituting the state equation (5.62) as follows

$$J = \int_0^T \sum_n \left[c_{xn}^+ X_n(t) + c_{un} u_n(t)\right] dt + c_T T = \sum_n c_{un} X_{Tn} + \int_0^T \sum_n c_{xn}^+ X_n(t) dt + c_T T .$$

Since the total production cost $\sum_n c_{un} X_{Tn}$ is a constant which does not influence the optimization, the objective of the problem takes the form

$$J = \int_0^T \sum_n c_{xn}^+ X_n(t) dt + c_T T \to \min . \tag{5.65}$$

The fact that the total production cost is independent of the control function is due to the fixed terminal inventory and linearity of the production cost function.

Dual Formulation

The Hamiltonian, H, resulting from the maximum principle is

$$H(t) = -\sum_n c_{xn}^+ X_n(t) + \sum_n \psi_n(t) u_n(t) = c_T . \tag{5.66}$$

The co-state variable, $\psi_n(t)$ in (5.66), satisfies the following co-state differential equation with no transversality condition

$$\dot{\psi}_n(t) = c_{xn}^+ . \tag{5.67}$$

Maximization of the Hamiltonian implies that

$$u_n(t) = \begin{cases} U_n, & \text{if } \psi_n(t) > 0 \text{ and } U_n \psi_n(t) > U_{n'} \psi_{n'}(t), \ \forall n' \neq n \\ 0, & \text{if } \psi_n(t) < 0, \\ b, \ b \in [0, U_n], & \text{if } \psi_n(t) \geq 0 \text{ and } U_n \psi_n(t) \geq U_{n'} \psi_{n'}(t), \ \forall n' \neq n. \end{cases} \tag{5.68}$$

Algorithm 5.6

Step 1:

Sort and re-number products in the non-decreasing order of $c_{xn}^{+}U_n$. Set

$t_0 = 0$ and calculate time points $t_n = \sum_{n'=1}^{n} \frac{X_{Tn'}}{U_{n'}}$ for n=1,...,N.

Step 2:

Assign $T = t_N$ and for n=1,...,N, set the optimal solution as

$u_n(t) = U_n$ for $t_{n-1} \le t < t_n$,

$u_n(t) = 0$ for $0 \le t < t_{n-1}$ and $t_n \le t \le T$;

STOP.

Discussion

From Algorithm 5.6, one can observe that the terminal constraints $X_n(T) = X_{Tn}$ dictate the required amount of the products to be carried out as fast as possible regardless of the production costs. The following lemma proves that the WLPR rule is optimal for the $1|M|u(t) \in [0, U], X(T) = X_T \| L_u, L_x, T$ problem which is, therefore, solvable in $O(N \log N)$ time.

LEMMA 5.18. (Correctness of Algorithm 5.6) Given problem (5.62)-(5.65) and that the products are sorted in the non-decreasing order of $c_{xn}^{+}U_n$, the optimal control is $u_n(t) = U_n$ for $t_{n-1} \le t < t_n$, $u_n(t) = 0$ for $0 \le t < t_{n-1}$ and $t_n \le t \le T$,

where $t_0 = 0$, $t_n = \sum_{n'=1}^{n} \frac{X_{Tn'}}{U_{n'}}$ for n=1,...,N, $T = t_N$.

PROOF. Consider time horizon $T = \sum_{n} \frac{X_{Tn}}{U_n}$ divided into intervals of $\frac{X_{Tn}}{U_n}$ by points $t_0 = 0$, $t_n = \sum_{n'=1}^{n} \frac{X_{Tn'}}{U_{n'}}$ for n=1,...,N. At each interval $t_{n-1} < t < t_n$, consider the following solution

$$u_{n'}(t) = 0,\ X_{n'}(t) = X_{Tn'},\ \psi_{n'}(t)U_{n'} < \psi_n(t)U_n \text{ for } n'=1,\dots,n\text{-}1;$$

$$u_n(t) = U_n,\ X_n(t) = U_n(t - t_{n-1}), \tag{5.69}$$

$$u_{n'}(t) = 0,\ X_{n'}(t) = 0,\ \psi_{n'}(t)U_{n'} < \psi_n(t)U_n \text{ for } n'=n+1,\dots,N.$$

$$\psi_{n'}(t) = \psi_{n'}(0) + c_{xn'}^{+}t \text{ for } n'=1,\dots,N.$$

From solution (5.69), it follows that

$$U_n\psi_n(t) > U_{n+1}\psi_{n+1}(t) \text{ for } 0 < t < t_n,$$

$$U_n\psi_n(t_n) = U_{n+1}\psi_{n+1}(t_n), \tag{5.70}$$

$$U_n\psi_n(t) < U_{n+1}\psi_{n+1}(t) \text{ for } t_n < t < T.$$

With respect to the co-state equation (5.67), conditions (5.70) can be feasible only if $U_n c_{xn}^+ \le U_{n+1} c_{xn+1}^+$, as required in the lemma. Finally, to prove that the constructed solution is optimal, we verify that

$$H(t) = -\sum_n c_{xn}^+ X_n(t) + \sum_n \left(\psi_n(0) + c_{xn}^+ t\right) u_n(t) = c_T \tag{5.71}$$

at every interval $t_{n-1} < t < t_n$. By substituting (5.69) into (5.71), we obtain

$$H(t) = -\sum_{n'=1}^{n-1} c_{xn'}^+ X_{Tn'} - c_{xn}^+ U_n (t - t_{n-1}) + \left(\psi_n(0) + c_{xn}^+ t\right) U_n = c_T$$

from where $\psi_n(0) = \frac{1}{U_n}\left(c_T + \sum_{n'=1}^{n-1} \frac{X_{Tn'}}{U_{n'}}\left(c_{xn'}^+ U_{n'} - c_{xn}^+ U_n\right)\right)$, that is (5.71) holds over the entire planning horizon.

■

Special Cases of Zero Costs

Since the optimal solution that Algorithm 5.6 finds is independent of the production and planning horizon cost coefficients, the $1|N|\,u(t) \in [0, U], X(T) = X_T \,\|\, L_x, T$ and $1|1|\,u(t) \in [0, U], X(T) = X_T \,\|\, L_u, L_x, T, c_T = 0$ problems are solvable by this algorithm with no modification.

The $1|N|\,u(t) \in [0, U], X(T) = X_T \,\|\, L_u, L_x, T, c_x^+ = 0$ problem is equivalent to the makespan problem, whose solution is trivial because the maximum completion time does not depend on a production sequence, $T = \sum_n \frac{X_{Tn}}{U_n}$.

6

PARALLEL MACHINE PROBLEMS

This chapter deals with scheduling of multiple machines and products under both limited and unlimited production capacity. Since the unlimited capacity problems can be decomposed into a number of one-machine, one-product problems, they are considered only in Section 6.1, for general demands over a finite planning horizon. Because the capacitated, multi-machine, scheduling problems are much harder to solve than the problems considered in the previous chapters, the objective of this chapter is to study the optimal behavior for only some special cases of scheduling parallel machines with a linear performance measure.

6.1 DEMAND OVER FINITE PLANNING HORIZON

This section addresses parallel machine, multiple product, scheduling problems characterized by demand specified over a finite planning horizon.

6.1.1 UNLIMITED CAPACITY

Backlogs

The first problem to be considered here is M-machine, N-product-type scheduling with backlogs and unlimited production capacity to minimize the total linear cost of inventory and production, that is, the $M|N|\, u(t) \in [0,\infty]\, |d(t),T|\, L_u, L_x$ problem which is stated as follows

$$J = \int_0^T \sum_n \left[c_{xn}^+ X_n^+(t) + c_{xn}^- X_n^-(t) + \sum_m c_{unm} u_{nm}(t) \right] dt \to \min, \tag{6.1}$$

$$X_n^+(t) = \max\{X_n(t), 0\}, \quad X_n^-(t) = \max\{-X_n(t), 0\}$$

subject to

$$\dot{X}_n(t) = \sum_m u_{nm}(t) - d_n(t), \quad X_n(0) = 0, \tag{6.2}$$

$$u_{nm}(t) \geq 0. \tag{6.3}$$

Dual Formulation

The Hamiltonian, H, for problem (6.1)-(6.3) is

$$H = -\sum_n \left[c_{xn}^+ X_n^+(t) + c_{xn}^- X_n^-(t) + \sum_m c_{unm} u_{nm}(t) \right] + \sum_n \psi_n(t) \left(\sum_m u_{nm}(t) - d_n(t) \right). \tag{6.4}$$

The co-state variables, $\psi_n(t)$ in (6.4), satisfy the following co-state differential equations with corresponding transversality conditions

$$\dot{\psi}_n(t) = \begin{cases} c_{xn}^+, \text{ if } X_n(t) > 0; \\ -c_{xn}^-, \text{ if } X_n(t) < 0; \\ a, \ a \in [-c_{xn}^-, c_{xn}^+], \text{ if } X_n(t) = 0; \end{cases} \qquad \psi_n(T) = 0. \tag{6.5}$$

Maximization of the Hamiltonian at every point of time with respect to admissible controls $u_{nm}(t)$ implies that

$$u_{nm}(t) = \begin{cases} b \geq 0, \ \text{if } \psi_n(t) = c_{unm}, \\ 0, \ \text{if } \psi_n(t) < c_{unm}. \end{cases} \tag{6.6}$$

The first condition in equation (6.6) defines the production regime for product n on machine m. The second condition defines the no-production regime for product n on machine m.

Algorithm 6.1

Step 1:

For each product n, find machine $m(n)$ such that its production cost is minimal, $c_{unm(n)} = \min_m c_{unm}$. Calculate $t_n = T - \dfrac{c_{unm(n)}}{c_{xn}^-}$ for n=1,...,N.

Step 2:

For each n, if $t_n \geq 0$, then set the optimal solution as

$u_{nm(n)}(t) = d_n(t)$ for $0 \leq t \leq t_n$,

$u_{nm(n)}(t) = 0$ for $t_n < t \leq T$,

$u_{nm}(t) = 0$ for $0 \leq t \leq T$, $m \neq m(n)$;

otherwise,

$u_{nm}(t) = 0$ for $0 \leq t \leq T$, $m=1,\ldots,M$.

STOP.

Discussion

Algorithm 6.1 solves the problem in $O(NM)$ time by assigning a product to the machine which produces the product with the minimum cost. Thus, Algorithm 6.1 decomposes the problem into at most N problems of uncapacitated, one-machine scheduling. Exactly the same decomposition is true for the other production environments considered in this book: due-date, demand over infinite planning horizon and variable planning horizon. The formal proof, which is similar to that presented below, is left for the reader. Once the problem is decomposed into one-machine problems, the corresponding algorithms from Sections 3.1.1, 3.2.1, 3.3.1 and 3.4.1 can be straightforwardly employed to solve them.

LEMMA 6.1. (Correctness of Algorithm 6.1) Given problem (6.1)-(6.3), the optimal solution is:

if $t_n = T - \dfrac{c_{unm(n)}}{c_{xn}^-} \geq 0$, then $u_{nm(n)}(t) = d_n(t)$ for $0 \leq t \leq t_n$, $u_{nm(n)}(t) = 0$ for $t_n < t \leq T$, $u_{nm}(t) = 0$ for $0 \leq t \leq T$ for $m \neq m(n)$,

otherwise, $u_{nm}(t) = 0$ for $m=1,\ldots,M$ for $0 \leq t \leq T$, where $m(n) = \arg\min_m c_{unm}$.

PROOF. Given the control function stated in the lemma, the corresponding inventory level is obtained by integrating state equation (6.2)

$$X_n(t) = 0 \text{ for } 0 \leq t \leq t_n, \text{ and } X_n(t) < 0 \text{ for } t_n < t \leq T.$$

Consequently, $\psi_n(t) = c_{unm(n)}$ for $0 \leq t \leq t_n$ and $\dot{\psi}_n(t) = -c_{xn}^-$ for $t_n < t \leq T$. Next, by integrating the last equation from $t_n = T - \dfrac{c_{unm(n)}}{c_{xn}^-}$ to T, we obtain $\psi_n(T) = 0$. Finally, one can observe that $\psi_n(t) \leq c_{unm}$ for $m \neq m(n)$, i.e. the optimality condition (6.6) is met. Thus, the solution stated in the lemma is optimal.

■

Special Cases of Zero Costs

Since cost coefficient c_{xn}^{+} is not involved in Algorithm 6.1, the solution of the $M|N|u(t)\in[0,\infty]|d(t),T|L_u,L_x,c_x^{+}=0$ problem is equivalent to the solution of the $M|N|u(t)\in[0,\infty]|d(t),T|L_u,L_x$ problem.

The algorithm for the $M|N|u(t)\in[0,\infty]|d(t),T|L_x$ problem is straightforwardly obtained by substituting $c_{unm}=0$ into the corresponding conditions and formulas of Algorithm 6.1. As a result, the solution becomes trivial because any machine can be selected to exactly track the demand of any set of products over the entire planning horizon

$$u_{nm(n)}(t)=d_n(t) \quad \text{for } 0\le t\le T,$$

where $m(n)$ is selected arbitrarily;

$$u_{nm}(t)=0 \quad \text{for } 0\le t\le T, \quad m\ne m(n).$$

No Backlog

When backlogs are disallowed, i.e. for the $M|N|u(t)\in[0,\infty],X(t)\ge 0|d(t),T|L_u,L_x$ problem, the solution algorithm can be obtained by an easy modification of Algorithm 6.1. Indeed, no backlog means having c_{xn}^{-} very large, that is the first step of Algorithm 6.1 will always result in $t_n=T$. This implies that the fractional production on the cheapest machine $m(n)$, $u_{nm(n)}(t)=d_n(t)$, is optimal along the entire planning horizon.

6.1.2 LIMITED CAPACITY

Backlogs

The next problem we consider is M-machine, N-product-type scheduling with backlogs and limited production rate to minimize the total linear cost of inventory and production, that is, the $M|N|u(t)\in[0,U]|d(t),T|L_u,L_x$ problem which is stated as follows

$$J=\int_0^T\sum_n\left[c_{xn}^{+}X_n^{+}(t)+c_{xn}^{-}X_n^{-}(t)+\sum_m c_{unm}u_{nm}(t)\right]dt\to\min, \tag{6.7}$$

$$X_n^{+}(t)=\max\{X_n(t),0\},\quad X_n^{-}(t)=\max\{-X_n(t),0\}$$

subject to

$$\dot{X}_n(t)=\sum_m u_{nm}(t)-d_n(t),\quad X_n(0)=0, \tag{6.8}$$

$$0 \le u_{nm}(t) \le U_{nm}, \tag{6.9}$$

$$\sum_{n} \frac{u_{nm}(t)}{U_{nm}} \le 1. \tag{6.10}$$

Dual Formulation

The Hamiltonian, H, and the co-state differential equations for problem (6.7)-(6.10) do not change. Maximization of the Hamiltonian implies that for each machine m

$$u_{nm}(t) = \begin{cases} U_{nm}, \text{ if } \psi_n(t) > c_{unm} \text{ and } U_{nm}(\psi_n(t) - c_{unm}) > U_{n'm}(\psi_{n'}(t) - c_{un'm}), \ \forall n' \ne n \\ 0, \text{ if } \psi_n(t) < c_{unm}, \\ b,\ b \in [0, U_{nm}], \text{ if } \psi_n(t) \ge c_{unm} \text{ and } U_{nm}(\psi_n(t) - c_{unm}) \ge U_{n'm}(\psi_{n'}(t) - c_{un'm}), \ \forall n' \ne n. \end{cases} \tag{6.11}$$

The first condition in equation (6.11) defines production of product n on machine m at the maximum rate (full production regime *FP*). The second condition defines the no-production regime, *NP*, of product n on machine m. The last condition in (6.11) presents the fractional (singular) production regime, *SP*, which allows production at an intermediate rate to track the demand.

Algorithm 6.2

Step 1:

Run Algorithm 6.1. If the solution found by Algorithm 6.1 is such that $\sum_{n} \frac{u_{nm(n)}(t)}{U_{nm(n)}} \le 1$ for each $m(n)$ assigned to produce product n, then this solution is optimal. Otherwise, go to the next step.

Step 2:

If $c_{unm} = c_{unm'}$ for all n, $m \ne m'$ and machines have uniform speeds in terms of Definition 6.1 below, then set $U_n = \sum_{m} U_{nm}$ and $c_{un} = c_{un1}$, and run Algorithm 5.1.

Step 3:

Set the optimal solution as

$$u_{nm}(t) = \frac{\beta_m}{\sum_{m'} \beta_{m'}} u_n(t) \text{ for all } n,m,\ 0 \le t \le T. \text{ STOP.}$$

Algorithm 6.3
Step 1:

If N=1, demand is $\bar{K}$-interval, single-mode, piece-wise constant and $\max_t d_1(t) \geq \sum_m U_{1m}$, then omit product index n, sort and re-number the machines with respect to their production cost coefficients

$$c_{u1} \leq c_{u2} \leq \ldots \leq c_{uM}.$$

Step 2:

Calculate time points $\bar{t}_{1,m}$ and $\bar{t}_{2,m}$, such that

$$d(t) < \sum_{m'=1}^{m} U_{m'} \text{ for } 0 \leq t < \bar{t}_{1,m} \text{ and } \bar{t}_{2,m} < t \leq T;$$

$$d(t) \geq \sum_{m'=1}^{m} U_{m'} \text{ for } \bar{t}_{1,m} \leq t \leq \bar{t}_{2,m}.$$

Set $\bar{t}_{1,0} = 0$ and $\bar{t}_{2,0} = T$.

Denote $2M$ time intervals $\tau_m = [\bar{t}_{1,m-2}, \bar{t}_{1,m-1}]$, m=2,...,M+1, and $\tau_{M+m} = [\bar{t}_{2,M-m+2}, \bar{t}_{2,M-m+1}]$, m=2,...,M+1. In addition, define τ_1 and τ_{2M+2} as the intervals which are degenerated to a point, $\tau_1 = \{0\}$ and $\tau_{2M+2} = \{T\}$. Assign K=2M+2.

Step 3 and Step 4 are identical to Steps 2 and 3 of Algorithm 3.5 respectively.

Subroutine 1 to Algorithm 6.3

(i) Set $t^n_{left} = \min\{t | t \in \tau_n\}$, $t^n_{right} = \max\{t | t \in \tau_n\}$,
$t^m_{left} = \min\{t | t \in \tau_m\}$, $t^m_{right} = \max\{t | t \in \tau_m\}$.

(ii) Find $\bar{k}_{n1}$, $\bar{k}_{m1}$, $\bar{k}_{n2}$ and $\bar{k}_{m2}$ such that
$\bar{k}_{n1} = \min\{k | t_k \geq t^n_{left}\}$, $\bar{k}_{n2} = \max\{k | t_k < t^n_{right}\}$,
$\bar{k}_{m1} = \min\{k | t_k \geq t^m_{left}\}$, $\bar{k}_{m2} = \max\{k | t_k < t^m_{right}\}$.

(iii) Assign $k_1 = \bar{k}_{n1}$, $k_2 = \bar{k}_{m1}$, $k_3 = \bar{k}_{n1} + 1$.

(iv) Set m_1, m_2 and m_3 such that $\sum_{m'=1}^{m_1-1} U_{m'} < d_{k1} \leq \sum_{m'=1}^{m_1} U_{m'}$,

$$\sum_{m'=1}^{m_2-1} U_{m'} < d_{k2} \le \sum_{m'=1}^{m_2} U_{m'}, \sum_{m'=1}^{m_3-1} U_{m'} < d_{k3} \le \sum_{m'=1}^{m_3} U_{m'}.$$

Solve the following system in three unknowns t_{in}, t_{out} and t_ψ

$$\sum_{m'=m_1+1}^{m_3} U_{m'}(t_\psi - t_{m'}) + U_{m_1}(t_\psi - t_{out}) = (t_{k1} - t_{out})d_{k1-1} + (t_\psi - t_{k3})d_{k3} + \sum_{k=k1}^{k3-1} d_k(t_{k+1} - t_k),$$

$$\sum_{m'=m_3}^{m_2+1} U_{m'}(t_{m'} - t_\psi) + U_{m_2}(t_{in} - t_\psi) = (t_{in} - t_{k2})d_{k2} + (t_{k3+1} - t_\psi)d_{k3} + \sum_{k=k3+1}^{k2-1} d_k(t_{k+1} - t_k),$$

$$c_x^+\left(t_\psi - t_{out}\right) = c_x^-\left(t_{in} - t_\psi\right),$$

where $t_{m_1+1} = t_{out} + \dfrac{c_{um_1+1} - c_{um_1}}{c_x^+}$, $t_{m'} = t_{m'-1} + \dfrac{c_{um'} - c_{um'-1}}{c_x^+}$,

$m' = m_1 + 2, \dots, m_3$,

$t_{m_2+1} = t_{in} - \dfrac{c_{um_2+1} - c_{um_2}}{c_x^+}$, $t_{m'} = t_{m'-1} - \dfrac{c_{um'} - c_{um'-1}}{c_x^+}$,

$m' = m_3, \dots, m_2 + 1$.

(v) If $t_{out} \in [t_{k_1-1}, t_{k_1}]$, $t_x \in [t_{k_2-1}, t_{k_2}]$, $t_{in} \in [t_{k_3-1}, t_{k_3}]$, then set $arc_{nm} = (\tau_n, \tau_m) \in A$ which is characterized by two weights $a(arc_{nm}) = t_{out}$, $b(arc_{nm}) = t_{in}$, STOP.
Otherwise go to (vi).

(vi) If $k_1 < \bar{k}_{2n}$, then assign $k_1 = k_1 + 1$ and go to (iv).
Otherwise, set $k_1 = \bar{k}_{n1}$. If $k_2 < \bar{k}_{2m} - 1$, then assign $k_2 = k_2 + 1$ and go to (iv).
Otherwise, set $k_2 = \bar{k}_{m1}$. If $k_3 < \bar{k}_{2m-1}$, then assign $k_3 = k_3 + 1$ and go to (iv).
Otherwise, no arc links nodes n and m; STOP.

Definition 6.1. Machines have *uniform speeds*, if there exist numbers β_m such that $U_{nm} = \beta_m U_{n1}$ for all *n*,*m*.

Example 6.1

Consider the $3|1|u(t) \in [0,U]\,|d(t),T|\,L_u, L_x$ problem with: the length of the planning horizon, *T*, set at 15hrs; the demand as depicted in Figure 6.1; and the production parameters shown in Table 6.1.

Table 6.1. Machine parameters and priorities

m	U_m	c_{um}	Optimal priority
1	10	1/2	2
2	15	2/3	3
3	20	1/10	1

In addition, inventory and backlog costs are c_x^+ = 0.025\$/unit.hr and c_x^-=0.5\$/unit.hr respectively.

Step 1 of Algorithm 6.3 sorts products with respect to the production cost coefficients as 3-1-2 (see the Optimal priority column in Table 6.1).

Three nodes of graph *G* are formed at Step 2 from the intervals at which the demand is constant as illustrated in Figure 6.1. The interval at which *d*(*t*)=50, does not form a node, since it exceeds the total machine capacity. Nodes 1 and 5 present the initial and terminal points of the planning horizon.

Ten pair-wise links between five nodes of graph *G* are to be checked at Step 3.

- Linking nodes 1 and 2. Since *X*(0)=0 already satisfies the condition of the fractional regime, the time of exiting node 0 is $t_{out}^1 = t_{in}^2 = 0$ and $\psi(0) = c_{u3} = 1/10$.
  Thus, if this link is included in the optimal solution, then, at the initial interval of time, the third machine will be functioning according to the fractional regime while the other machines are idle

$$u_1(t) = 0, \;\; u_2(t) = 0, \;\; u_3(t) = d_1 = 5, \;\; t = 0.$$

- Linking nodes 1 and 3.
  $\psi(t_{in}^3) = c_{u2}$=2/3 , $X(t_{in}^3) = 0$, $t_{out}^1 = 0$, where $\psi(t_{in}^3) = \psi(0) + c_x^+ t_{in}^3$,

$X(t_{in}^3) = (U_3 - d_1)\tau + (U_3 + U_1 - d_1)(t_1 - \tau) + (U_3 + U_1 - d_2)(t_{in}^3 - t_1) = 0$ and $c_x^+ \tau = c_{u1} - \psi(0)$.

Substituting the values of the machine and demand parameters, we obtain

$$\tau = 1\frac{7}{9},\ t_{in}^3 = 8\frac{4}{9},\ \psi(0) = \frac{41}{90}.$$

Thus, if this link is included in the optimal solution, then, at the initial interval of time, machines 3 and 1 are functioning at full capacity (machine 1 is idle before $\tau = 1\frac{7}{9}$). Machine 2, which is idle before $t_{in}^3 = 8\frac{4}{9}$, is on a fractional regime thereafter

$$u_1(t) = 0,\ t \in [0,\tau) \text{ and } u_1(t) = 10,\ t \in [\tau, t_{in}^3];$$

$$u_2(t) = 0,\ t \in [0, t_{in}^3) \text{ and } u_2(t) = d_2 - U_1 - U_3 = 5,\ t = t_{in}^3;$$

$$u_3(t) = 20,\ t \in [0, t_{in}^3].$$

- Linking nodes 2 and 3.
  $\psi(t_{in}^3) = c_{u2}$, $X(t_{in}^3) = 0$, where $\psi(t_{in}^3) = \psi(t_{out}^2) + c_x^+(t_{in}^3 - t_{out}^2)$ and $\psi(t_{out}^2) = c_{u3}$. Substituting the parameters of the machines, we obtain that $t_{in}^3 - t_{out}^2 = 22\frac{2}{3}$. However, $t_{in}^3 \in [t_1, t_2] = [2,10]$. Thus, even if t_{out}^2=0, then $t_{in}^3 = 22\frac{2}{3} > 10$, i.e. this link is not feasible.

- Linking nodes 3 and 4.
  $\psi(t_{in}^4) = c_{u2}$, $X(t_{in}^4) = 0$, where $\psi(t_{in}^4) = \psi(t_{out}^3) + c_x^+ \tau - c_x^-(t_{in}^4 - \tau - t_{out}^3)$,

  $\psi(t_{out}^3) = \psi(t_{in}^4) = c_{u2}$,

$$\begin{aligned} X(t_{out}^3 + \tau) = (t_2 - t_{out}^3)(U_1 + U_2 + U_3 - d_2) + \\ + (t_{out}^3 + \tau - t_2)(U_1 + U_2 + U_3 - d_3) = 0 \end{aligned},$$

$$\begin{aligned} X(t_{in}^4) = (t_3 - t_{out}^3 - \tau)(U_1 + U_2 + U_3 - d_3) + \\ + (t_{in}^4 - t_3)(U_1 + U_2 + U_3 - d_4) = 0 \end{aligned}$$

Substituting the values of the machine and demand parameters, we obtain

$$\tau = 2\frac{6}{7},\ t_{in}^{4} = 12\frac{1}{21},\ t_{out}^{3} = 9\frac{1}{21}.$$

Thus, if this link is included in the optimal solution, then from $t_{out}^{3} = 9\frac{1}{21}$ to $t_{in}^{4} = 12\frac{1}{21}$ all three machines function at the full production rate

$$u_1(t) = 10,\ \ u_2(t) = 15 \text{ and } u_3(t) = 20,\ \ t \in [t_{out}^{3}, t_{in}^{4}]\ .$$

- Linking nodes 4 and 5.
  $\psi(T) = 0$, where $\psi(T) = \psi(t_{out}^{4}) - c_x^{-}(T - t_{out}^{4})$, $\psi(t_{out}^{4}) = c_{u2}$, $t_{in}^{5} = T$.
  Substituting the values of the machine and demand parameters, we obtain $t_{out}^{4} = 13\frac{2}{3}$. In addition, points of time t_x and t_y, where the machines switch off are found from $\psi(t_x) = c_{u1} = \psi(t_{out}^{4}) - c_x^{-}(t_x - t_{out}^{4})$ and $\psi(t_y) = c_{u3} = \psi(t_{out}^{4}) - c_x^{-}(t_y - t_{out}^{4})$.
  Thus, if this link is included in the optimal solution, then, at the terminal interval of time, machines 2, 1 and 3 are respectively switched off at points $t_{out}^{4} = 13\frac{2}{3}$, $t_x = 14$ and $t_y = 14\frac{4}{5}$,

$$u_1(t) = 10,\ \ u_2(t) = 0 \text{ and } u_3(t) = 20,\ \ t \in [t_{out}^{4}, t_x);\ \ u_2(t) = 15,\ \ t < t_{out}^{4};$$

$$u_1(t) = 0,\ \ u_2(t) = 0 \text{ and } u_3(t) = 20,\ \ t \in [t_x, t_y);$$

$$u_1(t) = 0,\ \ u_2(t) = 0 \text{ and } u_3(t) = 0,\ \ t \in [t_y, T].$$

It is easy to verify that all the remaining links are not feasible. The resultant graph G is depicted in Figure 6.2.

At Step 4, the algorithm first colors all arcs gray and re-colors arcs (1,3), (3,4) and (4,5) black. Arc (1,2) becomes red. Thus, the feasible path is found as (1,3,4,5).

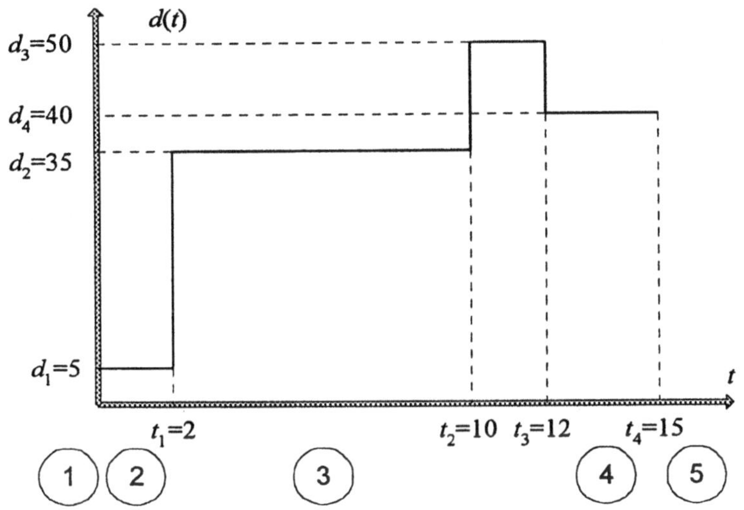

Figure 6.1. Demand rate for the example problem.

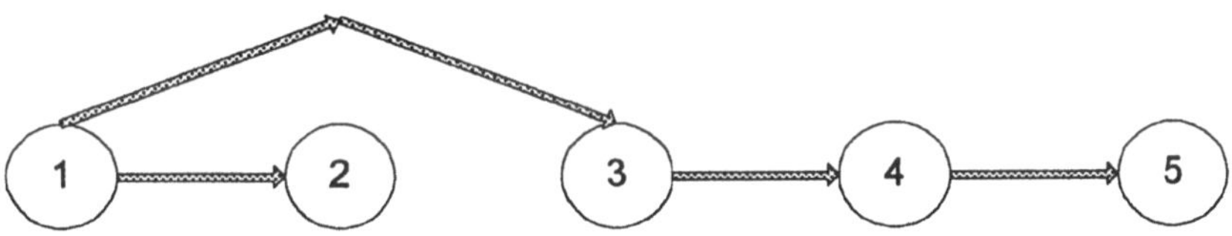

Figure 6.2. Graph G for the example problem.

Discussion

Algorithm 6.2 solves the problem in $O(NM)$ time for two special cases. Step 1 treats the problem as though it is uncapacitated and then verifies whether the capacity constraint is not violated. Step 2 solves the problem characterized by identical production costs and uniform speeds of the machines. The solution is based on coalescing the machines into a single machine with joined production capacity $U_n = \sum_m U_{nm}$, n=1,...,N. Then, Algorithm 5.1 for the single machine is executed and its solution is distributed proportionally among the machines.

Algorithm 6.3 solves the problem for machines producing a single product type, the demand for which is K-level, single-mode, piece-wise constant. The algorithm is similar to Algorithms 3.5 and 4.3 which reduce a continuous-time scheduling problem to a combinatorial search for a feasible path in a 2-weighted graph. Step 3 of Algorithm 6.3 invokes four subroutines intended to setting arcs on the graph. Since the subroutines are similar to those of Algorithm 3.5, we present only Subroutine 1. The others are left as an exercise for the reader.

LEMMA 6.2. (Correctness of Algorithm 6.2) Given problem (6.7)-(6.10), if the machines have uniform speeds (see Definition 6.1), $c_{unm} = c_{unm'}$ for all n, $m \neq m'$ and $u_n(t)$ is the optimal solution of the $1|N|u(t) \in [0,U]|d(t),T|L_u, L_x$ problem with $U_n = \sum_m U_{nm}$, $c_{un} = c_{un1}$, then the optimal solution is

$$u_{nm}(t) = \frac{\beta_m}{\sum_{m'} \beta_{m'}} u_n(t) \text{ for all } n,m,\ 0 \le t \le T.$$

PROOF. To prove the lemma, it is sufficient to show that solution $u_{nm}(t) = \frac{\beta_m}{\sum_{m'} \beta_{m'}} u_n(t)$ is feasible, i.e. it satisfies the capacity constraints (6.9) and (6.10). Since $u_n(t)$ is the optimal solution of the $1|N|u(t) \in [0,U]|d(t),T|L_u, L_x$ problem with

$$U_n = \sum_m U_{nm}, \tag{6.12}$$

then

$$u_n(t) \le U_n, \tag{6.13}$$

$$\sum_n \frac{u_n(t)}{U_n} \le 1. \tag{6.14}$$

From (6.13) and (6.12), it follows that

$$u_{nm}(t) = \frac{\beta_m}{\sum_{m'} \beta_{m'}} u_n(t) \le \frac{\beta_m}{\sum_{m'} \beta_{m'}} U_n = \frac{\beta_m}{\sum_{m'} \beta_{m'}} \sum_{m'} U_{nm'}.$$

Due to the uniform speeds of the machines, we obtain

$$u_{nm}(t) \le \frac{\beta_m U_{n1}}{\sum_{m'} \beta_{m'}} \sum_{m'} \beta_{m'} = \beta_m U_{n1} = U_{nm}, \tag{6.15}$$

that is, constraint (6.9) is met. Similarly, based on (6.14), we find

$$\sum_n \frac{u_{nm}(t)}{U_{nm}} = \sum_n \frac{\frac{\beta_m}{\sum_{m'} \beta_{m'}} u_n(t)}{\beta_m U_{n1}} = \sum_n \frac{u_n(t)}{\sum_{m'} \beta_{m'} U_{n1}} = \sum_n \frac{u_n(t)}{U_n} \leq 1,$$

that is, constraint (6.10) is met.

■

The next five lemmas prove the optimality of Algorithm 6.3 which solves the problem for N=1.

LEMMA 6.3. If a machine m' is on the SP regime at an interval of time, $\tau \subseteq [0,T]$, then sets I_1, I_2 and I_3 of the machines producing in accordance to the FP, SP and NP regimes respectively, are determined as follows

$$I_1 = \{m \mid c_{um} < c_{um'}\},\ I_2 = \{m \mid c_{um} = c_{um'}\},\ I_3 = \{m \mid c_{um} > c_{um'}\}. \tag{6.16}$$

PROOF. From the third line of (6.11), it follows that if a machine m' is on SP, then $\psi(t) = c_{um'}$ and all the other machines for which c_{um} is equal to $c_{um'}$ must also function according to the SP regime. Further, from the first and second lines of (6.11), we conclude that the machines for which c_{um} is less than $c_{um'}$ must be operating. The machines for which c_{um} is greater than $c_{um'}$ must be idle.

■

LEMMA 6.4. If I_1, I_2 and I_3 are the sets of the machines producing at an interval of time $\tau \subseteq [0,T]$, in accordance to the FP, SP and NP regimes respectively, then the optimal control is

$$u_{m'}(t) = U_{m'} \frac{d(t) - \sum_{m \in I_1} U_m}{\sum_{m \in I_2} U_{m'}},\ m' \in I_2,\ t \in \tau. \tag{6.17}$$

PROOF. Since the SP regime condition $\psi(t) = c_{um'}$ (see (6.11)) holds at interval $\tau \subseteq [0,T]$, $\dot{\psi}(t) = 0$ at this interval. By taking into account co-state equation (6.5), we conclude that to keep the SP regime, we need $X(t)$=0, $t \in \tau$. By differentiating the zero inventory condition, we find

$$\sum_m u_m(t) = d(t)\,. \tag{6.18}$$

By taking into account that $u_m(t) = U_m,\ \ m \in I_1$ and $u_m(t) = 0,\ \ m \in I_3$, equation (6.17) is immediately obtained from (6.18).
■

From Lemma 6.4, it follows that $X(t)$=0 and $\sum_m U_m \geq d(t)$, $t \in \tau$, are necessary for the SP regime to exist at τ.

Lemmas 6.3 and 6.4 establish a priority rule for the optimal ordering of the machines, $m_i = m_1, m_2, \ldots, m_M$ so that $c_{um_1} \leq c_{um_2} \leq \ldots \leq c_{um_M}$. In other words, the machines with higher production costs can start working only after the machines with lower costs are already working at full capacity. Henceforth, we assume that the machines are ordered according to the priority rule.

The next lemma determines the sequence of the regimes joining the optimal trajectory by studying the dynamics of the co-state variable.

LEMMA 6.5. Given demand as single-mode, the dynamics of $\psi(t)$ is also single-mode, i.e. there exists a time point $t_\psi \in [0, T]$ such that $\psi(t)$ does not decrease before t_ψ and does not increase after it.

PROOF (by contradiction). Let function $\psi(t)$ have more than one local maximum, and $t_1^{\max}$ and $t_2^{\max}$ be the points of adjacent local maximums. Let $t^{\min}$ be the point of a local minimum between $t_1^{\max}$ and $t_2^{\max}$ (see Figure 6.3). With respect to co-state equation (6.5), $X(t)$ must be zero at all three points $t_1^{\max}$, $t^{\min}$ and $t_2^{\max}$. By considering interval $[t_1^{\max}, t^{\min}]$, we conclude that

- $d(t_1^{\max} + 0) > \sum_m u_m(t_1^{\max} + 0)$ (since $X(t)$ decreases right of point $t_1^{\max}$),
- $\sum_m u_m(t_1^{\max} + 0) \geq \sum_m u_m(t^{\min} - 0)$ (since $\psi(t)$ decreases),
- $\sum_m u_m(t^{\min} - 0) > d(t^{\min} - 0)$ (since $X(t)$ increases left of point $t^{\min}$).

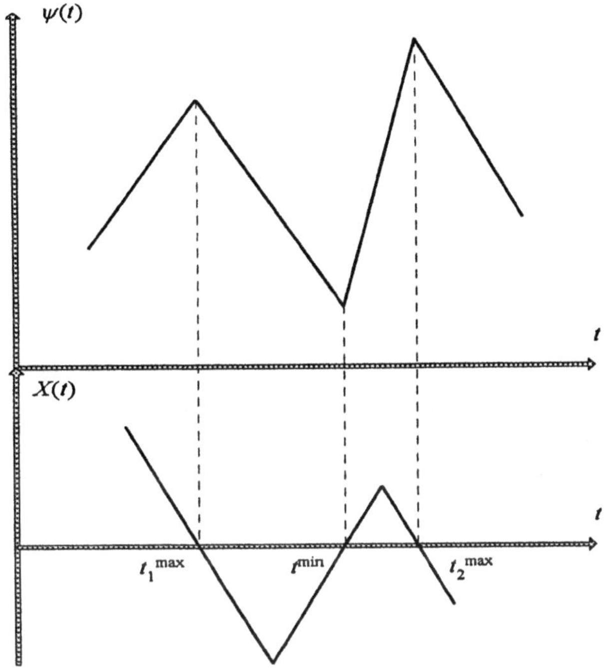

Figure 6.3. Multiple-mode dynamics of $\psi(t)$ and corresponding behavior of $X(t)$.

Therefore, demand $d(t)$ decreases at the interval $[t_1^{\max}, t^{\min}]$. Similarly, considering interval $[t^{\min}, t_2^{\max}]$, we conclude that demand $d(t)$ increases at this interval. Such a behavior contradicts the single-mode dynamics of the demand given in the lemma.
■

The single-mode $\psi(t)$ implies that the machines start working (in full production regime) one after another according to the priority order (see Lemma 6.3) and stop working (no-production regime) in the inverse order. Such a "wave" occurs only once over the planning horizon. However, there can be partial production regimes between full and no-production. Therefore, the next step in characterizing the optimal trajectory is to locate SP regimes along the planning horizon and link them. Lemma 6.6 below limits the number of possible SP regimes over the optimal trajectory.

LEMMA 6.6. Given K-level, single-mode, piece-wise demand, there are no more than K fractional production regimes over the optimal trajectory.

PROOF. To prove the lemma, we show that there can be no more than one SP regime at every interval $k = 1,\ldots,K$ where the demand is constant. Let there be a SP regime of machine m at interval $[t_k, t_{k+1})$ on which the demand is constant. Let t_{out} be a time of exiting the SP regime. Since at t_{out} the conditions of the SP regime hold, $X(t_{out}) = 0$ and $\psi(t_{out}) = c_{um}$.

If machine m re-switches on the full production regime, then both $X(t)$ and $\psi(t)$ increase after t_{out} (see equations (6.8) and (6.5)). Since the demand does not change until t_{k+1}, this increase will neither allow the inventory level to return to zero nor the co-state variable to return to c_{um}. Thus, another SP regime cannot start until t_{k+1}. Similarly, if machine m re-switches on the no-production regime, then both $X(t)$ and $\psi(t)$ decrease after t_{out}. Since the demand does not change until t_{k+1}, this decrease will not allow another SP regime to start until t_{k+1}. Thus, each interval $[t_k, t_{k+1})$, $k = 0,\ldots,K-1$, where the demand is constant, can contain no more than one SP regime.
■

If demands d_k and d_{k+1} over two adjacent intervals lie within the capacity of the same machines

$$\sum_{m'=1}^{m} U_{m'} \le d_k \le \sum_{m'=1}^{m+1} U_{m'} \text{ and } \sum_{m'=1}^{m} U_{m'} \le d_{k+1} \le \sum_{m'=1}^{m+1} U_{m'},$$

then we will not distinguish between these intervals, because they correspond to the SP regime of the same machine. Therefore, the maximum number of SP regimes over the optimal trajectory cannot exceed $2M$ (no more than M when $t<t_d$ and no more than M when $t>t_d$). Thus, the number of SP regimes is limited by $\min\{K,2M\}$.

Lemma 6.7 below determines quantitative conditions for linking SP regimes.

LEMMA 6.7. Let $u_m(t)$ be the optimal solution, SP_1 and SP_2 be adjacent fractional production regimes of machines m_1 and m_2 respectively, then the time t_{out} of exiting SP_1, and the time t_{in} of entering SP_2, are determined by solving no more than $M+1-\max\{m_1, m_2\}$ systems, each of which has three linear algebraic equations.

PROOF. Since the SP_1 and SP_2 regimes are adjacent, i.e. there are no SP regimes between them, all control functions $u_m(t)$ are determined at interval $[t_{out}, t_{in}]$ by co-state variable $\psi(t)$ as shown in Lemma 6.4. Since at t_{out} the conditions of the SP_1 regime are satisfied,

$$X(t_{out}) = 0 \text{ and } \psi(t_{out}) = c_{um_1} . \tag{6.19}$$

To enter the SP$_2$ regime, the conditions of this regime must also be satisfied at t_{in}

$$X(t_{in}) = 0 \text{ and } \psi(t_{in}) = c_{um_2} . \tag{6.20}$$

Now, we can integrate analytically the system of state and co-state equations (6.8) and (6.5) over interval $[t_{out}, t_{in}]$ with initial boundary conditions (6.19). For example, (see Figure 6.4), if there are two jumps of the demand at points t_{k1}, t_{k2}, $m_1 = m_2$, $|I_1| \le m_1$ and $|I_3| \le m_1$, then the integration results in three linear algebraic equations in three unknowns t_{out}, t_ψ and t_{in}

$$\left(t_\psi - t_{out}\right)\sum_{m'=1}^{m1} U_{m'} - (t_{k1} - t_{out})d_{k1-1} - (t_\psi - t_{k1})d_{k1} = 0,$$

$$\left(t_{in} - t_\psi\right)\sum_{m'=1}^{m1} U_{m'} - (t_{k2} - t_\psi)d_{k1} - (t_{in} - t_{k2})d_{k1+1} = 0, \tag{6.21}$$

$$c_x^+\left(t_\psi - t_{out}\right) = c_x^-\left(t_{in} - t_\psi\right).$$

The first equation follows from the integration of (6.8) over interval $[t_{out}, t_\psi]$ with boundary condition $X(t_\psi) = 0$. The last two equations are obtained by the integration of (6.8) and (6.5) over interval $[t_{out}, t_{in}]$ with boundary conditions (6.20).

Similarly, when the number of jumps in demand is greater than two and additional machines switch on (off) over interval $[t_{out}, t_{in}]$, the integration leads to three linear algebraic equations with the same three unknowns t_{out}, t_ψ and t_{in}. However, the number of terms involved in every equation of this algebraic system proportionally increases to the number of demand jumps and to the number of additional machines switched on (off) over interval $[t_{out}, t_{in}]$.

Finally, system (6.21) has been obtained with the condition that the number of machines involved in production at interval $[t_{out}, t_{in}]$ is given. Since this is not the case, we solve $M + 1 - \max\{m_1, m_2\}$ systems by starting without any additional machine (i.e. the number of working machines is $\max\{m_1, m_2\}$) and then going up to all M machines involved in the production.

■

Linking two SP regimes is readily formalized in the following way. Let τ_1 and τ_2 be intervals of time where SP_1 and SP_2 can occur. These intervals can then be linked if t_{out}, t_ψ and t_{in}, found in Lemma 6.7, are feasible, i.e. $t_{out} \in \tau_1$, $t_{in} \in \tau_2$ and $t_{out} \le t_\psi \le t_{in}$.

When linking the initial point of the planning horizon t=0 with the first SP regime, t_{out} is set at zero, and the system of three linear algebraic equations is solved with respect to $\psi(0)$, t_ψ and t_{in}. When linking the last SP regime with the terminal point of the planning horizon, t=T, t_{in} is set at T, and the system of three linear algebraic equations is solved with respect to t_{out}, t_ψ and $X(T)$.

THEOREM 6.1. If N=1 and demand is K-interval, single-mode, piece-wise constant, then problem (6.7)-(6.10) is solvable in $O\left(KM^2(\max\{K,2M\})^2\right)$ time.

PROOF. To estimate the algorithm's complexity, let us consider it step by step. At Step 1, in order to sort the machines, we need $O(M\log M)$ operations. At Step 2, we form graph nodes by comparing demand at each subsequent interval, where it is constant, with the machine capacities. This takes no more than MK operations. At Step 3, we solve no more than M times the system of three equations in three unknowns. The complexity of constructing and solving each system is $O(KM)$ (see Lemma 6.7). Thus, this step requires $O(KM^2)$ operations in linking any two nodes by an arc. Since the maximum number of such arcs is $\frac{(\min\{K,2M\}+2)^2}{2}$, the total complexity of this step is $O\left(KM^2(\min\{K,2M\})^2\right)$. The coloring procedure executed in the last step of the algorithm requires $O\left(K^3+M^3\right)$ (see Lemma 3.11). Thus, the algorithm totally requires $O\left(KM^2(\max\{K,2M\})^2\right)$ operations.

■

Special Cases of Zero Costs

The $M|N|\,u(t)\in[0,U]\,|d(t),T|\,L_u, L_x, c_x^+=0$ problem is straightforwardly solved by Algorithms 6.2 and 6.3 with no simplification. Zero inventory costs, however, extend the class of demand functions for which Algorithm 6.3 finds the optimal solution. Now, the single mode is no longer needed for the piece-wise constant demand because the co-state variable can only decrease, i.e. it is single-mode anyway (see Lemma 6.5).

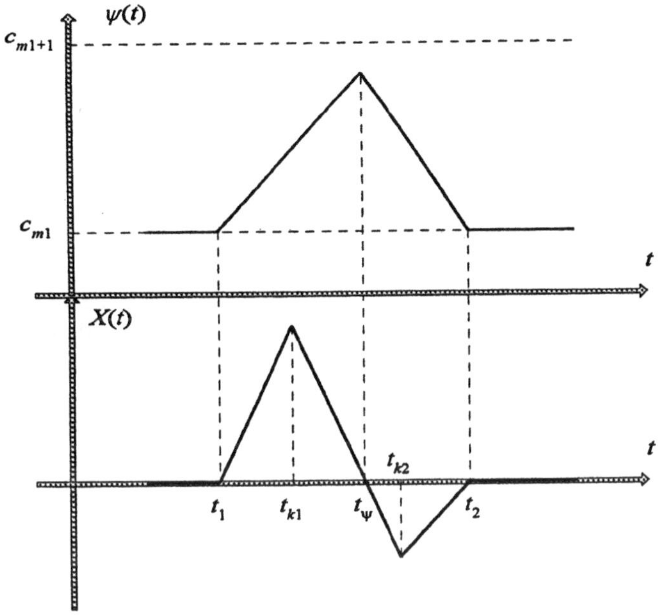

Figure 6.4. State and co-state variables evolution between two adjacent SP regimes.

Algorithm 6.2 for the $M|N|u(t) \in [0,U]|d(t),T| L_x$ problem is modified. Step 1 of Algorithm 6.2 cannot straightforwardly run Algorithm 6.1 to solve the problem as though it is uncapacitated. Rather, the problem with K-level, piece-wise constant demand can be transformed into a system of linear equations and inequalities in unknown u_{nmk} as follows

$$\sum_m u_{nmk} = d_{nk},\ n=1,...,N,\ k=1,...,K, \tag{6.22}$$

$$0 \le u_{nmk} \le U_{nm},\ n=1,...,N,\ k=1,...,K,\ m=1,...,M, \tag{6.23}$$

$$\sum_n \frac{u_{nmk}}{U_{nm}} \le 1,\ k=1,...,K,\ m=1,...,M. \tag{6.24}$$

If system (6.22)-(6.24) has a feasible solution, then the optimal solution of the $M|N|u(t) \in [0,U]|d(t),T| L_x$ problem is

$$u_{nm}(t) = u_{nmk} \text{ for } t_{k-1} \le t < t_k \text{ for all } n,m \text{ and } k. \tag{6.25}$$

The proof that this solution is optimal, immediately follows from the fact that $X_n(t) \equiv 0$, $\psi_n(t) \equiv 0$ and (6.25) is feasible along the entire planning horizon.

Note that a feasible solution of (6.22)-(6.24) can be found by linear programming with a fictitious objective, for example

$$\sum_m \sum_n \frac{u_{nmk}}{U_{nm}} \to \min, \qquad k=1,\ldots,K.$$

In contrast to Step 1 of Algorithm 6.2, Step 2 does not change and solves both the case of uniform speed machines and the case of a single product type solved by Algorithm 6.3. Indeed, $c_{unm} = 0$ implies that all production costs are equal and if N=1, the machines always have uniform speeds.

No Backlog

When backlogs are disallowed, i.e. for the $M|N|\,u(t) \in [0,\infty], X(t) \geq 0\,|d(t),T|\, L_u, L_x$ problem, the solution algorithm can be obtained by an easy modification of Algorithms 6.2 and 6.3. Indeed, Step 1 of Algorithm 6.2 will always result in $t_n = T$, because no backlog implies that c_{xn}^- is infinitely large. Steps 2 and 3 remain unchanged because the solution they find has no backlog.

Algorithm 6.3 also simplifies due to a point t^* after which the optimal solution is simply to track the demand. Clearly, the solution tends to reach this fractional regime as soon as possible because it gives no increment to the inventory-related term of the objective, while the production-related term is minimal if $\sum_m u_m(t) = d(t)$. Therefore, $t^* = t_k$, where k is such that $d_{k-1} > \sum_m U_m$ and $d_k \leq \sum_m U_m$. Before point t^*, however, the optimal solution is obtained by applying Algorithm 6.3 on the shortened planning horizon $[0, t^*]$. The proof of such modifications is left for the reader.

Algorithm 6.4

Step 1:

If N=1, demand is $\overline{K}$-interval, single-mode, piece-wise constant and $\max_t d_1(t) \geq \sum_m U_{1m}$, then omit product index n, sort and re-number the machines according to their production cost

$$c_{u1} \leq c_{u2} \leq \ldots \leq c_{uM}.$$

Step 2:

Define time point $t^* = t_k$, where k is such that $d_{k-1} > \sum_m U_m$ and $d_k \le \sum_m U_m$. Calculate time points $\bar{t}_{1,m} < t^*$ such that

$$d(t) < \sum_{m'=1}^{m} U_{m'} \text{ for } 0 \le t < \bar{t}_{1,m};$$

$$d(t) \ge \sum_{m'=1}^{m} U_{m'} \text{ for } \bar{t}_{1,m} \le t \le t^*. \text{ Set } \bar{t}_{1,0} = 0 \text{ and } \bar{t}_{1,m+1} = t^*.$$

Denote M time intervals $\tau_m = [\bar{t}_{1,m-2}, \bar{t}_{1,m-1}]$, m=2,...,M+1. In addition, define τ_1 and τ_{M+2} as the intervals which are degenerated to a point: $\tau_1 = \{0\}$ and $\tau_{M+2} = \{t^*\}$. Assign K=M+2.

Step 3 and Step 4 are identical to Steps 2 and 3 of Algorithm 3.5 respectively.

Step 5:

Define sets $I_1 = \left\{ m \middle| \sum_{m'=1}^{m} U_{m'} \le d(t) \right\}$, $I_2 = \{|I_1|+1\}$ and $I_3 = \{|I_2|+1, \ldots, M\}$. Set the optimal solution after t^* for $t_k < t \le t_{k+1}$ as

$$u_m(t) = U_m \text{ for } m \in I_1,$$

$$u_{m'}(t) = U_{m'} \frac{d(t) - \sum_{m \in I_1} U_m}{\sum_{m \in I_2} U_m}, \quad m' \in I_2,$$

$$u_m(t) = 0 \text{ for } m \in I_3.$$

STOP.

All subroutines of Algorithm 6.4 are identical to the corresponding subroutines of Algorithm 6.3.

6.2 DUE DATE

This section addresses parallel-machine, multiple product-type, capacitated scheduling problems characterized by demand concentrated at a due date.

Backlogs

Consider M-machine, N-product-type scheduling with backlogs and limited production capacity to minimize the total linear cost of inventory and production, that is, the $M|N|\, u(t)\in[0,U]\,|\,D,T\,|\,L_u, L_x$ problem which is stated as follows

$$J=\int_0^T \sum_n \left[c_{xn}^{+} Z_n^{+}(t) + c_{xn}^{-} Z_n^{-}(t) + \sum_m c_{unm} u_{nm}(t) \right] dt \to \min, \qquad (6.26)$$

$$Z_n^{+}(t)=\max\{0, X_n(t)-D_n(t)\},\ Z_n^{-}(t)=\max\{0, D_n(t)-X_n(t)\},$$

subject to

$$\dot{X}_n(t)=\sum_m u_{nm}(t), \quad X_n(0)=0, \qquad (6.27)$$

$$0 \le u_{nm}(t) \le U_{nm}, \qquad (6.28)$$

$$\sum_n \frac{u_{nm}(t)}{U_{nm}} \le 1, \qquad (6.29)$$

where

$$D_n(t)=\begin{cases} D_n, & \text{if } t \ge t_{dn}; \\ 0, & \text{otherwise}; \end{cases} \qquad (6.30)$$

and D_n is the amount of product-type n to be produced by the due date t_{dn} .

Dual Formulation

The Hamiltonian, H, for problem (6.26)-(6.30) is

$$H(t)=-\sum_n \left[c_{xn}^{+} Z_n^{+}(t) + c_{xn}^{-} Z_n^{-}(t) + \sum_m c_{unm} u_{nm}(t) \right] + \sum_n \psi_n(t) \sum_m u_{nm}(t). \qquad (6.31)$$

The co-state variable, $\psi_n(t)$ in (6.31), satisfies the following co-state differential equation with its transversality condition

$$\dot{\psi}_n(t) = \begin{cases} c_{xn}^+, \text{ if } X_n(t) > D_n(t); \\ -c_{xn}^-, \text{ if } X_n(t) < D_n(t); \\ a, \ a \in [-c_{xn}^-, c_{xn}^+], \text{ if } X_n(t) = D_n(t); \end{cases} \qquad \psi_n(T) = 0. \tag{6.32}$$

Maximization of the Hamiltonian with respect to admissible controls $u_{nm}(t)$ implies that for each machine m

$$u_{nm}(t) = \begin{cases} U_{nm}, \text{ if } \psi_n(t) > c_{unm} \text{ and } U_{nm}(\psi_n(t) - c_{unm}) > U_{n'm}(\psi_{n'}(t) - c_{un'm}), \ \forall n' \neq n \\ 0, \text{ if } \psi_n(t) < c_{unm}, \\ b, \ b \in [0, U_{nm}], \text{ if } \psi_n(t) \geq c_{unm} \text{ and } U_{nm}(\psi_n(t) - c_{unm}) \geq U_{n'm}(\psi_{n'}(t) - c_{un'm}), \ \forall n' \neq n. \end{cases} \tag{6.33}$$

Algorithm 6.5

Step 1:

If N=1, then omit product index n, sort and re-number the machines according to their production cost

$$c_{u1} \leq c_{u2} \leq \ldots \leq c_{uM}.$$

Step 2:

(i) Set k=1.

(ii) Solve the following system of two linear algebraic equations in t_1 and t_2

$$c_x^+(t_d - t_1) = c_x^-(t_2 - t_d),$$

$$(t_2 - t_1)\sum_{m=1}^{k} U_m - \frac{c_x^+ + c_x^-}{c_x^+ c_x^-}\sum_{m=1}^{k} U_m c_{um} = D.$$

(iii) If $t_1 \geq 0$, $c_{uk} \leq c_x^+(t_d - t_1) \leq c_{uk+1}$ and $t_2 \leq T$, then set the optimal solution as

$$u_m(t) = 0 \text{ for } 0 \leq t \leq t_1 + \frac{c_{um}}{c_x^+} \text{ and } t_2 - \frac{c_{um}}{c_x^-} \leq t \leq T, \ m \leq k;$$

$$u_m(t) = U_m \text{ for } t_1 + \frac{c_{um}}{c_x^+} < t < t_2 - \frac{c_{um}}{c_x^-}, \ m \leq k;$$

$u_m(t) = 0$ for $0 \le t \le T$, $m > k$. STOP.

Otherwise set k=k+1.
If $k > M$, then go to the next step, otherwise go to (ii).

Step 3:

(i) Set $k_1 = 0$ and assign $c_{u0} = -\infty$.

(ii) Set $k_2 = k_1$.

(iii) Solve the following system of two linear algebraic equations in $\psi(0)$ and t_2

$$\psi(0) + c_x^+ t_d = c_x^- (t_2 - t_d),$$

$$t_d \sum_{m=1}^{k_1} U_m + \sum_{m=k_1+1}^{k_2} \left(t_d - \frac{c_{um} - \psi(0)}{c_x^+} \right) U_m + \sum_{m=1}^{k_2} \left(t_2 - \frac{c_{um}}{c_x^-} - t_d \right) U_m = D.$$

(iv) If $c_{uk_1} \le \psi(0) < c_{uk_1+1}$, $c_{uk_2} \le \psi(0) + c_x^+ t_d \le c_{uk_2+1}$ and $t_2 \le T$, then set the optimal solution as

$$u_m(t) = 0 \text{ for } 0 \le t \le \frac{c_{um} - \psi(0)}{c_x^+}, \; k_1 < m \le k_2;$$

$$u_m(t) = 0 \text{ for } t_2 - \frac{c_{um}}{c_x^-} \le t \le T, \; 1 \le m \le k_2;$$

$$u_m(t) = U_m \text{ for } \frac{c_{um} - \psi(0)}{c_x^+} < t < t_2 - \frac{c_{um}}{c_x^-}, \; k_1 < m \le k_2;$$

$$u_m(t) = U_m \text{ for } 0 \le t < t_2 - \frac{c_{um}}{c_x^-}, \; m \le k_1;$$

$u_m(t) = 0$ for $0 \le t \le T$, $m > k_2$. STOP.

Otherwise set $k_2 = k_2 + 1$.

If $k_2 > M$, then set $k_1 = k_1 + 1$ and go to (v), otherwise go to (iii).

(v) If $k_1 > M$, then go to the next step, otherwise go to (ii).

Step 4:

(i) Calculate $t_1 = t_d - \dfrac{c_x^-(T-t_d)}{c_x^+}$ and set $c_{uM+1} = \infty$.

If $t_1 \geq 0$, then find k such that $c_{uk} \leq (t_d - t_1)c_x^+ < c_{uk+1}$ and set the optimal solution as

$$u_m(t) = 0 \text{ for } 0 \leq t \leq t_1 + \frac{c_{um}}{c_x^+} \text{ and } T - \frac{c_{um}}{c_x^-} \leq t \leq T,\ m \leq k;$$

$$u_m(t) = U_m \text{ for } t_1 + \frac{c_{um}}{c_x^+} < t < T - \frac{c_{um}}{c_x^-},\ m \leq k;$$

$$u_m(t) = 0 \text{ for } 0 \leq t \leq T,\ m > k. \text{ STOP.}$$

Otherwise go to the next step.

Step 5:

Set $c_{u0} = -\infty$ and $c_{uM+1} = \infty$. Calculate $\psi(0) = c_x^-(T-t_d) - c_x^+ t_d$ and find k_1 and k_2 such that $c_{uk_1} \leq \psi(0) < c_{uk_1+1}$ and $c_{uk_2} \leq (T-t_d)c_x^- < c_{uk_2+1}$.

Set the optimal solution as

$$u_m(t) = 0 \text{ for } 0 \leq t \leq \frac{c_{um} - \psi(0)}{c_x^+},\ k_1 < m \leq k_2;$$

$$u_m(t) = 0 \text{ for } T - \frac{c_{um}}{c_x^-} \leq t \leq T,\ 1 \leq m \leq k_2;$$

$$u_m(t) = U_m \text{ for } \frac{c_{um} - \psi(0)}{c_x^+} < t < T - \frac{c_{um}}{c_x^-},\ k_1 < m \leq k_2;$$

$$u_m(t) = U_m \text{ for } 0 \leq t < T - \frac{c_{um}}{c_x^-},\ m \leq k_1;$$

$$u_m(t) = 0 \text{ for } 0 \leq t \leq T,\ m > k_2. \text{ STOP.}$$

Algorithm 6.6

Step 1:

(i) Assign n=1.

(ii) Run Algorithm 6.5 and denote the time interval where $u_{nm}(t) = U_{nm}$ by (t^1_{nm}, t^2_{nm}).

(iii) If $n < N$, assign n=n+1 and go to (ii). Otherwise go to (iv).

(iv) Sort intervals (t^1_{nm}, t^2_{nm}). If they are disjoint for each machine m, $\bigcap_n (t^1_{nm}, t^2_{nm}) = \varnothing$, then the found solution is optimal. STOP.

Otherwise go to the next step.

Step 2:

If $c_{unm} = c_{unm'}$ for all n, $m \neq m'$ and machines have uniform speeds (see Definition 6.1), then set $U_n = \sum_m U_{nm}$, $c_{un} = c_{un1}$ and run Algorithm 5.4.

Step 3:

Set the optimal solution as

$$u_{nm}(t) = \frac{\beta_m}{\sum_{m'} \beta_{m'}} u_n(t) \text{ for all } n,m,\ 0 \le t \le T \text{. STOP.}$$

Example 6.2

Consider the $3|1|u(t) \in [0,U]\,|D,T|\, L_u, L_x$ problem with the length of the planning horizon, T, and the due date t_d set at 10hrs and 8hrs respectively. The demand required for the product is D=132 units and the production parameters are shown in Table 6.2.

Table 6.2. Machine parameters and priorities.

m	U_m	c_{um}	Optimal priority
1	10	1/2	2
2	15	2/3	3
3	20	1/10	1

In addition, inventory and backlog costs are c^+_x=0.1\$/unit.hr and c^-_x=0.5\$/unit.hr respectively.

Step 1 of Algorithm 6.5 sorts products with respect to the production cost coefficients as 3-1-2 (see the Optimal priority column in Table 6.2).

Solving the system of two linear equations at Step 2 for k=2 results in time points $t_1 = 2$ and $t_2 = 9.2$. Since $t_1 \geq 0$, $c_{u2} \leq c_x^+(t_d - t_1) = 0.6 \leq c_{u3}$ and $t_2 \leq T$, the optimal solution is set as

$u_3(t) = 0$ for $0 \leq t \leq 3$ and $9 \leq t \leq 10$, $u_3(t) = 20$ unit/hr. for $3 < t < 9$,

$u_1(t) = 0$ for $0 \leq t \leq 7$ and $8.2 \leq t \leq 10$, $u_1(t) = 10$ unit/hr. for $7 < t < 8.2$,

$u_2(t) = 0$ for $0 \leq t \leq 10$.

The found solution is depicted in Figure 6.5.

■

Discussion

Algorithm 6.5 deals with a special case which is characterized by a single product type. Generally, the algorithm solves it in $O(M^3)$ time due to Step 3, which is most time consuming. This step deals with a pressing due date and a loose planning horizon. However, if both the due date and the planning horizon are loose, Step 2 solves the problem in $O(M^2)$ time. Furthermore, when the planning horizon is pressing, either Step 4 or Step 5 solves the problem in $O(M)$ time which, together with sorting at Step 1, results in the total complexity of $O(M\log M)$ time.

Step 1 of Algorithm 6.6 treats the problem by decomposing it into N one-product sub-problems, each of which is solved by Algorithm 6.5 in $O(M^3)$ time. This decomposition is true if production intervals of the same machine do not overlap. This is verified by sorting in $O(MN\log N)$ time. Thus, the total complexity of Step 1 is $O(MN\max\{M^2,\log N\})$. If the problem cannot be decomposed, Steps 2 and 3 of Algorithm 6.6 solve the problem in $O(N\max\{M,N\})$ time for a special case characterized by equal production cost coefficients and uniform speed machines. The solution of Step 2 is based on coalescing the machines into a single machine with production capacity $U_n = \sum_m U_{nm}$, n=1,...,N. Then, Algorithm 5.4 for a single machine is executed and its solution is distributed proportionally among the machines.

LEMMA 6.8. (Correctness of Algorithm 6.5 Step 2) Let N=1 and machines be numbered in the non-decreasing order of the production costs $c_{u1} \leq c_{u2} \leq \ldots \leq c_{uM}$. If there exists machine k such that t_1 and t_2 satisfy the following system

$$c_x^+(t_d - t_1) = c_x^-(t_2 - t_d),$$

$$(t_2 - t_1)\sum_{m=1}^{k} U_m - \frac{c_x^+ + c_x^-}{c_x^+ c_x^-}\sum_{m=1}^{k} U_m c_{um} = D,$$

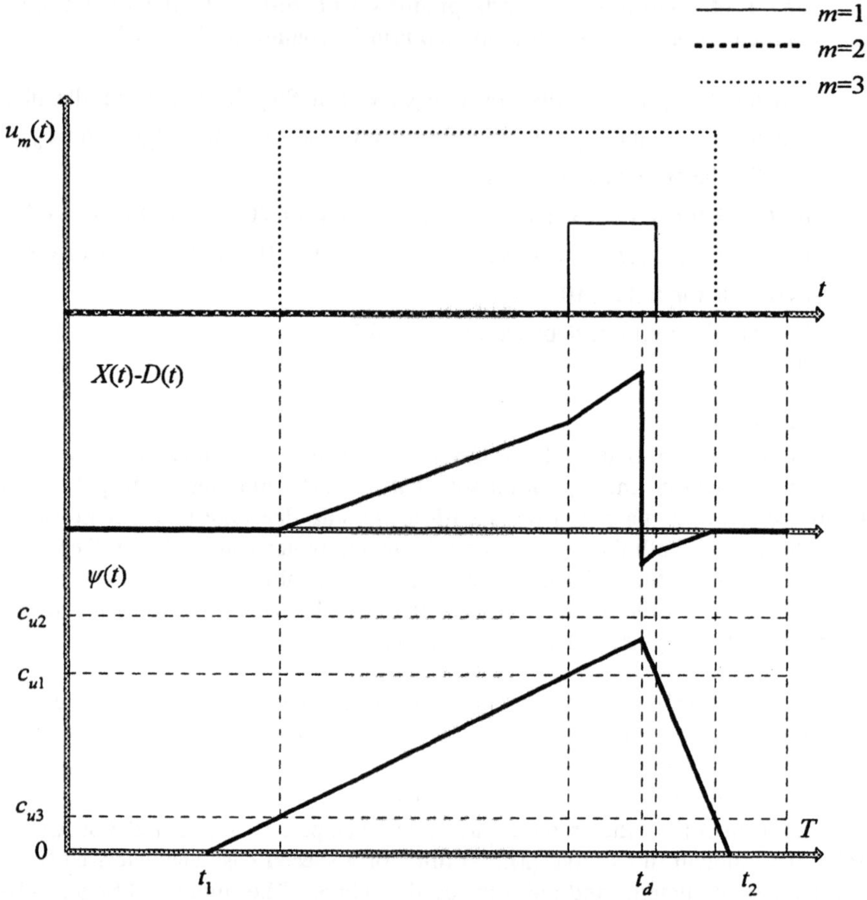

Figure 6.5. Optimal control, state and co-state variables for the $M|N|\, u(t) \in [0,U]\, |D,T|\, L_u, L_x$ problem of Example 6.2.

$$t_1 \geq 0\,,\ \ c_{uk} \leq c_x^+ (t_d - t_1) \leq c_{uk+1}\,,\ t_2 \leq T\,,$$

then the optimal solution is

$u_m(t) = 0$ for $0 \leq t \leq t_1 + \dfrac{c_{um}}{c_x^+}$ and $t_2 - \dfrac{c_{um}}{c_x^-} \leq t \leq T$, $m \leq k$;

$u_m(t) = U_m$ for $t_1 + \dfrac{c_{um}}{c_x^+} < t < t_2 - \dfrac{c_{um}}{c_x^-}$, $m \leq k$;

$u_m(t) = 0$ for $0 \leq t \leq T$, $m > k$.

PROOF. Given machines numbered in the non-decreasing order of the production cost coefficients, consider a solution with no production of machines $m=k+1,\dots,M$ over the entire planning horizon and full production of machines $m=1,\dots,k$ from point t_m^1 to t_m^2

$$u_m(t)=0 \text{ for } 0\le t\le T,\ m>k.$$

$$u_m(t)=U_m \text{ for } t_m^1<t<t_m^2,\ u_m(t)=0 \text{ for } 0\le t\le t_m^1,\ t_m^2\le t\le T,\ m\le k;$$

$$X(t)=X(t_m^1)+(t-t_m^1)\sum_{i=1}^{m}U_i,\ \text{ for } t_m^1\le t\le t_{m+1}^1,\ m<k;$$

$$X(t)=X(t_m^1)+(t-t_m^1)\sum_{i=1}^{m}U_i,\ \text{ for } t_m^1\le t\le t_m^2,\ m=k;$$

$$X(t)=X(t_{m+1}^2)+(t-t_{m+1}^2)\sum_{i=1}^{m}U_i,\ \text{ for } t_{m+1}^2\le t\le t_m^2,\ m<k;$$

$$X(t)=D(t) \text{ for } 0\le t\le t_1^1,\ t_1^2\le t\le T,$$

$$\psi(t)=c_x^+(t-t_1^1+\frac{c_{u1}}{c_x^+}) \text{ for } t_1^1-\frac{c_{u1}}{c_x^+}\le t\le t_d,$$

$$\psi(t)=c_x^-(t_1^2+\frac{c_{u1}}{c_x^-}-t) \text{ for } t_d\le t\le t_1^2+\frac{c_{u1}}{c_x^-},$$

$$\psi(t)=0 \text{ for } 0\le t\le t_1^1-\frac{c_{u1}}{c_x^+},\ t_1^2+\frac{c_{u1}}{c_x^-}\le t\le T.$$

By denoting $t_1=t_1^1-\frac{c_{u1}}{c_x^+}$ and $t_2=t_1^2+\frac{c_{u1}}{c_x^-}$ and taking into account boundary conditions $X(t_1^2)=D$ and $\psi(t_2)=0$, the last system of the state and co-state equations transforms into the two algebraic equations in t_1 and t_2 stated in the lemma. If the solution of these equations is feasible, i.e. $t_1\ge 0$, $t_2\le T$ and $c_{uk}\le\psi(t_d)\le c_{uk+1}$, then it is optimal.

■

The proof of the correctness for the remaining steps of Algorithm 6.5 is similar and is left as an exercise for the reader.

LEMMA 6.9. (Correctness of Algorithm 6.6 Step 1) Let one-product-type solutions be found by Algorithm 6.5 for each product separately. If time intervals (t^1_{nm}, t^2_{nm}) where $u_{nm}(t) = U_{nm}$ do not overlap, $\bigcap_n (t^1_{nm}, t^2_{nm}) = \varnothing$ for each m, then this solution is optimal.

PROOF. Condition $\bigcap_n (t_{nm}, t_{nm}) = \varnothing$ for each m ensures that the optimal production intervals found independently (see Lemma 6.8) for different products do not overlap and, thus, the capacity constraint of each machine is not violated.
■

LEMMA 6.10. (Correctness of Algorithm 6.6 Steps 2-3) If $c_{unm} = c_{unm'}$ for all n, $m \neq m'$, the machines have uniform speeds and $u_n(t)$ is the optimal solution of the $1|N|\,u(t) \in [0, U]\,|D,T|\,L_u, L_x$ problem with $U_n = \sum_m U_{nm}$ and $c_{un} = c_{un1}$, then the optimal solution is $u_{nm}(t) = \frac{\beta_m}{\sum_{m'} \beta_{m'}} u_n(t)$ for all n,m, $0 \le t \le T$.

PROOF. The proof is identical to that of Lemma 6.2. ■

THEOREM 6.2. If N=1, then problem (6.26)-(6.29) is solvable in $O\left(M^3\right)$ time.

PROOF. The complexity of Algorithm 6.5, which solves the problem for the case of a single product (see Lemma 6.8), is due to Step 3 which is most time-consuming because it involves $O(M)$ computations within a double loop.
■

Special Cases of Zero Costs

The zero inventory cost $M|N|\,u(t) \in [0, U]\,|D,T|\,L_u, L_x, c_x^+ = 0$ problem implies that the co-state variable $\psi(t)$ is a non-increasing function of time. As a result, the set of those machines, which produce before the due date, does not change. After the due date, this set changes because the machines switch off, as determined by Algorithm 6.5. Therefore, Algorithm 6.5 simplifies, as described below, and requires now $O(M^2)$ time. Algorithm 6.6 does not change because it does not operate with parameters c_{xn}^+.

Algorithm 6.7

Step 1:

If N=1, then omit product index n, sort and re-number the machines according to their production cost

$$c_{u1} \le c_{u2} \le \ldots \le c_{uM} .$$

Step 2:

(i) Set $k_1 = 0$ and assign $c_{u0} = -\infty$.

(ii) Solve the following two linear algebraic equations for $\psi(0)$ and t_2

$$\psi(0) = c_x^-(t_2 - t_d) ,$$

$$t_d \sum_{m=1}^{k_1} U_m + \sum_{m=1}^{k_1} \left(t_2 - \frac{c_{um}}{c_x^-} - t_d \right) U_m = D .$$

(iii) If $c_{uk_1} \le \psi(0) < c_{uk_1+1}$ and $t_2 \le T$, then set the optimal solution as

$u_m(t) = 0$ for $t_2 - \dfrac{c_{um}}{c_x^-} \le t \le T$, $m \le k_1$;

$u_m(t) = U_m$ for $0 \le t < t_2 - \dfrac{c_{um}}{c_x^-}$, $m \le k_1$;

$u_m(t) = 0$ for $0 \le t \le T$, $m > k_1$. STOP.

Otherwise, set $k_1 = k_1 + 1$.

(iv) If $k_1 > M$, then go to the next step, otherwise go to (ii).

Step 3:

Set $c_{u0} = -\infty$ and $c_{uM+1} = \infty$. Calculate $\psi(0) = c_x^-(T - t_d)$ and find k_1 such that $c_{uk_1} \le \psi(0) < c_{uk_1+1}$.

Set the optimal solution as

$u_m(t) = 0$ for $T - \dfrac{c_{um}}{c_x^-} \le t \le T$, $m \le k_1$;

$u_m(t) = U_m$ for $0 \le t < T - \dfrac{c_{um}}{c_x^-}$, $m \le k_1$;

$u_m(t) = 0$ for $0 \le t \le T$, $m > k_1$. STOP.

Next, we consider the zero production cost problem, $M|N|\, u(t) \in [0,U]\, |D,T|\, L_x$. By substituting $c_{unm} = 0$ in the formulas of Algorithm 6.5, one can observe that it becomes a special case of Steps 2 and 3 of Algorithm 6.6 which do not change. Therefore, the

$M|N|\,u(t)\in[0,U]\,|D,T|\,L_x$ problem is solvable in $O(N\max\{M,N\})$ time for the case of uniform speed machines. Finally, Step 1 of Algorithm 6.6 also simplifies to $O(NM)$ operations because it no longer needs sorting of the machines.

No Backlog

The solution algorithm (see Algorithm 6.8 below) for the $M|N|\,u(t)\in[0,U],X(t)\geq 0\,|D,T|\,L_u,L_x$ problem can be obtained by an easy modification of Algorithm 6.5. Indeed, since no backlog means that the demand for the product must be satisfied before the due date, the machines switch off at the due date.

As to Algorithm 6.6, it does not change. However, it now invokes the algorithms for solving the corresponding no-backlog problems (see No Backlog of Section 5.2).

Algorithm 6.8

Step 1:

If N=1, then omit product index n, sort and re-number the machines according to their production cost

$$c_{u1}\leq c_{u2}\leq\ldots\leq c_{uM}.$$

Step 2:

(i) Set k=1.

(ii) Solve the following linear algebraic equation in t_1

$$\sum_{m=1}^{k}(t_d-t_1-\frac{c_{um}}{c_x^+})U_m=D.$$

(iii) If $c_{uk}\leq c_x^+(t_d-t_1)\leq c_{uk+1}$ and $t_1\geq 0$, then set the optimal solution as

$u_m(t)=0$ for $0\leq t\leq t_1+\dfrac{c_{um}}{c_x^+}$ and $t_d\leq t\leq T$, $m\leq k$;

$u_m(t)=U_m$ for $t_1+\dfrac{c_{um}}{c_x^+}<t<t_d$, $m\leq k$;

$u_m(t)=0$ for $0\leq t\leq T$, $m>k$. STOP.

Otherwise, set k=k+1.

If $k>M$, then go to the next step; otherwise go to (ii).

Step 3:

(i) Set $k_1=0$ and assign $c_{u0}=-\infty$.

(ii) Set $k_2=k_1$.

(iii) Solve the following linear algebraic equation in $\psi(0)$

$$t_d \sum_{m=1}^{k_1} U_i + \sum_{m=k_1+1}^{k_2} \left(t_d - \frac{c_{um} - \psi(0)}{c_x^+} \right) U_m = D .$$

(iv) If $c_{uk_1} \le \psi(0) < c_{uk_1+1}$ and $c_{uk_2} \le \psi(0) + c_x^+ t_d \le c_{uk_2+1}$, then set the optimal solution as

$u_m(t) = 0$ for $0 \le t \le \dfrac{c_{um} - \psi(0)}{c_x^+}$, $k_1 < m \le k_2$;

$u_m(t) = 0$ for $t_d \le t \le T$, $1 \le m \le k_2$;

$u_m(t) = U_m$ for $\dfrac{c_{um} - \psi(0)}{c_x^+} < t < t_d$, $k_1 < m \le k_2$;

$u_m(t) = U_m$ for $0 \le t < t_d$, $m \le k_1$;

$u_m(t) = 0$ for $0 \le t \le T$, $m > k_2$. STOP.

Otherwise, set $k_2 = k_2 + 1$.

If $k_2 > M$, then set $k_1 = k_1 + 1$ and go to (v), otherwise go to (iii).

(v) If $k_1 > M$, then go to the next step; otherwise go to (ii).

6.3 DEMAND OVER INFINITE PLANNING HORIZON

This section addresses the scheduling problems characterized by periodic demand specified over an infinite planning horizon. To provide the periodicity of the solution, we consider the problem over demand period *T* with boundary condition $X_n(0) = X_n(T)$.

6.3.1 LINEAR PERFORMANCE MEASURE

Backlogs

The problem to be considered is *M*-parallel-machine, *N*-product-type scheduling with backlogs and limited production rate to minimize the linear cost of inventory and production over period *T*, that is, the $M|N|\, u(t) \in [0, U], X(0) = X(T)\, |d(t)|\, L_u, L_x$ problem which is stated as follows

$$J = \int_0^T \sum_n \left[c_{xn}^+ X_n^+(t) + c_{xn}^- X_n^-(t) + \sum_m c_{unm} u_{nm}(t) \right] dt \to \min, \qquad (6.34)$$

$$X_n^+(t) = \max\{X_n(t), 0\}, \quad X_n^-(t) = \max\{-X_n(t), 0\}$$

subject to

$$\dot{X}_n(t) = \sum_m u_{nm}(t) - d_n(t), \quad X_n(0) = X_n(T), \tag{6.35}$$

$$0 \le u_{nm}(t) \le U_{nm}, \tag{6.36}$$

$$\sum_n \frac{u_{nm}(t)}{U_{nm}} \le 1, \tag{6.37}$$

Dual Formulation

The Hamiltonian, H, resulting from the maximum principle is

$$H = -\sum_n \left[c_{xn}^+ X_n^+(t) + c_{xn}^- X_n^-(t) + \sum_m c_{unm} u_{nm}(t) \right] + \sum_n \psi_n(t) \left(\sum_m u_{nm}(t) - d_n(t) \right). \tag{6.38}$$

The co-state variables, $\psi_n(t)$ in (6.38), satisfy the following co-state differential equation with corresponding transversality condition

$$\dot{\psi}_n(t) = \begin{cases} c_{xn}^+, \text{ if } X_n(t) > 0; \\ -c_{xn}^-, \text{ if } X_n(t) < 0; \\ a, \ a \in [-c_{xn}^-, c_{xn}^+], \text{ if } X_n(t) = 0; \end{cases} \qquad \psi_n(T) = 0. \tag{6.39}$$

Maximization of the Hamiltonian implies that

$$u_{nm}(t) = \begin{cases} U_{nm}, \text{ if } \psi_n(t) > c_{unm} \text{ and } U_{nm}(\psi_n(t) - c_{unm}) > U_{n'm}(\psi_{n'}(t) - c_{un'm}), \ \forall n' \ne n \\ 0, \text{ if } \psi_n(t) < c_{unm}, \\ b, \ b \in [0, U_{nm}], \text{ if } \psi_n(t) \ge c_{unm} \text{ and } U_{nm}(\psi_n(t) - c_{unm}) \ge U_{n'm}(\psi_{n'}(t) - c_{un'm}), \ \forall n' \ne n. \end{cases} \tag{6.40}$$

Algorithm 6.9

Step 1:

For each product n, find machine $m(n)$ such that its production cost is minimal, $c_{unm(n)} = \min_m c_{unm}$. If $\sum_n \frac{d_n(t)}{U_{nm(n)}} \le 1$ for each $m(n)$ assigned to produce product n, then set the optimal solution as $u_{nm(n)}(t) = d_n(t)$ for $0 \le t \le T$,

$u_{nm}(t)=0$ for $0 \le t \le T$, $m \ne m(n)$. STOP.

Step 2:

If $c_{unm}=c_{unm'}$ for all n, $m \ne m'$ and machines have uniform speeds (see Definition 6.1), then set $U_n = \sum_m U_{nm}$ and $c_{un}=c_{un1}$ and run Algorithm 5.5.

Step 3:

Set the optimal solution as

$$u_{nm}(t)=\frac{\beta_m}{\sum_{m'} \beta_{m'}} u_n(t) \text{ for all } n,m,\ 0 \le t \le T. \text{ STOP.}$$

Algorithm 6.10

Step 1:

If N=1, then omit product index n, sort and re-number the machines according to their production costs

$$c_{u1} \le c_{u2} \le \ldots \le c_{uM}.$$

If $d(t)=\begin{cases} d_1, & \text{if } 0 \le t \le \bar{t}, \\ d_2, & \text{if } \bar{t} < t \le T, \end{cases}$ and $d_1\bar{t}+d_2(T-\bar{t}) \le T\sum_{m=1}^{M} U_m$, then go to the next step. Otherwise STOP.

Step 2:

(i) Find k_1 and k_2 such that $\sum_{m=1}^{k_1-1} U_m < d_1 \le \sum_{m=1}^{k_1} U_m$ and $\sum_{m=1}^{k_2-1} U_m < d_2 \le \sum_{m=1}^{k_2} U_m$ respectively.

(ii) Solve the following two systems of two linear algebraic equations in unknowns t_1, t_2, t_3 and t_4

$$c_x^+(t_2-t_1)=c_{uk_2}-c_{uk_1},$$

$$d_1(\bar{t}-t_1)+d_2(t_2-\bar{t})=(t_2-t_1)\sum_{m=1}^{k_1} U_m + \sum_{m=k_1+1}^{k_2-1} U_m \frac{c_{uk_2}-c_{um}}{c_x^+},$$

$$c_x^-(t_4 - t_3) = c_{uk_2} - c_{uk_1},$$

$$d_1(t_4 - T) + d_2(T - t_3) = (t_4 - t_3)\sum_{m=1}^{k_1} U_m + \sum_{m=k_1+1}^{k_2-1} U_m \frac{c_{uk_2} - c_{um}}{c_x^-}.$$

(iii) If $t_1 \geq t_4 - T$ and $t_2 \leq t_3$, then set the optimal solution as

$$u_m(t) = U_m \text{ for } t_2 - \frac{c_{uk_2} - c_{um}}{c_x^+} \leq t \leq t_3 + \frac{c_{uk_2} - c_{um}}{c_x^-}, \quad k_1 \leq m < k_2.$$

$$u_m(t) = 0 \text{ for } t_3 + \frac{c_{uk_2} - c_{um}}{c_x^-} \leq t \leq t_2 - \frac{c_{uk_2} - c_{um}}{c_x^+} + T, \quad k_1 < m < k_2.$$

$$u_{k_1}(t) = d_1 - \sum_{m=1}^{k_1-1} U_m \text{ for } t_4 - T \leq t \leq t_1,$$

$$u_{k_2}(t) = d_2 - \sum_{m=1}^{k_2-1} U_m \text{ for } t_2 \leq t \leq t_3,$$

$$u_{k_2}(t) = 0 \text{ for } t_3 \leq t \leq T + t_2,$$

$$u_m(t) = U_m \text{ for } 0 \leq t \leq T, \ m < k_1;$$

$u_m(t) = 0$ for $0 \leq t \leq T$, $m > k_2$. STOP.

Otherwise go to the next step.

Step 3:

(i) Set $k_1 = 0$.

(ii) Set $k_2 = k_1$.

(iii) Set $k_3 = k_1$.

(iv) Solve the following system of four linear algebraic equations in $X(0)$, $\psi(0)$, t_1 and t_2

$$c_x^-(T - t_2 + t_1) = c_x^+(t_2 - t_1),$$

$$X(0)+(T-t_2)\sum_{m=1}^{k_1}U_m+\sum_{m=k_1+1}^{k_2}\left(T-\frac{c_{um}-\psi(0)}{c_x^-}-t_2\right)U_m=d_2(T-t_2),$$

$$(t_2-t_1)\sum_{m=1}^{k_3}U_m+\sum_{m=k_3+1}^{k_2}\frac{c_{um}-\psi(0)+t_1c_x^-}{c_x^+}U_m=d_2(t_2-\bar{t})+d_1(\bar{t}-t_1),$$

$$X(0)+t_1\sum_{m=1}^{k_3}U_m+\sum_{m=k_3+1}^{k_1}\frac{\psi(0)-c_{um}}{c_x^-}U_m=d_1t_1,$$

(v) If $c_{uk_1}\le\psi(0)<c_{uk_1+1}$, $c_{uk_2}\le\psi(0)+c_x^-(T-t_2)\le c_{uk_2+1}$, $c_{uk_3}\le\psi(0)-c_x^-t_1\le c_{uk_3+1}$ and $t_1\le t_2$, then set the optimal solution as

$u_m(t)=U_m$ for $t_1+\dfrac{c_x^-t_1-\psi(0)+c_x^-t_1}{c_x^+}\le t\le T+\dfrac{\psi(0)-c_{um}}{c_x^-}$, $k_3<m\le k_1$;

$u_m(t)=0$ for $\dfrac{\psi(0)-c_{um}}{c_x^-}\le t\le t_1+\dfrac{c_x^-t_1-\psi(0)+c_x^-t_1}{c_x^+}$, $k_3<m\le k_1$;

$u_m(t)=U_m$ for $t_2-\dfrac{c_x^-(T-t_2)-c_{um}+\psi(0)}{c_x^+}<t<T-\dfrac{c_{um}-\psi(0)}{c_x^-}$,

$k_1<m\le k_2$;

$u_m(t)=0$ for $T-\dfrac{c_{um}-\psi(0)}{c_x^-}<t<T+t_2-\dfrac{c_x^-(T-t_2)-c_{um}+\psi(0)}{c_x^+}$,

$k_1<m\le k_2$;

$u_m(t)=U_m$ for $0\le t\le T$, $m\le k_3$;

$u_m(t)=0$ for $0\le t\le T$, $m>k_2$. STOP.

Otherwise, set $k_3=k_3-1$
If $k_3<0$, then set $k_2=k_2+1$ and go to (vi), otherwise go to (iv).

(vi) If $k_2>M$, then set $k_1=k_1+1$ and go to (vii), otherwise go to (iii).

(vii) If $k_1>M$, then go to the next step, otherwise go to (ii).

Step 4:

(i) Find k_1 such that $\sum_{m=1}^{k_1-1} U_m < d_1 \le \sum_{m=1}^{k_1} U_m$ and set $k_2 = k_1$.

(ii) Solve the following system of three linear algebraic equations in unknowns t_1, t_2 and t_3

$$c_x^+(t_2 - t_1) = c_x^-(t_3 - t_2),$$

$$d_1(\bar{t} - t_1) + d_2(t_2 - \bar{t}) = (t_2 - t_1)\sum_{m=1}^{k_1} U_m + \sum_{m=k_1+1}^{k_2} U_m(t_2 - t_1 - \frac{c_{um} - c_{uk_1}}{c_x^+}),$$

$$d_1(t_3 - T) + d_2(T - t_2) = (t_3 - t_2)\sum_{m=1}^{k_1} U_m + \sum_{m=k_1+1}^{k_2} U_m(t_3 - t_2 - \frac{c_{um} - c_{uk_1}}{c_x^-}).$$

(iii) If $t_1 \ge t_3 - T$ and $c_{uk_2} \le c_{uk_1} + c_x^+(t_2 - t_1) \le c_{uk_2+1}$ then set the optimal solution as

$u_m(t) = U_m$ for $t_2 - \dfrac{c_{uk_1} + c_x^+(t_2 - t_1) - c_{um}}{c_x^+} \le t \le t_2 + \dfrac{c_{uk_1} + c_x^+(t_2 - t_1) - c_{um}}{c_x^-}$,

$k_1 \le m \le k_2$.

$u_m(t) = 0$ for $t_2 + \dfrac{c_{uk_1} + c_x^+(t_2 - t_1) - c_{um}}{c_x^-} \le t \le t_2 - \dfrac{c_{uk_1} + c_x^+(t_2 - t_1) - c_{um}}{c_x^+} + T$,

$k_1 < m \le k_2$.

$u_{k_1}(t) = d_1 - \sum_{i=1}^{k_1-1} U_i$ for $t_3 - T \le t \le t_1$,

$u_m(t) = U_m$ for $0 \le t \le T$, $m < k_1$;

$u_m(t) = 0$ for $0 \le t \le T$, $m > k_2$. STOP.

Otherwise, set $k_2 = k_2 + 1$.

If $k_2 > M$, then go to the next step, otherwise go to (ii).

Step 5:

(i) Find k_2 such that $\sum_{m=1}^{k_2-1} U_m < d_2 \le \sum_{m=1}^{k_2} U_m$ and set $k_1 = k_2$.

(ii) Set $k_3 = k_1$.

(iii) Solve the following system of five linear algebraic equations in $X(0)$, $\psi(0)$, t_1, t_2 and t_3

$$c_x^- t_1 + c_x^+ (t_2 - t_1) = c_{uk_2} - \psi(0),$$

$$c_x^- (T - t_3) = c_{uk_2} - \psi(0),$$

$$-X(0) + (T - t_3) \sum_{m=1}^{k_1} U_m + \sum_{m=k_1+1}^{k_2} \left(T - \frac{c_{um} - \psi(0)}{c_x^-} \right) U_m = d_2 (T - t_2),$$

$$(t_2 - t_1) \sum_{m=1}^{k_3} U_i + \sum_{m=k_3+1}^{k_2} \frac{c_{um} - \psi(0) + t_1 c_x^-}{c_x^+} U_m = d_2 (t_2 - \bar{t}) + d_1 (\bar{t} - t_1),$$

$$X(0) + t_1 \sum_{m=1}^{k_3} U_m + \sum_{m=k_3+1}^{k_1} \frac{\psi(0) - c_{um}}{c_x^-} U_m = d_1 t_1,$$

(iv) If $c_{uk_1} \le \psi(0) < c_{uk_1+1}$, $c_{uk_3} \le \psi(0) - c_x^- t_1 \le c_{uk_3+1}$ and $0 \le t_1 \le t_2 \le t_3 \le T$, then set the optimal solution as

$u_m(t) = U_m$ for $t_1 + \dfrac{c_x^- t_1 - \psi(0) + c_x^- t_1}{c_x^+} \le t \le T + \dfrac{\psi(0) - c_{um}}{c_x^-}$, $k_3 < m \le k_1$;

$u_m(t) = 0$ for $\dfrac{\psi(0) - c_{um}}{c_x^-} \le t \le t_1 + \dfrac{c_x^- t_1 - \psi(0) + c_x^- t_1}{c_x^+}$, $k_3 < m \le k_1$;

$u_m(t) = U_m$ for $t_2 - \dfrac{c_x^- (T - t_2) - c_{um} + \psi(0)}{c_x^+} < t < T - \dfrac{c_{um} - \psi(0)}{c_x^-}$, $k_1 < m \le k_2$;

$$u_m(t) = 0 \text{ for } T - \frac{c_{um} - \psi(0)}{c_x^-} < t < T + t_2 - \frac{c_x^-(T - t_2) - c_{um} + \psi(0)}{c_x^+},$$

$k_1 < m < k_2$;

$$u_{k_2}(t) = d_2 - \sum_{m=1}^{k_2-1} U_m \text{ for } t_2 \le t \le t_3,$$

$u_m(t) = U_m$ for $0 \le t \le T$, $m \le k_3$;

$u_m(t) = 0$ for $0 \le t \le T$, $m > k_2$. STOP.

Otherwise, set $k_3 = k_3 - 1$.

If $k_3 < 0$, then set $k_1 = k_1 - 1$ and go to (iii), otherwise go to (iv).

Example 6.3

Consider the $3|1|u(t) \in [0, U], X(0) = X(T)\,|d(t)|\, L_u, L_x$ problem with the length of the planning horizon, T, set at 10hrs and the demand

$$d(t) = \begin{cases} 15, & \text{if } 0 \le t \le 6, \\ 25, & \text{if } 6 < t \le 10. \end{cases}$$

The production parameters are shown in Table 6.3.

Table 6.3. Machine parameters and priorities

m	U_m	c_{um}	Optimal priority
1	10	1/2	2
2	15	2/3	3
3	20	1/10	1

In addition, inventory and backlog costs are c_x^+=0.1$/unit.hr and c_x^-=0.5$/unit.hr respectively.

Step 1 of Algorithm 6.5 sorts products with respect to the production cost coefficients as 3-1-2 (see the Optimal priority column in Table 6.3) and verifies that the total capacity is sufficient for a periodic solution to exist.

Step 2 determines k_1=3 and k_2=1 and solves the two systems of two linear equations as t_1=4, t_2=8, t_3=9.6 and t_4=10.4. Since $t_1 \ge t_4 - T$=0.4 and $t_2 \le t_3$, the optimal solution is set as

$u_3(t) = 20$ unit/hr for $4 \le t \le 10.4$,

$u_3(t) = 15$ unit/hr for $0.4 \le t \le 4$,
$u_1(t) = 5$ unit/hr for $8 \le t \le 9.6$,
$u_1(t) = 0$ for $0 \le t < 8$ and $9.6 < t \le 10$,
$u_2(t) = 0$ for $0 \le t \le 10$.

The found solution is depicted in Figure 6.6.

■

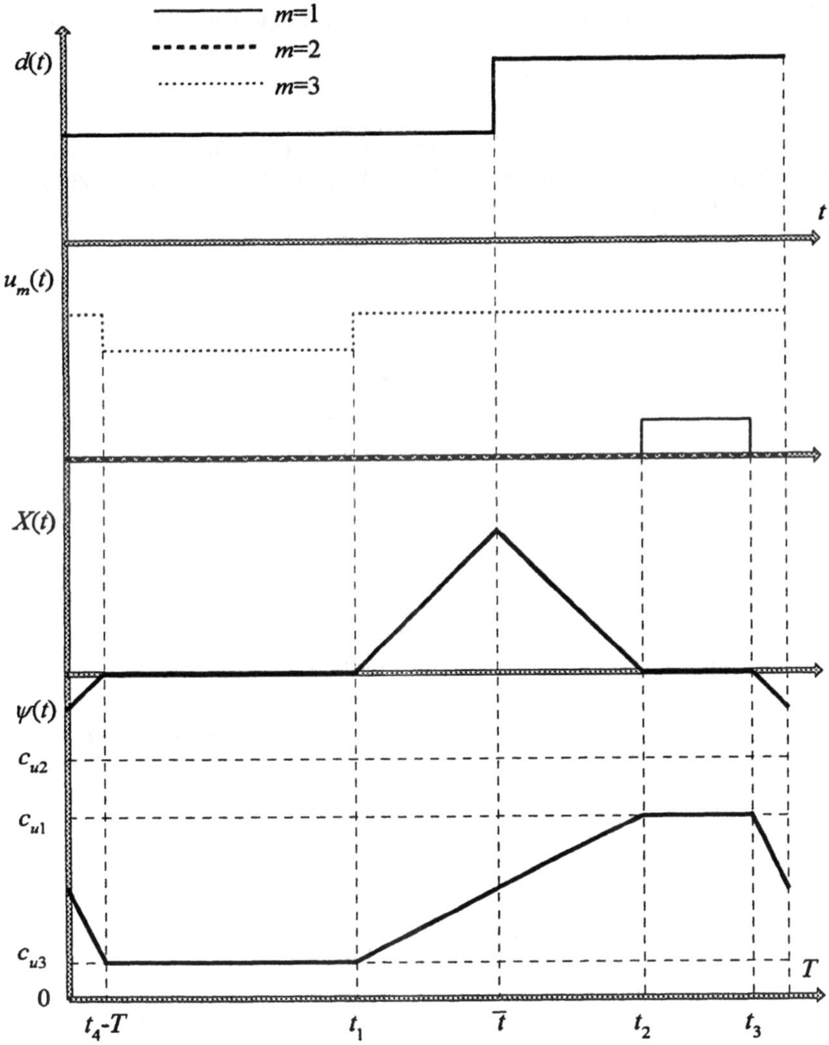

Figure 6.6. Optimal control, state and co-state variables for the $M|N|\,u(t) \in [0, U], X(0) = X(T)\,|d(t)|\, L_u, L_x$ problem of Example 6.3.

Discussion

This section presents three solvable cases. Step 1 of Algorithm 6.9 solves the problem in $O(NM)$ time by assigning a product to the machine which produces the product with minimal cost. This implies that Step 1 decomposes the problem into at most M problems of uncapacitated, one-machine scheduling. Steps 2 and 3 of Algorithm 6.9 solve the problem in $O(N\max\{M,N\})$ time for a special case characterized by identical production costs and uniform-speed machines. The solution of Step 2 is based on coalescing the machines into a single machine with aggregate production capacity $U_n = \sum_m U_{nm}$, $n=1,\ldots,N$. Then, Algorithm 5.5 for a single machine is executed and its solution is distributed proportionally among the machines.

Algorithm 6.10 deals with a single product type and a two-level, piece-wise constant, periodic demand. The algorithm solves the problem if the demand period is loose in either $O(M\log M)$ time via Steps 1 and 2 or in $O(M^2)$ time via Steps 1 and 4. Alternatively, if the demand period is pressing, then either Step 3 solves the problem in $O(M^4)$ or Step 5 does in $O(M^3)$ time.

LEMMA 6.11. (Correctness of Algorithm 6.9 Step 1) Given problem (6.34)-(6.37), and $m(n) = \arg\min_m c_{unm}$, if $\sum_n \frac{d_n(t)}{U_{nm(n)}} \le 1$ for each $m(n)$, then the optimal solution is

$$u_{nm(n)}(t) = d_n(t) \text{ for } 0 \le t \le T,\ \ u_{nm}(t) = 0 \text{ for } 0 \le t \le T \text{ for } m \ne m(n).$$

PROOF. Since the demand does not exceed the capacity, $\sum_n \frac{d_n(t)}{U_{nm(n)}} \le 1$ for $0 \le t \le T$, the following solution

$$\psi_n(t) = c_{unm(n)},\ \ u_{nm(n)}(t) = d_n(t),\ \ u_{nm}(t) = 0 \text{ for } m \ne m(n),\ X_n(t) = 0 \text{ for } 0 \le t \le T,$$

is always feasible and satisfies optimality conditions (6.40).
■

The next lemma proves the correctness of Step 2 of Algorithm 6.9, while the proof of the other steps is left as an exercise for the reader.

LEMMA 6.12. (Correctness of Algorithm 6.9 Step 2) If N=1, machines are numbered in the non-decreasing order of their production costs, $c_{u1} \le c_{u2} \le \ldots \le c_{uM}$ and $d(t)=\begin{cases} d_1, \text{ if } 0 \le t \le \bar{t}, \\ d_2, \text{ if } \bar{t} < t \le T, \end{cases}$ then the optimal solution is

$$u_m(t) = U_m \text{ for } t_2 - \frac{c_{uk_2} - c_{um}}{c_x^+} \le t \le t_3 + \frac{c_{uk_2} - c_{um}}{c_x^-}, \quad k_1 \le m < k_2;$$

$$u_m(t) = 0 \text{ for } t_3 + \frac{c_{uk_2} - c_{um}}{c_x^-} \le t \le t_2 - \frac{c_{uk_2} - c_{um}}{c_x^+} + T, \quad k_1 < m < k_2;$$

$$u_{k_1}(t) = d_1 - \sum_{m=1}^{k_1-1} U_m \text{ for } t_4 - T \le t \le t_1; \quad u_{k_2}(t) = d_2 - \sum_{m=1}^{k_2-1} U_m \text{ for } t_2 \le t \le t_3;$$

$u_m(t) = U_m$ for $0 \le t \le T$, $m < k_1$; $u_m(t) = 0$ for $0 \le t \le T$, $m > k_2$,
where t_1, t_2, t_3, t_4, $t_1 \ge t_4 - T$ and $t_2 \le t_3$ satisfy the following system of linear algebraic equations

$$c_x^+(t_2 - t_1) = c_{uk_2} - c_{uk_1},$$

$$c_x^-(t_4 - t_3) = c_{uk_2} - c_{uk_1},$$

$$d_1(\bar{t} - t_1) + d_2(t_2 - \bar{t}) = (t_2 - t_1)\sum_{m=1}^{k_1} U_m + \sum_{m=k_1+1}^{k_2-1} U_m \frac{c_{uk_2} - c_{um}}{c_x^+},$$

$$d_1(t_4 - T) + d_2(T - t_3) = (t_4 - t_3)\sum_{m=1}^{k_1} U_m + \sum_{m=k_1+1}^{k_2-1} U_m \frac{c_{uk_2} - c_{um}}{c_x^-},$$

and k_1 and k_2 are such that $\sum_{m=1}^{k_1-1} U_m < d_1 \le \sum_{m=1}^{k_1} U_m$ and $\sum_{m=1}^{k_2-1} U_m < d_2 \le \sum_{m=1}^{k_2} U_m$ respectively.

PROOF. Consider a solution containing two fractional regimes of machines k_1 and k_2 over intervals $[t_4 - T, t_1]$ and $[t_2, t_3]$ respectively

$$\psi(t) = c_{uk_1}, X(t)=0 \text{ for } t_4 - T \le t \le t_1; \quad \psi(t) = c_{uk_2}, X(t)=0 \text{ for } t_2 \le t \le t_3.$$

Then, the behavior of the state and co-state variables between these regimes is described as follows

$$\psi(t) = c_{uk_1} + c_x^+(t - t_1),\ X(t) > 0 \ \text{ for } t_1 < t < t_2; \tag{6.41}$$

$$\psi(t) = c_{uk_2} - c_x^-(t - t_3),\ X(t) < 0 \ \text{ for } t_3 < t < t_4. \tag{6.42}$$

By substituting boundary conditions $\psi(t_2) = c_{uk_2}$ and $\psi(t_4) = c_{uk_1}$ into (6.41) and (6.42) respectively, we obtain the first two equations in unknowns t_1, t_2, t_3, t_4, as stated in the lemma. Given $\psi(t)$, the corresponding behavior of the control functions stated in the lemma is immediately obtained from the first two conditions of (6.40), if the machines are ordered as $c_{u1} \le c_{u2} \le \ldots \le c_{uM}$.

The remaining two equations for unknowns t_1, t_2, t_3, t_4 are obtained by integrating the state equation over intervals $t_1 \le t \le t_2$ and $t_3 \le t \le t_4$ with boundary conditions $X(t_1) = X(t_2) = X(t_3) = X(t_4)$.

The described solution is feasible, if $t_1 \ge t_4 - T$, $t_2 \le t_3$, and k_1, k_2 are such that $\sum_{m=1}^{k_1-1} U_m < d_1 \le \sum_{m=1}^{k_1} U_m$ and $\sum_{m=1}^{k_2-1} U_m < d_2 \le \sum_{m=1}^{k_2} U_m$ respectively.

■

Special Cases of Zero Costs

The $M|N|u(t) \in [0, U], X(0)=X(T)|d(t)|\, L_u, L_x, c_x^+ = 0$ problem is straightforwardly solved by Algorithm 6.9 with no simplification. The zero inventory costs, however, simplify Algorithm 6.10 and extend the class of the demand for which the problem is solvable. Now the two-level, piece-wise constant demand mode is no longer needed because the co-state variable can only decrease. This, due to the periodicity of the solution, implies that the co-state variable must be constant for an arbitrary demand function.

Algorithm 6.11

Step 1:

If N=1, then omit product index n, sort and re-number the machines according to their production costs

$$c_{u1} \le c_{u2} \le \ldots \le c_{uM}.$$

If $\int_0^T d(t)dt \le T \sum_{m=1}^{M} U_m$, then go to the next step. Otherwise STOP, a periodic solution does not exist.

Step 2:

Find k_1 such that $T\sum_{i=1}^{k_1-1} U_i < \int_0^T d(t)dt \le T\sum_{i=1}^{k_1} U_i$ and choose any

$$X(0) = X(T) \ge -\min_t \left\{ \frac{t}{T} \int_0^T d(\tau)d\tau - \int_0^t d(\tau)d\tau \right\}.$$

Set the optimal solution for $0 \le t \le T$ as

$$u_{k_1}(t) = \frac{1}{T} \int_0^T d(\tau)d\tau - \sum_{m=1}^{k_1-1} U_m ,$$

$u_m(t) = U_m$ for $m < k_1$,

$u_m(t) = 0$ for $m > k_1$. STOP.

From Step 2 of Algorithm 6.11, one can observe that the optimal control is constant in time, providing $\psi(t) = c_{uk_1}$ along the period. This solution is feasible if the initial inventory is chosen so that there are no shortages along the period, i.e. $X(0) \ge -\min_t \left\{ \frac{t}{T} \int_0^T d(\tau)d\tau - \int_0^t d(\tau)d\tau \right\}$.

The algorithm for the $M|N|\, u(t) \in [0, U]$,$X(0)$=$X(T)|d(t)|\, L_x$ problem is also modified. Step 1 of Algorithm 6.9 cannot be straightforwardly implemented to solve the problem as though it is uncapacitated. Rather, the problem with a K-level, piece-wise constant demand can be transformed into a system of linear equations and inequalities in unknowns u_{nmk}, as was the case with the $M|N|\, u(t) \in [0, U]\, |d(t), T|\, L_x$ problem,

$$\sum_m u_{nmk} = d_{nk} ,$$

$$0 \le u_{nmk} \le U_{nm} ,$$

$$\sum_n \frac{u_{nmk}}{U_{nm}} \le 1 .$$

If the above system has a feasible solution, then the optimal solution of the $M|N|\, u(t) \in [0, U]$,$X(0)$=$X(T)|d(t)|\, L_x$ problem is

$$u_{nm}(t) = u_{nmk} \text{ for } t_{k-1} \leq t < t_k \text{ for all } n,m \text{ and } k. \tag{6.43}$$

The proof that this solution is optimal immediately follows from the fact that $X_n(t) \equiv 0$, $\psi_n(t) \equiv 0$ and (6.43) is feasible along the entire planning horizon.

In contrast to Step 1 of Algorithm 6.9, Step 2 by not changing, solves both the case of uniform-speed machines and the case of a single product type treated before by Algorithm 6.10. Indeed, $c_{unm} = 0$ implies that all production costs are equal and if N=1, the machines always have uniform speeds.

No Backlog

The solution of the $M|N|\,u(t) \in [0,U]$, $X(t) \geq 0$,$X(0)$=$X(T)|d(t)|\,L_u, L_x$ problem coincides with the solution of the $M|N|\,u(t) \in [0,U]$,$X(0)$=$X(T)|d(t)|\,L_u, L_x, c_x^+ = 0$ problem, which has no shortages as explained in the previous sub-section.

6.3.2 MAINTAINING STABLE INVENTORY

Backlogs

The problem to be considered is M-machine, N-product-type control with limited production rate to maintain stable inventory under an unknown demand function, that is the $M|N|\,u(t) \in [0,U]\,|d(t)|\,\overline{X}$ problem which is stated as follows

$$\sup_{0 \leq t < \infty} |X_n(t)| < \infty, \tag{6.44}$$

$$\dot{X}_n(t) = \sum_m u_{nm}(t) - d_n(t), \tag{6.45}$$

$$0 \leq u_{nm}(t) \leq U_{nm}, \tag{6.46}$$

$$\sum_n \frac{u_{nm}(t)}{U_{nm}} \leq 1. \tag{6.47}$$

Feedback Rule 6.1
Proportional, multi-machine, multi-product, feedback control

$$\text{set } u_{nm}(X_1(t),\ldots,X_N(t)) = \begin{cases} U_{nm}\dfrac{G_n(\overline{X}_n - X_n(t))}{\sum_{n'} G_{n'}(\overline{X}_{n'} - X_{n'}(t))}, & \text{if } X_n(t) < \overline{X}_n, \\ 0, \text{ if } X_n(t) \geq \overline{X}_n, & \end{cases} \quad \text{at every } t.$$

Feedback Rule 6.2
Critical-ratio feedback control:

(i) Denote by A_n a set of machines which produce product n and assign $A_n = \varnothing$ for all n.

(ii) Select product n and machine m such that

$$T_{nm}(t) = \min_{m' \notin \bigcup_i A_i} \max_{n'} \frac{\overline{X}_{n'} - X_{n'}(t)}{\sum_{j \in A_{n'}} U_{n'j} + U_{n'm'}}.$$

(iii) If $X_n(t) \leq \overline{X}_n$, then set $u_{nm}(t) = U_{nm}$ and add machine m to set A_n, $u_{n'm}(t) = 0$ for all $n' \neq n$;
otherwise, set $u_{nm}(t) = 0$ for all $m \in A_n$, assign $A_n = \varnothing$ and go to (ii).

Discussion

Feedback Rule 6.1 is based on the proportional policy and Feedback Rule 6.2 is based on the bang/bang policy. The rules are of $O(MN)$ and $O(M^2N)$ complexity respectively at every time t. Specifically, Rule 6.1, similar to Rule 5.1 suggested for a single machine, traces the demand simultaneously for all products.

Rule 6.2 switches the machines from product n which has reached the desired level $\overline{X}_n$ to the product with minimal critical ratio. This implies that the product with the longest remaining processing time is assigned to the fastest machine with respect to this product.

Bounded evolution of inventories under Feedback Rules 6.1 and 6.2 is very similar to that illustrated for Feedback Rules 5.1 and 5.2 shown in Figure 5.4 for those cases when most of the time the demand is weak. The following lemmas identify the demand-related conditions sufficient for the inventory to be bounded for the two respective feedback rules.

LEMMA 6.13. (Stability of Feedback Rule 6.1) Given problem (6.44)-(6.47), if Feedback Rule 6.1 is applied with $\overline{X}_n \geq 0$ and demand is such that $\lim_{t\to\infty} \frac{1}{t} \int_0^t \sum_n \frac{d_n(\tau)}{\sum_m U_{nm}} d\tau < 1$, then $X_n(t) \leq \overline{X}_n$ for all t and $\inf_{0\leq t<\infty} X_n(t) > -\infty$.

PROOF. First note, that if $X_n(0) \leq \overline{X}_n$, the inventory level can never exceed $\overline{X}_n$. This is due to the proportional feedback control, in which once $X_n(t)$ reaches $\overline{X}_n$, production stops and demand drains inventory again below $\overline{X}_n$, i.e.

$$X_n(t) \leq \overline{X}_n. \tag{6.48}$$

To prove that the inventory is bounded from below, we can substitute the first condition of the rule into the state equation (6.45)

$$\dot{X}_n(t) = \sum_m U_{nm} \frac{G_n(\overline{X}_n - X_n(t))}{\sum_{n'} G_{n'}(\overline{X}_{n'} - X_{n'}(t))} - d_n(t), \quad X_n(0) = 0. \tag{6.49}$$

Next, by summing equations (6.49) for $n=1,\ldots,N$, we find

$$\sum_n \frac{\dot{X}_n(t)}{\sum_m U_{nm}} = 1 - \sum_n \frac{d_n(t)}{\sum_m U_{nm}}. \tag{6.50}$$

From (6.50), it follows that the inventories $X_n(t)$ are bounded from below if

$$\lim_{t\to\infty} \frac{1}{t} \int_0^t \sum_n \frac{d_n(\tau)}{\sum_m U_{nm}} d\tau < 1. \tag{6.51}$$

■

LEMMA 6.14. (Stability of Feedback Rules 6.2) Given problem (6.44)-(6.47), if Feedback Rules 6.2 is applied with $\overline{X}_n \geq 0$ and demand is such that $\lim_{t\to\infty} \frac{1}{t} \int_0^t \sum_n \frac{d_n(\tau)}{\min_m U_{nm}} d\tau < M$, then $X_n(t) \leq \overline{X}_n$ for all t and $\inf_{0\leq t<\infty} X_n(t) > -\infty$.

PROOF. According to Rule 6.2, if $X_n(0) \le \overline{X}_n$, the inventory level can never exceed $\overline{X}_n$. To prove that the inventory is bounded from below, we sum up N state equations (6.45) divided by $\min_m U_{nm}$

$$\sum_n \frac{\dot{X}_n(t)}{\min_m U_{nm}} = \sum_n \sum_{m'} \frac{u_{nm'}(t)}{\min_m U_{nm}} - \sum_n \frac{d_n(t)}{\min_m U_{nm}}. \tag{6.52}$$

Next, we consider the worst case, when the machine capacity is fully utilized. Then, by substituting capacity inequality constraint (6.47) as equality, we find

$$\sum_n \frac{\dot{X}_n(t)}{\min_m U_{nm}} \ge M - \sum_n \frac{d_n(t)}{\min_m U_{nm}}. \tag{6.53}$$

From (6.53), it follows that the inventories $X_n(t)$ are bounded from below if

$$\lim_{t\to\infty} \frac{1}{t} \int_0^t \sum_n \frac{d_n(\tau)}{\min_m U_{nm}} d\tau < M.$$

■

No Backlog

For the $M|N|\,u(t) \in [0, U], X(t) \ge 0\,|d(t)|\,\overline{X}$ problem, additional conditions are required to ensure no backlogs. Specifically, if $X_n(0) = \overline{X}_n$ and demand is weak at every point of time, i.e. $\sum_n \frac{d_n(t)}{\sum_m U_{nm}} < 1$ and $\sum_n \frac{d_n(t)}{\min_m U_{nm}} < M$ for Feedback Rules 6.1 and 6.2 respectively, the desired inventory level is strictly maintained, $X_n(t) \equiv \overline{X}_n$. Thus, these rules are evidently stable and provide no backlog for such a case.

6.4 VARIABLE PLANNING HORIZON

This section addresses the scheduling problems characterized by amount of products to be completed by the end of a planning horizon. The length of the planning horizon, which is unknown *a priori*, is to be minimized along with the total cost of inventory and production.

We consider M-machine, N-product-type scheduling with limited production rate to minimize the total linear cost of inventory, production and planning horizon, that is, the $M|N|\, u(t) \in [0,U], X(T) = X_T \,||\, L_u, L_x, T$ problem which is stated as follows

$$J = \int_0^T \sum_n \left[c_{xn}^+ X_n(t) + \sum_m c_{unm} u_{nm}(t) \right] dt + c_T T \to \min, \tag{6.54}$$

subject to

$$\dot{X}_n(t) = \sum_m u_{nm}(t), \quad X_n(0) = 0, \quad X_n(T) = X_{Tn}, \tag{6.55}$$

$$0 \le u_{nm}(t) \le U_{nm}, \tag{6.56}$$

$$\sum_n \frac{u_{nm}(t)}{U_{nm}} \le 1. \tag{6.57}$$

Note, the above statement is characterized by non-negative inventory along the overall planning horizon because the planning horizon ends as soon as the required amount of the products, $X_n(T) = X_{Tn}$, is completed. If the production cost is identical for all machines, i.e. $c_{unm} = c_{un}$ for all m, the objective can be simplified by substituting the state equation (6.55),

$$J = \int_0^T \sum_n \left[c_{xn}^+ X_n(t) + \sum_m c_{unm} u_{nm}(t) \right] dt + c_T T = \sum_n c_{un} X_{Tn} + \int_0^T \sum_n c_{xn}^+ X_n(t) dt + c_T T .$$

Since the total production cost, $\sum_n c_{un} X_{Tn}$, is a constant which does not influence the optimization, the objective of the problem in such a case takes the form

$$J = \int_0^T \sum_n c_{xn}^{+} X_n(t)dt + c_T T \to \min. \tag{6.58}$$

which is identical to that of the corresponding one-machine problem (see (5.65)).

Dual Formulation

The Hamiltonian, H, resulting from the maximum principle is

$$H = -\sum_n \left[c_{xn}^{+} X_n(t) + \sum_m c_{unm} u_{nm}(t) \right] + \sum_n \psi_n(t) \sum_m u_{nm}(t) = c_T. \tag{6.59}$$

The co-state variable, $\psi_n(t)$ in (6.59), satisfies the following co-state differential equation with no transversality condition

$$\dot{\psi}_n(t) = c_{xn}^{+}. \tag{6.60}$$

Maximization of the Hamiltonian implies that

$$u_{nm}(t) = \begin{cases} U_{nm}, \text{ if } \psi_n(t) > c_{unm} \text{ and } U_{nm}(\psi_n(t) - c_{unm}) > U_{n'm}(\psi_{n'}(t) - c_{un'm}), \ \forall n' \neq n \\ 0, \text{ if } \psi_n(t) < c_{unm}, \\ b, \ b \in [0, U_{nm}], \text{ if } \psi_n(t) \geq c_{unm} \text{ and } U_{nm}(\psi_n(t) - c_{unm}) \geq U_{n'm}(\psi_{n'}(t) - c_{un'm}), \ \forall n' \neq n. \end{cases} \tag{6.61}$$

Algorithm 6.12

Step 1:

If N=1, then omit product index n, sort and re-number the machines according to their production costs

$$c_{u1} \leq c_{u2} \leq \ldots \leq c_{uM}.$$

Step 2:

(i) Set $k_1 = 1$.

(ii) Set $k_2 = k_1$.

(iii) Solve the following system of two linear algebraic equations in $\psi(0)$ and T

$$\psi(0) \sum_{m=1}^{k_1} U_m = c_T + \sum_{m=1}^{k_1} c_{um} U_m,$$

$$T\sum_{m=1}^{k_1} U_m + \sum_{m=k_1+1}^{k_2}\left(T - \frac{c_{um} - \psi(0)}{c_x^+}\right)U_m = X_T .$$

(iv) If $c_{uk_1} \le \psi(0) \le c_{uk_1+1}$ and $c_{uk_2} \le \psi(0) + c_x^+ T \le c_{uk_2+1}$, then set the optimal solution as

$u_m(t) = U_m$ for $m \le k_1$, $0 \le t \le T$,

$u_m(t) = 0$ for $m > k_2$, $0 \le t \le T$,

$u_m(t) = 0$ for $0 \le t < \dfrac{c_{um} - \psi(0)}{c_x^+}$, $k_1 < m \le k_2$,

$u_m(t) = U_m$ for $\dfrac{c_{um} - \psi(0)}{c_x^+} \le t \le T$, $k_1 < m \le k_2$. STOP.

Otherwise, set $k_2 = k_2 + 1$.
If $k_2 > M$, then set $k_1 = k_1 + 1$ and go to (ii), otherwise go to (iii).

Algorithm 6.13
Step 1:

(i) If $c_T = 0$, then assign n=1.
(ii) Run Algorithm 6.12 and denote the time interval where $u_{nm}(t) = U_{nm}$ by (t_{nm}, T_n).
(iii) If $n < N$, assign n=n+1 and go to (ii). Otherwise go to (iv).
(iv) If every machine m produces at most one product, then set $T = \max_n T_n$ and shift each time interval (t_{nm}, T_n) by $T - T_n$. The found solution is optimal. STOP.
Otherwise go to the next step.

Step 2:

If $c_{unm} = c_{unm'}$ for all n, $m \ne m'$ and machines have uniform speeds (see Definition 6.1), then set $U_n = \sum_m U_{nm}$, $c_{un} = c_{un1}$ and run Algorithm 5.6.

Step 3:

Set the optimal solution as

$$u_{nm}(t) = \frac{\beta_m}{\sum_{m'} \beta_{m'}} u_n(t) \text{ for all } n, m, \ 0 \le t \le T \text{. STOP.}$$

Example 6.4
Consider the $3|1|u(t) \in [0,U], X(T) = X_T \,\|\, L_u, L_x, T$ problem with the cost of the planning horizon, c_T=1\$/hr and the required amount of the product X_T=115unit. The production parameters are shown in Table 6.3. In addition, inventory and backlog costs are c_x^+=0.1\$/unit.hr and c_x^-=0.5\$/unit.hr respectively.

Step 1 of Algorithm 6.5 sorts products with respect to the production cost coefficients as 3-1-2 (see the Optimal priority column in Table 6.3).

Step 2 determines k_1=3 and k_2=1 and solves the system of two linear equations as $\psi(0) = 0.15$ and T=5hr. Since $c_{u3} \le \psi(0) \le c_{u1}$ and $c_{u1} \le \psi(0) + c_x^+ T \le c_{u2}$, the optimal solution is set as
$u_3(t) = 20$ unit/hr for $0 \le t \le 5$,
$u_1(t) = 0$ for $0 \le t \le 3.5$,
$u_1(t) = 10$ unit/hr for $3.5 < t \le 5$,
$u_2(t) = 0$ for $0 \le t \le 5$.
The found solution is depicted in Figure 6.7.
■

Discussion
Algorithms 6.12 and 6.13 solve three special cases of the problem. Specifically, Algorithm 6.12 solves one-product, M-machine scheduling problem in $O(M^3)$ time. With respect to the algorithm, the machines switch on according to the order of their production cost, produce at the maximum rate and do not preempt until the required amount of the product $X_n(T) = X_{Tn}$ is completed.

Step 1 of Algorithm 6.13 treats the problem by decomposing it into N one-product sub-problems, each of which is solved by Algorithm 6.12 in $O(M^3)$ time. This decomposition is true if each machine produces only one product type that is verified in $O(MN)$ time. Thus, the total complexity of Step 1 is $O(NM^3)$. If the problem cannot be decomposed, Steps 2 and 3 of Algorithm 6.13 solve the problem in $O(N\max\{M,\log N\})$ time for the case of identical production costs and uniform-speed machines. The solution of Step 2 is based on coalescing the machines into a single machine with aggregate production capacity $U_n = \sum_m U_{nm}$, n=1,...,N. Then, Algorithm 5.6 for a single machine is executed and its solution is distributed proportionally among the machines.

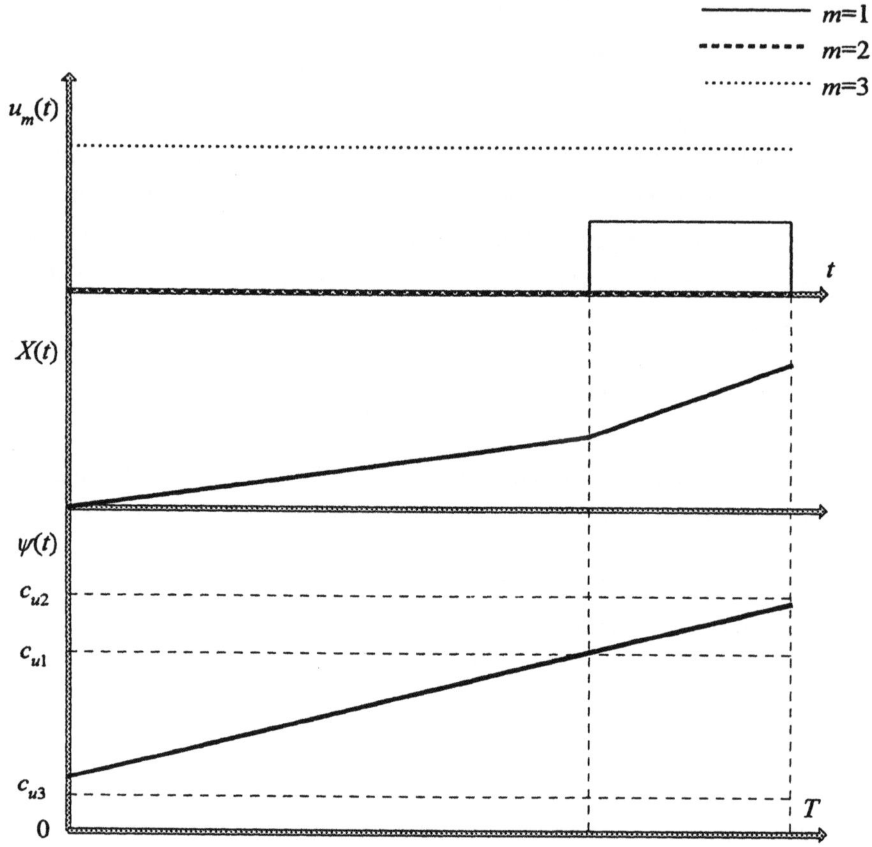

Figure 6.7. Optimal control, state and co-state variables for the $M|N|u(t)\in[0,U], X(T)=X_T \,||\, L_u, L_x, T$ problem of Example 6.4.

LEMMA 6.15. (Correctness of Algorithm 6.12) Given problem (6.54)-(6.57), if N=1, machines are numbered in the non-decreasing order of their production costs, $c_{u1} \le c_{u2} \le \ldots \le c_{uM}$, then the optimal solution is

$u_m(t)=U_m$ for $m \le k_1$, $0 \le t \le T$,

$u_m(t)=0$ for $m>k_2$, $0 \le t \le T$, $u_m(t)=0$ for $0 \le t < \dfrac{c_{um}-\psi(0)}{c_x^+}$, $k_1 < m \le k_2$,

$u_m(t)=U_m$ for $\dfrac{c_{um}-\psi(0)}{c_x^+} \le t \le T$, $k_1 < m \le k_2$, where $\psi(0)$ and T satisfy the following system of two linear algebraic equations

$$\psi(0)\sum_{m=1}^{k_1} U_m = c_T + \sum_{m=1}^{k_1} c_{um} U_m ,$$

$$T\sum_{m=1}^{k_1} U_m + \sum_{m=k_1+1}^{k_2}\left(T - \frac{c_{um} - \psi(0)}{c_x^+}\right)U_m = X_T ,$$

and k_1 and k_2 are such that $c_{uk_1} \le \psi(0) \le c_{uk_1+1}$ and $c_{uk_2} \le \psi(0) + c_x^+ T \le c_{uk_2+1}$ respectively.

PROOF. Consider the following solution for the co-state variable

$$\psi(t) = \psi(0) + c_x^+ t \text{ for } 0 \le t \le T ,$$

$$c_{uk_1} \le \psi(0) \le c_{uk_1+1} \text{ and } c_{uk_2} \le \psi(0) + c_x^+ T \le c_{uk_2+1} . \tag{6.62}$$

Then, according to (6.61), the optimal control is determined as stated in the lemma. By substituting this control into the state equation (6.55), we find the following behavior of the state variable along the planning horizon

$$X(t) = t\sum_{m'=1}^{k_1} U_{m'} + \sum_{m'=k_1+1}^{m}\left(t - \frac{c_{um'} - \psi(0)}{c_x^+}\right)U_{m'} ,\quad \frac{c_{um} - \psi(0)}{c_x^+} \le t \le \frac{c_{um+1} - \psi(0)}{c_x^+} . \tag{6.63}$$

By substituting (6.62) and (6.63) into the boundary conditions $X(T) = X_T$ and $H(0) = c_T$, we obtain two equations in unknowns $\psi(0)$ and T, as stated in the lemma.

The described solution is evidently feasible if the machines are ordered as $c_{u1} \le c_{u2} \le \ldots \le c_{uM}$.

■

LEMMA 6.16. (Correctness of Algorithm 6.13 Step 1) Given problem (6.54)-(6.57) and one-product solutions found by Algorithm 6.12, $u_{nm}(t) = U_{nm}$ at (t_{nm}, T_n), if every machine produces no more than one product, then $T = \max_n T_n$ and the optimal solution is obtained by shifting each time interval (t_{nm}, T_n) by $T - T_n$.

PROOF. The condition that every machine produces at most one product ensures that the optimal production intervals found independently (see Lemma 6.15) for different products do not overlap and thus the capacity constraint of each machine is not violated.

■

LEMMA 6.17. (Correctness of Algorithm 6.13 Steps 2-3) If $c_{unm} = c_{unm'}$ for all n, $m \neq m'$, machines have uniform speeds and $u_n(t)$ is the optimal solution of the $1|N|\, u(t) \in [0,U], X(T) = X_T \,||\, L_u, L_x$ problem with $U_n = \sum_m U_{nm}$ and $c_{un} = c_{un1}$, then the optimal solution is $u_{nm}(t) = \frac{\beta_m}{\sum_{m'} \beta_{m'}} u_n(t)$ for all n,m, $0 \le t \le T$.

PROOF. The proof is similar to that of Lemma 6.2. ■

THEOREM 6.3. If N=1, then problem (6.54)-(6.57) is solvable in $O\left(M^3\right)$ time.

PROOF. The complexity of Algorithm 6.12, which solves the problem for the case of a single product (see Lemma 6.15), is due to Step 2. This step is the most time-consuming because procedure (iii) of $O(M)$ complexity is repeated at most $\sum_{k_1=1}^{M} \left(M - k_1\right)$ times.

■

Special Cases of Zero Costs

Given zero production costs, if only one product is produced, N=1, then the machines have uniform speeds. Therefore, Algorithm 6.12 becomes a special case of Steps 2 and 3 of Algorithm 6.13 which do not change. Thus, the $M|N|\, u(t) \in [0,U], X(T) = X_T \,||\, L_x$ problem is solvable in $O(N\max\{M,\log N\})$ time for the case of uniform-speed machines. Step 1 of Algorithm 6.13, however, never results in a feasible solution. This implies that the zero production cost problem cannot be decomposed into one-product problems.

The $M|N|\, u(t) \in [0,U], X(T) = X_T \,||\, L_u, L_x, T, c_x^+ = 0$ problem can be straightforwardly reformulated into a linear programming problem because it does not depend on sequencing of the products on the machines. Therefore, it is polynomially solvable and is presented as follows

$$c_T T + \sum_{nm} c_{unm} p_{nm} U_{nm} \rightarrow \min$$

subject to

$$\sum_{m} p_{nm} U_{nm} = X_{Tn} \text{ for all } n, \tag{6.64}$$

$$\sum_{n} p_{nm} \leq T \text{ for all } m.$$

where unknown p_{nm} denotes the time interval at which machine m produces product n.

Note that if both $c_{xn}^{+} = 0$ and $c_{unm} = 0$, then formulation (6.64) simplifies to the well-known problem of preemptive scheduling of parallel machines to minimize the makespan.

7

SERIAL MACHINE PROBLEMS

This book is devoted to classifying and solving scheduling problems in the production systems of simple machine configurations. Although such configurations as job-shop, open-shop, and re-entrant lines are not within the book's scope, the maximum principle applied to scheduling problems in complex machine configurations may still result in efficient combinatorial algorithms. This chapter deals with scheduling of multiple tandem machines with limited production capacity and demand concentrated at a due date. Among the various scheduling problems, this one is chosen to illustrate the solution methodology suggested in this book for those cases when configuration of the production system is not as simple as a parallel-machine cell.

Consider M-serial-machine, one-product-type scheduling with limited production capacity to minimize the linear inventory cost. The product requires M processing stages which are carried out by M tandem machines. A buffer is placed after each machine m to store product units on completing the m-th stage. The units stored in the m-th buffer are then processed by the $(m+1)$-th machine. The M-th buffer contains the finished product which can be backlogged. Shortages of the non-finished units in buffers $m=1,\ldots,M-1$ are disallowed. The problem is stated as follows

$$J = \int_0^T \left[\sum_{m=1}^{M-1} c_{xm}^{+} X_m(t) + c_{xM}^{+} Z_M^{+}(t) + c_{xM}^{-} Z_M^{-}(t) \right] dt \to \min, \tag{7.1}$$

$$Z_M^{+}(t) = \max\{0, X_M(t) - D(t)\}, Z_M^{-}(t) = \max\{0, D(t) - X_M(t)\},$$

subject to

$$\dot{X}_m(t) = u_m(t) - u_{m+1}(t), \quad X_m(0) = 0, \quad m=1,\ldots,M-1, \tag{7.2}$$

$$\dot{X}_M(t) = u_M(t), \quad X_M(0) = 0, \tag{7.3}$$

$$X_m(t) \geq 0, \quad m=1,\ldots,M\text{-}1, \tag{7.4}$$

$$0 \leq u_m(t) \leq U_m, \tag{7.5}$$

where

$$D(t) = \begin{cases} D, \text{ if } t \geq t_d; \\ 0, \text{ otherwise;} \end{cases} \tag{7.6}$$

and D is the amount of the product to be produced by the due date t_d .

Dual Formulation

The Hamiltonian, H, for problem (7.1)-(7.5) is

$$H(t) = -\sum_{m=1}^{M-1} c_{xm}^+ X_m(t) - c_{xM}^+ Z_M^+(t) - c_{xM}^- Z_M^-(t) + \\ + \sum_{m=1}^{M-1} \psi_m(t)(u_m(t) - u_{m+1}(t)) + \psi_M(t) u_M(t). \tag{7.7}$$

The co-state variable, $\psi_m(t)$ in (7.7), satisfies the following co-state differential equation with its transversality condition

$$d\psi_m(t) = c_{xm}^+ dt - d\mu_m(t), \quad \psi_m(T+0) = 0, \quad m=1,\ldots,M\text{-}1, \tag{7.8}$$

$$\dot{\psi}_M(t) = \begin{cases} c_{xM}^+, \text{ if } X_M(t) > D(t); \\ -c_{xM}^-, \text{ if } X_M(t) < D(t); \\ a, \ a \in [-c_{xM}^-, c_{xM}^+], \text{ if } X_M(t) = D(t); \end{cases} \qquad \psi_M(T) = 0. \tag{7.9}$$

Lagrange multiplier $d\mu_m(t) \geq 0$ corresponds to the no-shortage constraint (7.4) and satisfies the complementary slackness condition

$$\int_0^T X_m(t) d\mu_m(t) = 0. \tag{7.10}$$

Maximization of the Hamiltonian with respect to admissible controls $u_m(t)$ implies that for each machine m

$$u_m(t)=\begin{cases}U_m, \text{ if } \psi_m(t)-\psi_{m-1}(t)>0,\ m=2,\dots,M \text{ and } \psi_m(t)>0,\ m=1 \text{(production regime - PR)};\\ u\in[0,U_m], \text{ if } \psi_m(t)-\psi_{m-1}(t)=0,\ m=2,\dots,M \text{ and } \psi_m(t)=0,\ m=1 \text{ (singular regime - SR)};\\ 0, \text{ if } \psi_m(t)-\psi_{m-1}(t)<0,\ m=2,\dots,M \text{ and } \psi_m(t)<0,\ m=1 \text{(idle regime - IR)}.\end{cases} \quad (7.11)$$

Algorithm 7.1

Step 1:

(i) If $c^+_{1x}\le c^+_{2x}\le\ldots\le c^+_{Mx}$, then define array $p[1,\dots,M]$. Assign $p[M]=M$, $m=M-1$, $\overline{U}=U_M$ and $m''=M$.

(ii) If $U_m<\overline{U}$, then set $\overline{U}=U_m$, $p[m]=m$, $p[m']=m''$ for $m'=m+1,\dots,m''$. Assign $m''=m$.

(iii) If $m>1$, then set $m=m-1$ and go to (ii), otherwise go to the next step.

Step 2:

Select machines m for which $p[m]=m$ and re-number them as $m=1,\dots,R$. Solve the following system of $2R$ linear algebraic equations in $2R$ time points t^a_m and t^b_m, $m=1,\dots,R$

$$U_m(t^a_{m+1}-t^a_m)=(U_{m+1}-U_m)(t^b_m-t^a_{m+1}),\ m=1,\dots,R-1,$$

$$U_R(t^b_{R-1}-t^a_R)+\sum_{m=2}^{R-1}U_m(t^b_{m-1}-t^b_m)+U_1(t^b_R-t^b_1)=D,$$

$$c^+_{xm}(t^b_m-t^a_m)+\sum_{m'=1}^{m-1}c^+_{xm'}(t^a_{m'+1}-t^a_{m'})=c^-_{xR}(t^b_R-t^b_m),\ m=1,\dots,R-1, \quad (7.12)$$

$$c^+_{xR}(t_d-t^a_R)=c^-_{xR}(t^b_R-t_d),$$

and set the optimal solution as

$u_m(t)=0$ for $0\le t<t^a_m$ and $t^b_R\le t\le T$, $m=1,\dots,R$,

$u_m(t)=U_m$ for $t^a_m\le t<t^b_{m-1}$, $m=2,\dots,R-1$,

$u_R(t)=U_R$ for $t^a_R\le t<t^b_{R-1}$,

$u_m(t)=U_{m'}$ for $t^b_{m'}\le t<t^b_{m'-1}$, $m'=2,\dots,R-1$, $m=m'+1,\dots,R$,

$u_m(t)=U_1$ for $t^b_1\le t<t^b_R$, $m=2,\dots,R$.

$u_1(t) = U_1$ for $t_1^a \le t < t_R^b$.

Step 3:

Restore the original indexes of machines m=1,...,R and set the optimal control for the other machines as

$u_m(t) = u_{p[m]}(t)$, for $0 \le t \le T$. STOP.

Example 7.1

Consider an instance of one-product, five-serial-machine scheduling problem with the length of the planning horizon, T, and the due date t_d set at 10hrs and 8hrs respectively. The demand required for the product is D=70units and the production parameters are shown in Table 7.1.

Table 7.1. Machine parameters

m	U_m	c_{xm}^{+}	$p[m]$
1	10	0.1	1
2	15	0.1	3
3	12	0.2	3
4	20	0.3	5
5	15	0.4	5

In addition, the backlog cost of the finished product is $c_x^- = 1$\$/unit.hr.

Step 1 of Algorithm 7.1 fills in array $p[m]$, m=1,...,5 as shown in the last column of Table 7.1. At Step 2, the algorithm selects machines m=1,3,5 which restrict downstream production, $p[m]$=m. Since the number of restricting machines is R=3, the system of six linear equations is solved (in original indexes) as

$$t_1^a = 2.42\,,\ t_3^a = 3.48,\ t_5^a = 4.45,\ t_1^b = 8.78,\ t_3^b = 8.34\,,\ t_5^b = 9.42\,.$$

Then the optimal solution for the restricting machines is set as

$u_1(t) = 0$ for $0 \le t < 2.42$ and $9.42 \le t \le 10$,

$u_3(t) = 0$ for $0 \le t < 3.48$ and $9.42 \le t \le 10$,

$u_5(t) = 0$ for $0 \le t < 4.45$ and $9.42 \le t \le 10$,

$u_1(t) = 10$ unit/hr for $2.42 \le t < 8.78$,

$u_3(t) = 12$ unit/hr for $3.48 \le t < 8.78$,

$u_5(t) = 15$ unit/hr for $4.45 \le t < 8.34$,

$u_5(t) = 12$ unit/hr for $8.34 \le t < 8.78$,

$u_1(t) = u_3(t) = u_5(t) = 10$ unit/hr for $8.78 \le t < 9.42$.

At Step 3, the optimal solution for the non-restricting machines is set as $u_2(t) = u_3(t)$ and $u_4(t) = u_5(t)$ for $0 \leq t \leq 10$.

The found solution is depicted in Figure 7.1.

■

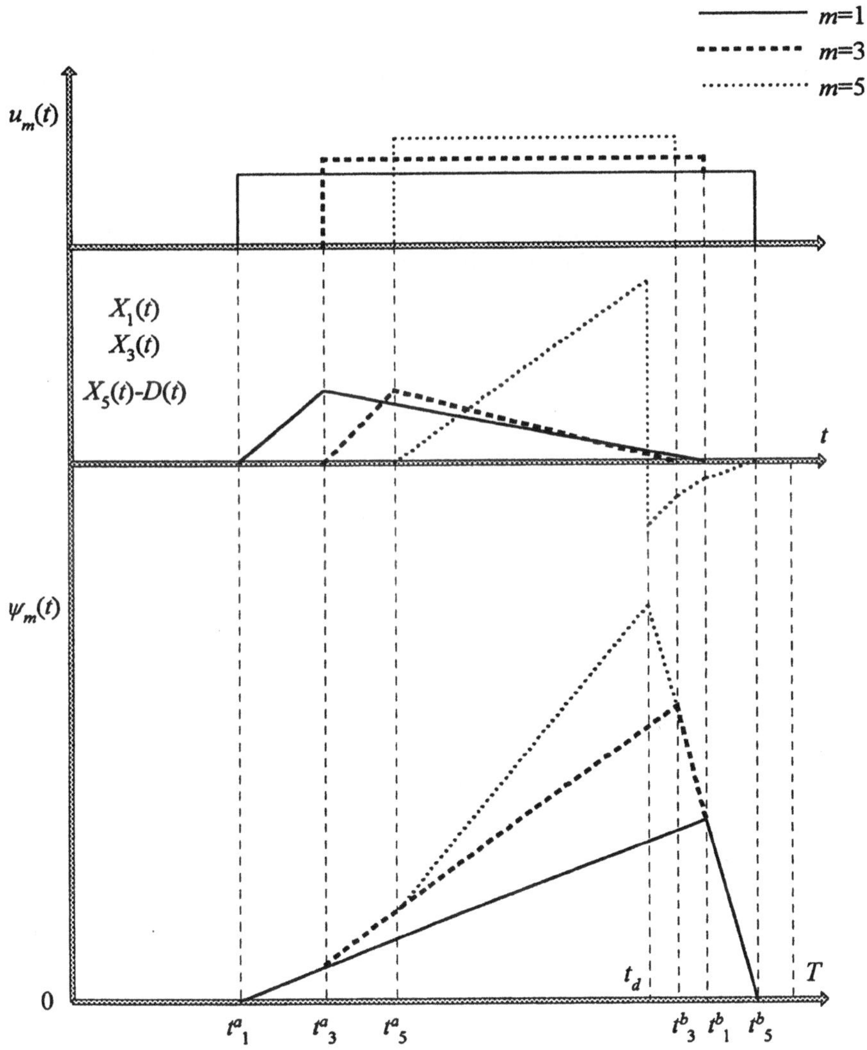

Figure 7.1. Optimal control, state and co-state variables for the restricting machines in Example 7.1.

Definition 7.1. Machine m' is *restricting* if either $m'=M$, or $U_{m'}<U_m$ for all $m>m'$.

Discussion

Algorithm 7.1 solves the serial-machine scheduling problem in $O(M^3)$ time, if the inventory costs on later production stages increase. The complexity is due to Step 2, where a system of at most $2M$ equations is solved. The algorithm utilizes the fact that only machine m', which restricts the downstream production in terms of Definition 7.1, influences the optimal solution. The optimal solution of a non-restricting machine is just to copy the optimal control of the closest downstream restricting machine. Note that the utmost restricting machine is the bottleneck machine.

The optimal behavior of the restricting machines is characterized by consecutive switching the downstream machines on the full production regime before the due date. After the due date, the restricting machines consecutively decrease their production rate to copy the upstream machine. At a final time interval, all restricting machines work with the maximal production rate of the bottleneck machine.

LEMMA 7.1. (Correctness of Algorithm 7.1 Step 2) Given problem (7.1)-(7.5), if $c_{1x}^+ \le c_{2x}^+ \le \ldots \le c_{Mx}^+$ and $U_1 < U_2 < \ldots < U_M$, then the optimal solution is

$u_m(t)=0$ for $0 \le t < t_m^a$ and $t_M^b \le t \le T$, m=1,...,M,

$u_m(t)=U_m$ for $t_m^a \le t < t_{m-1}^b$, m=2,...,M-1, $u_M(t)=U_M$ for $t_M^a \le t < t_{M-1}^b$,

$u_m(t)=U_{m'}$ for $t_{m'}^b \le t < t_{m'-1}^b$, m'=2,...,M-1, $m=m'+1,\ldots,M$,

$u_m(t)=U_1$ for $t_1^b \le t < t_R^b$, m=2,...,M, $u_1(t)=U_1$ for $t_1^a \le t < t_M^b$,

where points t_m^a and t_m^b, m=1,...,M satisfy (7.12) with R=M.

PROOF. Consider a solution for which

$$X_m(t)=0 \text{ for } 0 \le t < t_m^a \text{ and } t_m^b \le t \le T, \quad X_m(t)>0 \text{ for } t_m^a < t < t_m^b, m<M,$$

$$X_M(t)=0 \text{ for } 0 \le t < t_M^a,\ X_M(t)>0 \text{ for } t_M^a < t < t_M^b \text{ and } X_M(t)=D \text{ for } t_M^b \le t \le T. \tag{7.13}$$

Then, the control function for m>1 consists of the no-production regime at interval $0 \le t < t_m^a$, the full production regime at $t_m^a \le t < t_{m-1}^b$, the fractional regime at $t_{m-1}^b \le t < t_M^b$ and again the no-production regime at $t_M^b \le t \le T$. The bottleneck machine, m=1, does not have the fractional regime. Consequently, by integrating the state equations (7.2) and (7.3) with boundary conditions from (7.13), $X_m(t_m^b)=0$, m=1,...,M-1 and $X_M(t_M^b)=D$, we obtain the first two equations from (7.12), as stated in the lemma. The last two equations of (7.12) are found by integration of the co-state equations (7.8) and (7.9) with boundary conditions $\psi_m(t_1^a) = \psi_m(t_M^b)$, m=1,...,M. Since the co-state variables are piece-wise linear functions of time, the found solution is feasible only if $\dot{\psi}_1 \le \dot{\psi}_2 \le \ldots \le \dot{\psi}_M$, i.e. $c_{1x}^+ \le c_{2x}^+ \le \ldots \le c_{Mx}^+$.

■

LEMMA 7.2. (Correctness of Algorithm 7.1 Step 3) Let $p[m]$ be the closest to m, downstream restricting machine and $u_{p[m]}(t)$ be the optimal solution for the set of the restricting machines, $p[m]$=m. If $c_{1x}^+ \le c_{2x}^+ \le \ldots \le c_{Mx}^+$, then the optimal solution for the non-restricting machines is $u_m(t) = u_{p[m]}(t)$, for $0 \le t \le T$.

PROOF. Consider the primal solution for the non-restricting machine m

$$X_m(t) = 0,\ u_m(t) = u_{p[m]}(t) \text{ for } 0 \le t \le T.$$

Then, the corresponding dual solution depends on the closest to machine m, upstream restricting machine m'. If m' exists, then

$$\psi_m(t) = \psi_{m'}(t) \text{ for } 0 \le t \le T,$$

otherwise

$$\psi_m(t) = 0 \text{ for } 0 \le t \le T.$$

Such solution is evidently feasible if $c_{(m-1)x}^+ \le c_{mx}^+$ and $U_m \ge U_{m+1}$.

■

INDEX

Applied Optimization

1. D.-Z. Du and D.F. Hsu (eds.): *Combinatorial Network Theory.* 1996
ISBN 0-7923-3777-8
2. M.J. Panik: *Linear Programming: Mathematics, Theory and Algorithms.* 1996
ISBN 0-7923-3782-4
3. R.B. Kearfott and V. Kreinovich (eds.): *Applications of Interval Computations.* 1996 ISBN 0-7923-3847-2
4. N. Hritonenko and Y. Yatsenko: *Modeling and Optimimization of the Lifetime of Technology.* 1996 ISBN 0-7923-4014-0
5. T. Terlaky (ed.): *Interior Point Methods of Mathematical Programming.* 1996
ISBN 0-7923-4201-1
6. B. Jansen: *Interior Point Techniques in Optimization.* Complementarity, Sensitivity and Algorithms. 1997 ISBN 0-7923-4430-8
7. A. Migdalas, P.M. Pardalos and S. Storøy (eds.): *Parallel Computing in Optimization.* 1997 ISBN 0-7923-4583-5
8. F.A. Lootsma: *Fuzzy Logic for Planning and Decision Making.* 1997
ISBN 0-7923-4681-5
9. J.A. dos Santos Gromicho: *Quasiconvex Optimization and Location Theory.* 1998
ISBN 0-7923-4694-7
10. V. Kreinovich, A. Lakeyev, J. Rohn and P. Kahl: *Computational Complexity and Feasibility of Data Processing and Interval Computations.* 1998
ISBN 0-7923-4865-6
11. J. Gil-Aluja: *The Interactive Management of Human Resources in Uncertainty.* 1998
ISBN 0-7923-4886-9
12. C. Zopounidis and A.I. Dimitras: *Multicriteria Decision Aid Methods for the Prediction of Business Failure.* 1998 ISBN 0-7923-4900-8
13. F. Giannessi, S. Komlósi and T. Rapcsák (eds.): *New Trends in Mathematical Programming.* Homage to Steven Vajda. 1998 ISBN 0-7923-5036-7
14. Ya-xiang Yuan (ed.): *Advances in Nonlinear Programming.* Proceedings of the '96 International Conference on Nonlinear Programming. 1998 ISBN 0-7923-5053-7
15. W.W. Hager and P.M. Pardalos: *Optimal Control.* Theory, Algorithms, and Applications. 1998 ISBN 0-7923-5067-7
16. Gang Yu (ed.): *Industrial Applications of Combinatorial Optimization.* 1998
ISBN 0-7923-5073-1
17. D. Braha and O. Maimon (eds.): *A Mathematical Theory of Design: Foundations, Algorithms and Applications.* 1998 ISBN 0-7923-5079-0

Applied Optimization

18. O. Maimon, E. Khmelnitsky and K. Kogan: *Optimal Flow Control in Manufacturing.* Production Planning and Scheduling. 1998 ISBN 0-7923-5106-1
19. C. Zopounidis and P.M. Pardalos (eds.): *Managing in Uncertainty: Theory and Practice.* 1998 ISBN 0-7923-5110-X
20. A.S. Belenky: *Operations Research in Transportation Systems:* Ideas and Schemes of Optimization Methods for Strategic Planning and Operations Management. 1998 ISBN 0-7923-5157-6
21. J. Gil-Aluja: *Investment in Uncertainty.* 1999 ISBN 0-7923-5296-3
22. M. Fukushima and L. Qi (eds.): *Reformulation: Nonsmooth, Piecewise Smooth, Semismooth and Smooting Methods.* 1999 ISBN 0-7923-5320-X
23. M. Patriksson: *Nonlinear Programming and Variational Inequality Problems.* A Unified Approach. 1999 ISBN 0-7923-5455-9
24. R. De Leone, A. Murli, P.M. Pardalos and G. Toraldo (eds.): *High Performance Algorithms and Software in Nonlinear Optimization.* 1999 ISBN 0-7923-5483-4
25. A. Schöbel: *Locating Lines and Hyperplanes.* Theory and Algorithms. 1999 ISBN 0-7923-5559-8
26. R.B. Statnikov: *Multicriteria Design.* Optimization and Identification. 1999 ISBN 0-7923-5560-1
27. V. Tsurkov and A. Mironov: *Minimax under Transportation Constrains.* 1999 ISBN 0-7923-5609-8
28. V.I. Ivanov: *Model Development and Optimization.* 1999 ISBN 0-7923-5610-1
29. F.A. Lootsma: *Multi-Criteria Decision Analysis via Ratio and Difference Judgement.* 1999 ISBN 0-7923-5669-1
30. A. Eberhard, R. Hill, D. Ralph and B.M. Glover (eds.): *Progress in Optimization.* Contributions from Australasia. 1999 ISBN 0-7923-5733-7
31. T. Hürlimann: *Mathematical Modeling and Optimization.* An Essay for the Design of Computer-Based Modeling Tools. 1999 ISBN 0-7923-5927-5
32. J. Gil-Aluja: *Elements for a Theory of Decision in Uncertainty.* 1999 ISBN 0-7923-5987-9
33. H. Frenk, K. Roos, T. Terlaky and S. Zhang (eds.): *High Performance Optimization.* 1999 ISBN 0-7923-6013-3
34. N. Hritonenko and Y. Yatsenko: *Mathematical Modeling in Economics, Ecology and the Environment.* 1999 ISBN 0-7923-6015-X
35. J. Virant: *Design Considerations of Time in Fuzzy Systems.* 2000 ISBN 0-7923-6100-8

Applied Optimization

36. G. Di Pillo and F. Giannessi (eds.): *Nonlinear Optimization and Related Topics*. 2000 ISBN 0-7923-6109-1
37. V. Tsurkov: *Hierarchical Optimization and Mathematical Physics*. 2000 ISBN 0-7923-6175-X
38. C. Zopounidis and M. Doumpos: *Intelligent Decision Aiding Systems Based on Multiple Criteria for Financial Engineering*. 2000 ISBN 0-7923-6273-X
39. X. Yang, A.I. Mees, M. Fisher and L.Jennings (eds.): *Progress in Optimization*. Contributions from Australasia. 2000 ISBN 0-7923-6175-X
40. D. Butnariu and A.N. Iusem: *Totally Convex Functions for Fixed Points Computation and Infinite Dimensional Optimization*. 2000 ISBN 0-7923-6287-X
41. J. Mockus: *A Set of Examples of Global and Discrete Optimization*. Applications of Bayesian Heuristic Approach. 2000 ISBN 0-7923-6359-0
42. H. Neunzert and A.H. Siddiqi: *Topics in Industrial Mathematics*. Case Studies and Related Mathematical Methods. 2000 ISBN 0-7923-6417-1
43. K. Kogan and E. Khmelnitsky: *Scheduling: Control-Based Theory and Polynomial-Time Algorithms*. 2000 ISBN 0-7923-6486-4

KLUWER ACADEMIC PUBLISHERS – DORDRECHT / BOSTON / LONDON

MIX
Papier aus verantwortungsvollen Quellen
Paper from responsible sources
FSC® C105338

If you have any concerns about our products,
you can contact us on
ProductSafety@springernature.com

In case Publisher is established outside the EU,
the EU authorized representative is:
Springer Nature Customer Service Center GmbH
Europaplatz 3, 69115 Heidelberg, Germany

Printed by Libri Plureos GmbH
in Hamburg, Germany